Advances in Experimental Medicine and Biology

Volume 1481

Advances in Experimental Medicine and Biology provides a platform for scientific contributions in the main disciplines of the biomedicine and the life sciences. This series publishes thematic volumes on contemporary research in the areas of microbiology, immunology, neurosciences, biochemistry, biomedical engineering, genetics, physiology, and cancer research. Covering emerging topics and techniques in basic and clinical science, it brings together clinicians and researchers from various fields.

Advances in Experimental Medicine and Biology has been publishing exceptional works in the field for over 40 years, and is indexed in SCOPUS, Medline (PubMed), EMBASE, Reaxys, EMBiology, the Chemical Abstracts Service (CAS), and Pathway Studio.

2023 CiteScore: 5.9

Christopher D. Gregory • Will Wood
Kodi Ravichandran
Editors

Apoptosis and Other forms of Cell Death in the Regulation of Inflammation in Health and Disease

Springer

Editors
Christopher D. Gregory
Centre for Inflammation Research, Institute for Regeneration and Repair
University of Edinburgh
Edinburgh, UK

Will Wood
Centre for Inflammation Research, Institute for Regeneration and Repair
University of Edinburgh
Edinburgh, UK

Kodi Ravichandran
Department of Pathology and Immunology
Washington University in St. Louis
St. Louis, Missouri, MO, USA

ISSN 0065-2598 ISSN 2214-8019 (electronic)
Advances in Experimental Medicine and Biology
ISBN 978-3-031-92784-3 ISBN 978-3-031-92785-0 (eBook)
https://doi.org/10.1007/978-3-031-92785-0

This Springer imprint is published by the registered company Springer Nature Switzerland AG
The registered company address is: Gewerbestrasse 11, 6330 Cham, Switzerland

"Dedicated to Ruth, Sam, John, Adam, Simla and Rosie"

Foreword

Elephants in the Room

Like people, and cells, research fields go through "ages." Shakespeare (speaking through Jacques in As You Like It) counted seven of these for men and women. In the field of cell death, there was a time when, like Shakespeare's Soldier, we were "jealous in honour, sudden and quick in quarrel, seeking the bubble reputation." Our vibrant meetings were replete with arguments, contradictions, and exorbitant claims, only some of which stood the test of time. Our field moved into the age of Justice, "Full of wise saws and modern instances." And despite predictions of the demise of this field (which are made every ten years or so), we have not reached that age where our once big voice turns again to childish treble and ends "sans teeth, sans eyes, sans taste, sans everything." Cell death research, as one sees in perusing the pages of the book you hold (or stream), continues to reveal the deep mysteries of this fundamental cellular process and how it interfaces with the physiology of the individual.

Thinking about scientific fields puts me in mind of elephants. Not research on elephants (which I am sure is very interesting), but instead, that parable of the blind people and the elephant, which is claimed by many cultures. I'm sure you know it: a group of blind people touching an elephant and proclaiming they have found a tree, a large leaf, a rope, a snake. But as a metaphor for science, I think it gets one thing wrong; there isn't only one elephant, but many. Each elephant is a "big question." Some elephants are surrounded by blind scientists, each trying to touch their small part, with one proclaiming that "my hypothesis is that the third eyelash on the left eye resembles the third eyelash on the right." Other elephants stand quietly alone, with perhaps only a few blind scientists hanging onto the tails. One elephant there was thought to be too old for attention, but new technologies have revived intense interest in it.

And the elephant that is "how do cells die?," this elephant of ours continues to surprise us. In this volume, the focus is not on different modes of cell death, but rather how dying cells influence inflammation and thus the whole organism. When a cell has gone to that "undiscovered country, from whose bourn no traveller returns" (same author, different play), the cell is actively cleared by phagocytosis, a process we now call efferocytosis (cell burial). Several chapters herein describe how efferocytosis impacts the inflammatory response and wound healing, with insights from a variety of experimental models. These chapters should (probably) not be read as the final word on this process, but rather as placeholders of our current understanding in a rapidly evolving research area.

I find it useful to think of the modes of active cell death as falling into three general categories. The first is suicide, cells killing themselves. This can be in response to cell stress, toxic agents, or extracellular ligands that signal via receptors. Apoptosis, discussed in detail in the first chapter (among others), is an example of cell suicide. Others include necroptosis and pyroptosis, discussed in the penultimate chapter. These are evolutionarily conserved mechanisms that, when triggered, cause the cell to die. A second category is sabotage, where a normal cellular mechanism is subverted, often by a chemical agent, causing the cell to die. Ferroptosis (also discussed in Chapter 8), cuproptosis, and disulfidptosis, among others, likely fall into this category. We still regard this as active cell death, since the normal cellular mechanism must be "on" for the disruption to occur (much as derailment will "kill" a train that is moving, but not one that is stopped). The third category is "murder," where the active cell death mechanism is imparted to the cell, often by another cell. Cytotoxic lymphocytes actively murder target cells. The phagocytosis of a living cell (say, by opsonization) is similarly murder. These categories blur, however, when we think of entosis, the subject of the final chapter. Entosis appears to be a murder of a cell by another cell that engulfs it. But this is only appearances. In entosis, a cell effectively "kills itself" by burrowing into another cell, and therefore it might be considered suicide. The engulfed cell can then experience apoptosis as it is deprived of growth factors (again, suicide), and if apoptosis is blocked, the cell can escape and survive. So is entosis suicide or murder? I leave it to the reader to decide.

Department of Immunology
St Jude Children's Hospital
Memphis, TN, USA

Douglas R. Green

Preface

Apoptosis and Other Forms of Cell Death in the Control of Inflammation in Health and Disease was conceived as a follow-up to its successful predecessor, *Apoptosis in cancer pathogenesis and anti-cancer therapy* (Adv Exp Med Biol 930, 2016). Naturally, over the course of a decade, much significant progress is made in any active research field worth its salt and the cell death field is no exception. Ever since the proposition in 1972 by John Kerr, Andrew Wyllie, and Alastair Currie that apoptosis is an "inherently programmed phenomenon…. initiated or inhibited by a variety of environmental stimuli, both physiological and pathological," it has gradually emerged that the host cell *responses* to apoptosis, especially those that govern inflammation and immunity, lie at the heart of many normal physiological processes in multiple organ systems undergoing tissue turnover. Indeed, as John Savill and Valerie Fadok pointed out at the beginning of the century, the phagocytes that clear dying cells essentially provide meaning to apoptosis, such as resolving inflammation or preventing autoimmune disease. Since publication of the previous volume in 2016, clarity on, and interest in, the communication between dying cells and their responders has continued to increase, a particularly memorable example being the surprising effects of secreted apoptotic cell metabolites on bacteria in their vicinity, as established by Kodi Ravichandran and C.J. Anderson, two of the many outstanding scientists we have been fortunate to count among the contributors to this book. Thus, as this collection attests, the capacity of dying cells to influence tissues and organisms is broader than previously realized and is often underappreciated, not only by those working outside the cell death field but also those within!

From the outset, the design and ambition of the current volume was not simply to provide an update to its predecessor's cancer theme, but rather to extend the principles underlying the communicative activities of dying cells to a raft of pathophysiological processes that are rooted in inflammation. The regulatory consequences of cell death affect many inflammatory processes that are commonly encountered throughout biology and medicine, from normal embryonic development and tissue homeostasis to diverse diseases and tissue challenges, including cancer, autoimmunity, infection, and tissue damage. At a time when it is increasingly difficult to keep pace with the shifting sands of change in one's own field, let alone those of complementary areas, it is hoped that this book will prove a unique and timely learning aid and reference resource for the enthusiastic biomedical sciences student and for the specialist cell death researcher alike.

Many engaging topics are discussed herein, starting with a detailed overview of the apoptosis machinery at the beginning of the book and concluding with the curious case of entosis—a "cell-in-cell" death program with intriguing and unique properties—at the end. Along the way the reader encounters a variety of engaging contributions that provide in-depth insights into cell death processes and their consequences for tissues and organisms from many perspectives. In addition to the common consideration of mammalian models, fundamental and ancient aspects of apoptosis and its intercellular communicative properties are explored in the fruit fly, *Drosophila melanogaster*. We learn how the fly lends itself appealingly and rigorously to cell death research through its powerful genetics in combination with in vivo real-time imaging capabilities. Notably, these approaches have proven potency in establishing the molecular principles of apoptosis-induced proliferation and apoptosis-induced apoptosis, among other evolutionarily conserved signaling pathways originating from dying cells and hold much promise for rapidly advancing knowledge in the field. Throughout the book, the molecular cell biology of efferocytosis—the term given to describe the phagocytic clearance of apoptotic cells by macrophages and other cell types—is evaluated from many angles in relation to its far-reaching roles in controlling inflammation: from the programming of the phagocyte phenotype to the metabolic mechanisms and consequences of digestion of apoptotic targets, and from its close connections to tissue homeostasis, wound repair, autoimmune disease, fibrosis, and cancer. Apoptotic cell-dependent communication pathways, whether associated with efferocytosis or not, feature in many chapters and include detailed discussions of basic underlying principles, of the biology of germinal centers and their malignant counterparts and of the important interactions between apoptotic cells and microbes in tissues like the lung and gut which face non-sterile environments. Other cell death modalities are also considered throughout the book in comparison with apoptosis in a multitude of contexts, including a compelling, encyclopedic account of necrotic cell death processes provided in the penultimate chapter.

Apoptosis and Other Forms of Cell Death in the Control of Inflammation in Health and Disease is intended to enlighten, educate and inspire readers at all stages in their careers, be they academic staff or students, clinical or nonclinical, or industry-based scientists, either within this exciting field or simply curious about it. Enjoy!

Edinburgh, UK Christopher D. Gregory

Contents

Molecular Cell Biology of Apoptosis in Health and Disease

Asma Ahmed and Stephen W. G. Tait

Abstract

Apoptotic cell death is fundamental to the life of multicellular organisms, making central contributions to processes ranging from embryonic development to tissue homeostasis. Two distinct apoptosis pathways have been defined: extrinsic apoptosis and mitochondrial apoptosis. As we discuss, apoptosis is an evolutionary conserved process that is, unsurprisingly, tightly regulated. Inhibition of apoptosis can promote cancer, whereas inappropriate apoptosis has been associated with various neurodegenerative diseases. At its core, apoptosis is initiated and executed by proteases called caspases that, once activated, rapidly dismantle dying cells, ensuring that apoptosis is immunologically silent. In this chapter, we discuss the molecular mechanisms of apoptosis and its evolutionary conservation. Secondly, we highlight the emerging concept that apoptosis signalling can be engaged at non-lethal levels with diverse biological effects. Finally, we provide an overview of how apoptosis can impact health and disease, discussing ways in which apoptosis can be therapeutically targeted.

Keywords

Cell death · Extrinsic apoptosis · Mitochondrial apoptosis · MOMP · BCL-2 protein family · Caspases

A. Ahmed · S. W. G. Tait (✉)
Cancer Research UK Scotland Institute, School of Cancer Sciences, University of Glasgow, Glasgow, UK
e-mail: stephen.tait@glasgow.ac.uk

C. D. Gregory et al. (eds.), *Apoptosis and Other forms of Cell Death in the Regulation of Inflammation in Health and Disease*, Advances in Experimental Medicine and Biology 1481, https://doi.org/10.1007/978-3-031-92785-0_1

Abbreviations

AML	Acute myeloid leukaemia
APAF-1	Apoptotic protease-activating factor 1
ATP	Adenosine triphosphate
BAD	BCL-2 antagonist of cell death
BAK	BCL-2 antagonist or killer
BAX	BCL-2-associated X protein
BCL-2	B cell lymphoma-2
BH	BCL-2 homology
BID	BH3-interacting domain death agonist
BIM	BCL-2-interacting mediator of cell death
BOK	BCL-2-related ovarian killer protein
CAD	Caspase-activated DNase
CAR	Chimeric antigen receptor
CED	Cell death abnormal
c-FLIP	Cellular FLICE-inhibitory protein
cGAS	Cyclic GMP-AMP synthase
cIAP	Cellular inhibitor of apoptosis protein
CICD	Caspase-independent cell death
CLL	Chronic lymphocytic leukaemia
DAMP	Danger-associated molecular pattern
DARK	Death-associated APAF-1-related killer
Dcp-1	Death caspase-1
DD	Death domain
DED	Death effector domain
DIAP1	Death-associated inhibitor of apoptosis 1
DISC	Death-inducing signalling complex
DR	Death receptor
DrICE	Death-related ICE-like caspase
EGL-1	Egg-laying defective
ER	Endoplasmic reticulum
FADD	Fas-associated death domain
Fas L	Fas-ligand
FDA	Food and Drug Administration
GSDM	Gasdermin
HMGB1	High mobility group box 1
IMS	Intermembrane space
LDH	Lactate dehydrogenase
LUBAC	Linear ubiquitin chain assembly complex
MCL1	Myeloid cell leukaemia sequence 1
miMOMP	Minority MOMP
MLKL	Mixed lineage kinase-like
MOM	Mitochondrial outer membrane
MOMP	Mitochondrial outer membrane permeabilization

mtDNA	Mitochondrial DNA
NF-κB	Nuclear factor kappa-light-chain-enhancer of activated B cells
NINJ1	Ninjurin-1
PMR	Plasma membrane rupture
PUMA	p53-upregulated modulator of apoptosis
RIPK1	Receptor-interacting serine/threonine-protein kinase 1
ROS	Reactive oxygen species
SMAC	Second mitochondria-derived activator of caspase
STING	Stimulator of interferon genes
tBID	Truncated BID
TKO	Triple knockout
TNF	Tumour necrosis factor
TNFR1	Type 1 TNF receptor
TRADD	TNFR-associated death domain protein
TRAF2	TNF-receptor-associated factor 2
TRAIL	TNF-related apoptosis-inducing ligand
TRAs	TRAIL receptors agonists
VDAC2	Voltage-dependent anion channel 2
XIAP	Caspase X-linked inhibitor of apoptosis protein

Introduction

Regulated cell death is fundamental to the life of multicellular organisms, playing a key role in processes including embryonic development, homeostasis, and disease control. Until recently, cell death was primarily thought to occur in two distinct forms: regulated cell death through apoptosis which is typically considered non-inflammatory or unregulated necrosis, a passive-type of cell death which is associated with inflammation. However, additional forms of regulated cell death have recently been identified including pyroptosis and necroptosis—these have been extensively reviewed elsewhere [35].

Apoptosis is the best characterized form of regulated cell death. Kerr et al. coined the term apoptosis, defining distinct morphological changes in cells undergoing apoptosis [59]. These changes include chromatin condensation, cytoplasmic shrinkage, followed by nuclear fragmentation and membrane blebbing, and the formation of apoptotic bodies. Apoptotic bodies are then taken up by phagocytes and degraded within lysosomes without causing an immune reaction [4, 59] (Fig. 1).

Caspases: Central Mediators of Apoptosis

Apoptotic cell death is initiated and executed by cysteine-aspartyl proteases called caspases that are essential for the biochemical and morphological hallmarks of apoptosis. Caspases—*cysteine aspartate-dependent proteases*—are so-called because their active site contains a cysteine residue, and they cleave substrates after

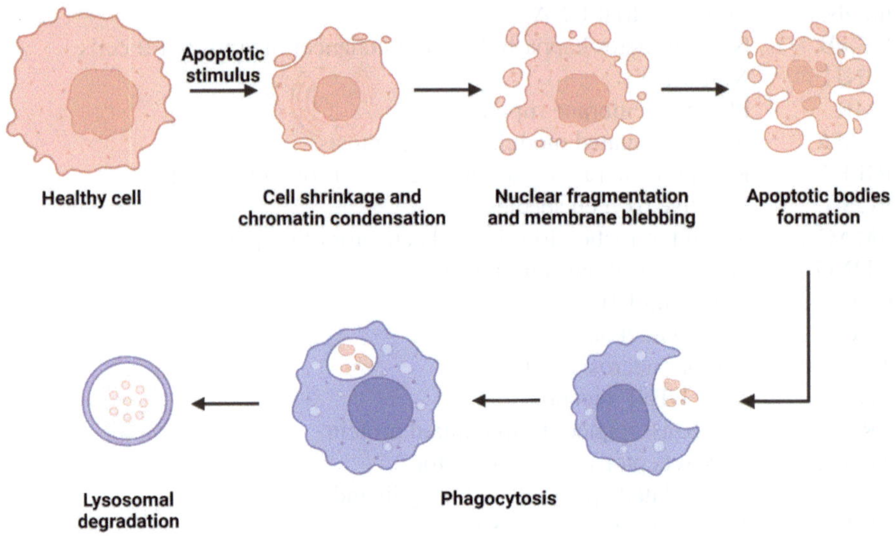

Fig. 1 Morphological features of apoptosis. Apoptotic dying cells display distinct morphological changes including chromatin condensation, cytoplasmic shrinkage, followed by nuclear fragmentation and membrane blebbing, and the formation of apoptotic bodies that are then taken up by phagocytes and degraded within lysosomes

aspartate residues. Besides cell death, caspases have been shown to play important roles in inflammation. Human caspases are broadly classified into three groups based on their biological functions: (i) apoptotic initiator caspases; caspase-2, caspase-8, caspase-9, and caspase-10, (ii) apoptotic executioner caspases; caspase-3, caspase-6, and caspase-7, and (iii) inflammatory caspases; caspase-1, caspase-4, and caspase-5 (caspase-11 in mouse) [49]. Caspases are synthesized as inactive zymogens, thus require activation; initiator and inflammatory caspases exist as inactive monomers in cells and are activated by dimerization. In comparison, executioner caspases exist as inactive dimers and are instead activated through cleavage (most often via initiator caspases) [75].

The biochemical and morphological characteristics of apoptosis described earlier are mediated by apoptotic executioner caspases [119]. Caspases induce these characteristic features of apoptosis by cleaving several hundreds of caspase substrates, rapidly destroying the cell and leading to the exposure of "find-me" and "eat-me" signals that stimulate the clearance of dead cells (a process called efferocytosis) by professional phagocytes such as macrophages and to a lesser extent monocytes and dendritic cells [4]. Thus, caspases play an important role in ensuring rapid cell death and removal of the dying cells, helping keep apoptosis immunologically silent [1].

Gasdermin Cleavage by Caspases and the Role of NINJ1 in Plasma Membrane Rupture

Gasdermins (GSDMs) encompass a protein family that, upon activation (usually via cleavage), leads to plasma membrane pore formation during cell death. This is best evidenced during pyroptosis, where inflammatory caspases (−1, −4, and −5 in humans and −1 and −11 in mice) cleave GSDMD, generating an active N-terminal fragment that oligomerizes, forming pores in the plasma membrane [71, 108]. Moreover, during apoptosis, caspase-8 can directly cleave GSDMD [91], whereas caspase-3 and caspase-8 can cleave GSDME, generating a GSDME N-terminal fragment that, via plasma membrane rupture (PMR), mediates the progression of apoptosis to pyroptosis or secondary necrosis [100, 124]. GSDME has been shown to act as a tumour suppressor as evidenced by inhibition of tumour growth upon overexpression of GSDME. GSDME-mediated PMR can induce an antitumour response mediated by the activation of tumour infiltrating cytotoxic killer cells ($CD8^+$ T cells and natural killer cells) and enhanced tumour cells phagocytosis by macrophages [133]. Interestingly, co-treatment of cancer cells with key mediators of killer cells cytotoxicity, namely perforin and granzyme B, can directly activate GSDME in a caspase-independent manner [133]. Strikingly, even induction of as low as 15% of tumour cell pyroptosis using nanoparticles can induce robust antitumour immunity and synergizes with anti-PD1 blockade for effective tumour control [125]. Intriguingly, GSDME and GSDMD N-terminal fragments can also directly permeabilize the mitochondrial membrane and thus augment caspase-3 activation during apoptosis, highlighting the crosstalk between pyroptosis and apoptosis in response to inflammasome activation [101].

Until recently, PMR was considered a passive event driven by osmotic pressure culminating from lytic cell death that leads to the release of intracellular proteins such as high mobility group box 1 (HMGB1) which is a danger-associated molecular pattern (DAMP) molecule and lactate dehydrogenase (LDH), a standard measure of PMR, leading to inflammation. However, it has been shown that PMR can be an active process mediated by a cell-surface protein called Ninjurin-1 (NINJ1) after induction of pyroptosis, necroptosis, apoptosis, and ferroptosis [54, 95]. In line with the essential role of NINJ1 in mediating PMR, NINJ1-deficient macrophages showed impaired PMR and failed to release HMGB1 and LDH in response to the induction of these cell death pathways [54]. How NINJ1 is activated remains unclear; caspase activity per se is not required for its activation given that NINJ1 also mediates PMR in caspase-independent forms of cell death such as pyroptosis [6]. Structural studies revealed that during lytic cell death, NINJ1 monomers oligomerize in amphipathic filaments in the membrane of dying cells and that ruptures the plasma membrane via NINJ1 hydrophilic side. Under steady state, homodimeric NINJ1 is autoinhibited to prevent PMR and ensure cell survival [23, 24, 94].

Apoptotic Pathways

Broadly speaking, there are two distinct pathways of apoptosis, originally named depending on the origin of the cell death-inducing signal: extrinsic (also called death receptor) and intrinsic (also called mitochondrial) apoptosis; both pathways converge at the point of activation of executioner caspases, which leads to cell demise [38].

Extrinsic Apoptosis

Extrinsic apoptosis is initiated by the binding of extracellular ligands to their cognate death receptors (DRs) that are members of the tumour necrosis factor (TNF) receptor super family, such as the binding of TNF, Fas ligand (Fas-L) or TNF-related apoptosis-inducing ligand (TRAIL) to type 1 TNF receptor (TNFR1), Fas/CD95 receptor or TRAIL receptor, respectively [132]. Ligand binding triggers the formation of muti-protein complexes that lead to cell death. Stimulation of CD95 and TRAIL receptors initiates the recruitment of the adaptor protein Fas-associated death domain (FADD) through homotypic interactions whereby the death domain (DD) of the receptor interacts with the adaptor's DD. FADD then recruits initiator caspase-8 and the regulatory protein cellular FLICE (FADD-like IL-1β-converting enzyme)-inhibitory protein (c-FLIP) using its death effector domain (DED) to form the death-inducing signalling complex (DISC) [26, 66]. The DISC enables the dimerization of caspase-8 leading to its activation. Active caspase-8 cleaves and activates executioner caspase-3 and -7 through cleavage. Interestingly, another secondary cytosolic complex was reported to form following CD95 receptor stimulation. This complex does not contain CD95 and only contains pro-caspase-8, c-FLIP, and FADD. This complex has been shown to play an important role in the amplification of caspase-8 activation [66] (Fig. 2).

In contrast to TRAIL and Fas-L apoptotic functions, TNF plays pleiotropic roles in tissue homeostasis, cell death, and survival. TNF is a master proinflammatory cytokine that regulates tissue homeostasis and immunity and thus has been linked to inflammation and cancer. Binding of TNF to TNFR1 can induce apoptosis or the activation of the transcription factor nuclear factor kappa-light-chain-enhancer of activated B cells (NF-κB) that leads to inflammation and cell survival. Inflammation is the predominant response following TNF signalling, indicating an active repression of TNF-induced apoptosis under normal conditions [123].

This life-or-death decision is regulated by downstream checkpoints following TNFR1 stimulation [2, 77]. A key checkpoint molecule in TNF signalling is the kinase receptor-interacting serine/threonine-protein kinase 1 (RIPK1). RIPK1 is an important regulator of cell survival, apoptosis, and necroptosis [2, 127]. The DD in RIPK1 mediates TNF-induced NF-κB pro-survival signal. Consequently, cells that lack RIPK1 fail to activate NF-κB and become highly sensitive to TNF-mediated cell death, and homozygous loss of RIPK1 induces early postnatal lethality in mice, supporting the notion that RIPK1 is essential for survival [58]. NF-κB counteracts

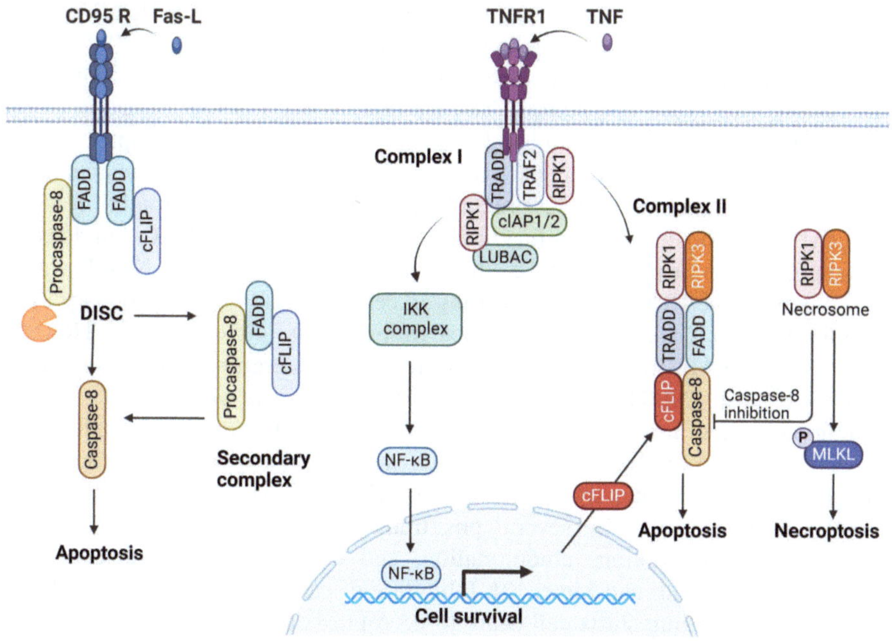

Fig. 2 Death receptor signalling. Binding of death receptor ligands such as tumour necrosis factor (TNF) or Fas/CD95 to their cognate death receptors, which are members of the TNF receptor super family, initiates the recruitment of various adaptor proteins through homotypic death domain interactions. Signalling from CD95 leads to the recruitment of Fas-associated death domain (FADD). FADD then recruits procaspase-8 and the regulatory protein cellular FLICE (FADD-like IL-1β-converting enzyme)-inhibitory protein (c-FLIP) to form the death-inducing signalling complex (DISC) or an alternative secondary cytoplasmic complex containing procaspase-8, FADD, and cFLIP. Both complexes enable caspase-8 activation leading to apoptosis. In contrast, TNF signalling can lead to survival, apoptosis, or necroptosis. Binding of TNF to TNF receptor 1 (TNFR1) leads to the formation of complex I that contains TNFR1, adaptor TNFR-associated death domain protein (TRADD), receptor-interacting serine/threonine-protein kinase 1 (RIPK1), TNF-receptor-associated factor 2 (TRAF2), E3 ubiquitin ligase cellular inhibitor of apoptosis protein 1 (cIAP1) and cIAP2, and linear ubiquitin chain assembly complex (LUBAC). Complex I triggers the recruitment of the IKK complex leading to the activation of NF-κB resulting in cell survival. Alternatively, a secondary cytoplasmic complex (complex II) can form following the dissociation of TRADD, RIPK1, and TRAF2 from TNFR1. This complex activates FADD and caspase-8 leading to apoptosis. Inhibition of caspase-8 in complex II shifts cell death to necroptosis resulting from the activation of RIPK3 by RIPK1 and the formation of the necrosome. Activated RIPK3 phosphorylates the pseudokinase mixed lineage kinase-like (MLKL) leading to necroptosis

cell death by enhancing the expression of potent anti-apoptotic proteins such as cFLIP [78] and maintaining cFLIP abundance by preventing its degradation by the proteasome [85].

c-FLIP is a catalytically inactive homolog of caspase-8. It has been shown to suppress apoptosis induced by TNF, Fas-L, and TRAIL, acting as a master regulator for apoptosis. cFLIP binds to caspase-8 forming a cFLIP/caspase-8 heterodimer that inactivates caspase-8 and consequently apoptosis. Hence, cFLIP deletion sensitizes mice to cell death and results in perinatal lethality [93, 103].

TNFR1 activation induces the formation of two distinct complexes: complex I and complex II. Upon stimulation, TNFR1 rapidly triggers the formation of the plasma membrane bound complex I that contains TNFR1, the adaptor TNFR-associated death domain protein (TRADD), RIPK1, TNF-receptor-associated factor 2 (TRAF2), the E3 ubiquitin ligases cellular inhibitor of apoptosis protein 1 (cIAP1) and cIAP2, and the linear ubiquitin chain assembly complex (LUBAC) but does not contain FADD or caspase-8. This complex triggers the activation of NF-κB via the recruitment of the IKK complex resulting in cell survival [77]. A secondary cytoplasmic complex (complex II) is formed following the dissociation of TRADD, RIPK1, and TRAF2 from TNFR1 as the receptor undergoes endocytosis, hence complex II lacks TNFR1. The DD of TRADD then interacts with the DD of FAAD leading to the recruitment and activation of FADD and caspase-8 to induce cell death (Fig. 2). Failure of complex I to activate NF-κB triggers the activation of complex II and consequently cell death [77]. The cytotoxic function of RIPK1 in complex II is regulated by several post-translational modification checkpoints, including phosphorylation, ubiquitylation, and caspase-mediated cleavage, as reviewed in Annibaldi and Meier [2]. Inhibition of caspase-8 in complex II downstream TNF signalling shifts cell death to necroptosis resulting from activation of RIPK3 by RIPK1 and the formation of the necrosome. Activated RIPK3 phosphorylates the pseudokinase, mixed lineage kinase-like (MLKL). Once phosphorylated, MLKL permeabilizes the plasma membrane leading to necroptotic cell death [40, 127] (Fig. 2).

Mitochondrial Apoptosis

Mitochondrial apoptosis is initiated by a vast array of intracellular perturbations including DNA damage, reactive oxygen species (ROS), endoplasmic reticulum (ER) stress, and nutrient deprivation. Mitochondrial outer membrane permeabilization (MOMP) is the seminal event that initiates mitochondrial apoptosis. MOMP typically represents the point of no return after which cells are committed to die irrespective of downstream caspase activity. This is primarily due to the fact that typically all mitochondria undergo rapid MOMP during apoptosis and this results in widespread adenosine triphosphate (ATP) loss and progressive cellular dysfunction [5]. Given its role in controlling life and death, mitochondrial apoptosis is tightly regulated by the B cell lymphoma-2 (BCL-2) family proteins that control mitochondrial outer membrane integrity [22].

Cellular stress such as DNA damage induces the recruitment and oligomerization of the pro-apoptotic proteins BCL-2-associated X protein (BAX) and BCL-2 antagonist or killer (BAK) at the mitochondrial membrane leading to MOMP induction [19, 126]. BAX and BAK are essential for mitochondrial cell death as demonstrated by the resistance of cells from BAX/BAK double knockout mice to cell death induced by different intrinsic stimuli [126]. MOMP allows the release of soluble intermembrane space (IMS) proteins, some of which promote caspase activation. Cytochrome *c* released into the cytosol binds to apoptotic protease-activating factor 1 (APAF-1) monomer and induces its oligomerization and the formation of the apoptosome which is a multi-subunit complex composed of heptameric oligomer of APAF-1 bound to cytochrome *c* in the presence of ATP. This leads to the activation of APAF-1 and causes conformational changes that cause the exposure of the caspase-activated domain in APAF-1, which allows the recruitment and the activation of caspase-9. Once activated, caspase-9 cleaves and activates executioner caspases, caspase-3 and caspase-7, to cleave cellular substrates, causing rapid cell death [5] (Fig. 3). MOMP also leads to the release of additional mitochondrial proteins that promote apoptosis. For instance, OMI (HTRA2) and second

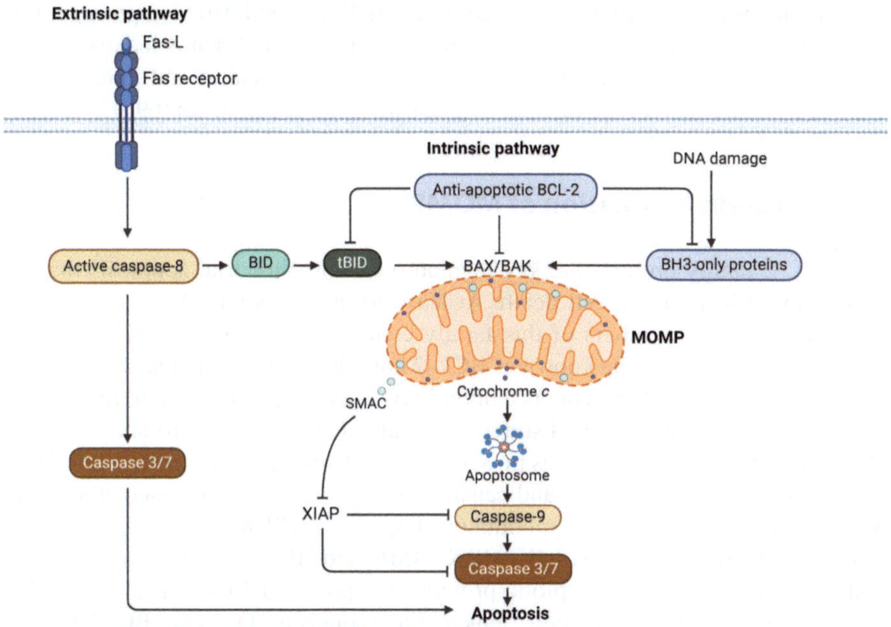

Fig. 3 Interplay between extrinsic and intrinsic apoptotic pathways. There are two distinct pathways of apoptosis: extrinsic and intrinsic. Binding of death receptor ligands to their cognate death receptors, for example, the binding of Fas-L to Fas receptors, initiates the activation of caspase-8. Active caspase-8 cleaves and activates the executioner caspases 3 and 7, leading to cell demise and rapid cell death. Various intrinsic stimuli such as DNA damage or nutrient deprivation activate the intrinsic (mitochondrial) apoptosis. Mitochondrial apoptosis is controlled by the balance between pro-apoptotic and anti-apoptotic B cell lymphoma 2 (BCL-2) protein family. Anti-apoptotic BCL2 proteins inhibit apoptosis, whereas BH3-only proteins activate the pro-apoptotic proteins BAX and BAK, leading to mitochondrial outer membrane permeabilization (MOMP) and the release of intermembrane space proteins such as cytochrome *c* and SMAC into the cytosol. Once released, cytochrome *c* binds to apoptotic peptidase activating factor 1 (APAF1) and pro-caspase-9 to form the apoptosome. This leads to the activation of caspase-9, which cleaves and activates executioner caspases 3 and 7, leading to apoptosis. SMAC release blocks the caspase inhibitor XIAP and thereby induces apoptosis. Extrinsic and intrinsic pathways crosstalk by caspase-8-mediated cleavage of BID to generate tBID that activates BAX and BAK leading to apoptosis

mitochondria-derived activator of caspase (SMAC, also called DIABLO) bind to and inhibit the caspase X-linked inhibitor of apoptosis protein (XIAP), allowing rapid and efficient activation of caspases [29, 122] (Fig. 3). Live single-cell imaging allowed the characterization of the kinetics and the sequence of the events that follow MOMP and has shown the rapid and early release of cytochrome *c* upon apoptosis initiation, despite the variability in onset. Remarkably, all cytochrome *c* is released from all mitochondria in individual cells within 5 min post MOMP irrespective of the strength or the time post the apoptotic stimuli [39]. Exactly how MOMP is coordinated in such a rapid manner across a cell remains unclear, but may, in part, be due to the ability of BAX and BAK to propagate their own activation. Crosstalk between the extrinsic and mitochondrial apoptotic pathways occurs by caspase-8-mediated cleavage of the BH3-interacting domain death agonist (BID) to generate the active form truncated BID (tBID). Full-length BID is a pro-apoptotic BCL-2 family member and a specific substrate for caspase-8 that is localized in the cytosol. Upon cleavage, tBID translocates to the mitochondria and induces cytochrome *c* release and the downstream signals that lead to cell death [69, 73] (Fig. 3).

BCL-2 Protein Regulation of MOMP

The BCL-2 family proteins are key regulators of mitochondrial apoptosis that can either inhibit or promote cell death. At least 20 members have been identified in mammalian cells. Members of this family contain at least one of four conserved sequence homology motifs known as BCL-2 homology (BH) domains, BH1 to BH4 [22]. In humans, based on sequence homology, structure, and function, BCL-2 proteins are classified into three distinct subgroups: (1) Initiators: pro-apoptotic proteins called BH3-only proteins as most of them only contain the BH3 domain, these are subdivided into activators and sensitizers; BCL-2-interacting mediator of cell death (BIM), p53 upregulated modulator of apoptosis (PUMA), BCL-2 antagonist of cell death (BAD), NOXA, BIK, HRK, BMF, and BID. (2) Guardians: multidomain pro-survival or anti-apoptotic proteins composed of BCL-2 itself, BCL-XL, BCL-W, MCL1 (myeloid cell leukaemia sequence 1), and BCL2A1. (3) Executioners: multidomain pro-apoptotic proteins; BAX, BAK, and BOK (BCL-2-related ovarian killer protein), another potential pore-forming protein. Guardians and executioners share four BH domains, a C-terminal transmembrane domain which is essential for membrane insertion and a hydrophobic surface groove that mediates the interactions between the BH3 domains. BCL-2 family proteins interact with each other through their BH3 domain, and these interactions dictate the cell fate [22] (Fig. 4).

It is well known that some BH3-only proteins selectively bind to a subset of pro-survival BCL-2 proteins, while others can neutralize all pro-apoptotic proteins. This discrepancy in selectivity is attributed to the differences in the BH3-only proteins' BH3 domains and in the grooves of their pro-survival binding partners. BIM, PUMA, and tBID bind with high affinity all pro-survival proteins and potently neutralize them. BAD, however, selectively binds BCL-2, BCL-XL, and BCL-W but

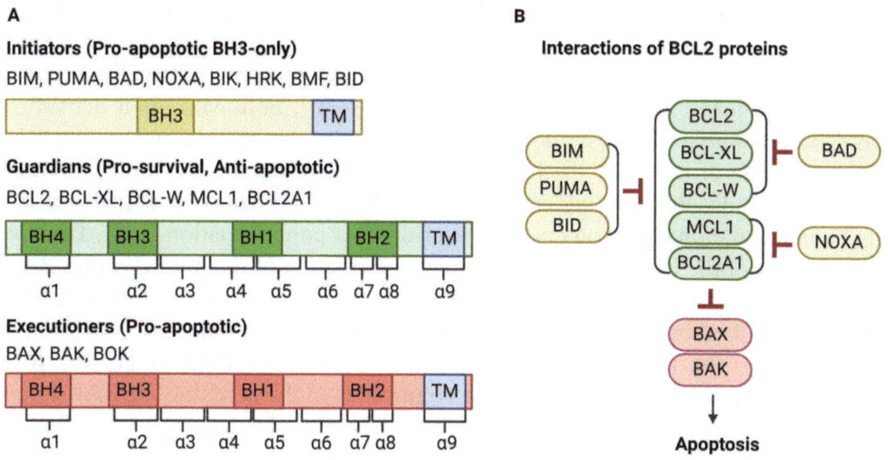

Fig. 4 BCL-2 family proteins. The BCL-2 family proteins are key regulators of mitochondrial apoptosis. (**a**) BCL-2 family members are subdivided based on their function into initiators, guardian, and executioners. They contain at least one of four conserved sequence homology motifs known as BCL-2 homology (BH) domains, BH1 to BH4. Initiator pro-apoptotic proteins are called BH3-only proteins as most of them only contain the BH3 domain; these are subdivided into activators and sensitizer. Guardians (pro-survival) and executioners (pro-apoptotic) share four BH domains. Most members of the BCL-2 family harbour a transmembrane (TM) domain in their C terminus, which is crucial for membrane insertion. (**b**) Interactions between initiators BH3 domains and guardians harbour varying levels of selectivity due to differences in their BH3 domains and in the grooves of their pro-survival binding partners. BIM, PUMA, and BID bind with high affinity to all pro-survival proteins and potently neutralize them. In contrast, BAD can only block BCL-2, BCL-XL, and BCL-W, while NOXA inhibits MCL1 and BCL2A1. Pro-survival proteins can also bind activated BAX and BAK and thereby sequester their pro-apoptotic function

binds weakly to BCL2A1 and has low affinity to MCL1, while NOXA selectively binds MCL-1 and BCL2A1 [11, 51, 64, 76] (Fig. 4). Interestingly, pro-survival proteins can also bind activated BAX and BAK and thereby sequester their pro-apoptotic function (Fig. 4). All of the pro-survival proteins can antagonize BAX [84, 90, 107], while BAK is selectively antagonized by MCL1, BCL2A1, and BCL-XL [109, 130]. Remarkably, inactivation of BCL-XL and MCL1 by BH3-mimetics induces the association of BAX and BAK with the mitochondrial outer membrane (MOM) causing their activation and subsequently MOMP in the absence of canonical pro-apoptotic BH3-only proteins suggesting that BCL2 proteins neutralization alone is sufficient to drive cell death [89].

While hugely disparate stimuli can trigger mitochondrial apoptosis, the main effectors that initiate MOMP are the multidomain proteins BAX and BAK. BH3-only proteins promote the activation of BAX and BAK. During apoptosis, BAX and BAK oligomerize on the mitochondrial outer membrane leading to MOMP [19]. In healthy conditions, BAX continuously cycles between the mitochondria and cytosol, through rounds of retrotranslocation [30]. BAX is kept as an inactive monomer in the cytosol with its C-terminal α9 helix occupying the dimerization pocket formed by BH1-3 domains [22, 116]. In contrast, BAK is an integral membrane protein with its C-terminal α9 helix constitutively inserted in MOM [12, 126]. Upon apoptosis, BH3-only proteins stimulate BAX to undergo stepwise structural reorganization that leads to its mitochondrial targeting [36]. Following their activation, BAX/BAK proteins form homodimers through the mutual binding of the BH3 domain of one protein into the hydrophobic groove of another protein. Individual homodimers subsequently undergo linking and multimerization eventually resulting in the formation of oligomers that are critical for pore formation [104]. The lipid composition of the MOM is important for the formation of the pore-forming BAX/BAK oligomeric assemblies. Cardiolipins enhances the association of distinct BAX dimers and thereby promote the formation of pore-forming BAX oligomers. Consequently, the lack of cardiolipins results in inactive BAX oligomers [65]. Likewise, lipid binding to BAK dimers contributes to the assembly of BAK oligomers [20]. Super resolution imaging revealed the organization of BAX and BAK oligomers into assemblies of varying sizes and shapes, including lines, arcs and rings, with the ability of arcs and rings to form pores [41, 104]. The expansion of these pores is key for the regulation of the release of cytochrome *c* in the early phase of apoptosis and the release of mitochondrial DNA (mtDNA) enabled by mitochondrial inner membrane permeabilization at the later stages of apoptosis [74, 97].

BAX and BAK are often the main effectors of MOMP, however, BOK can also cause MOMP independent of BAX and BAK. BOK is constitutively active and normally kept in check by proteasomal degradation, therefore, proteasome inhibition or the induction of ER stress induces BOK-mediated apoptosis [9, 72]. BOK also regulates non-apoptotic processes such as mitochondrial dynamics and metabolism [106, 112].

Intriguingly, although BOK shares a similar structure to BAX and BAK as shown by similar multi-BH domain and α-helical structural organization and induces similar ring-shaped oligomeric pores, it is regulated differently. Unlike BAX and BAK, BOK pro-apoptotic activity is not regulated by BCL2 proteins. Instead, it is limited by subcellular mitochondrial localization and regulated by proteasomal turnover [72]. Of note, BAX/BAK/BOK triple knockout (TKO) mice exhibited more severe developmental defects than BAX/BAK double knockout mice indicating an overlapping role for BOK with BAX and BAK during developmental cell death [56]. Interestingly, deletion of BID in these TKO mice further exacerbates the developmental anomalies as compared to TKO mice suggesting a new effector role for BID in apoptosis in parallel to BAX, BAK, and BOK, in keeping with the ability of tBID to induce MOMP independent of BAX and BAK [34, 57].

Intriguingly, voltage-dependent anion channel 2 (VDAC2) has been shown to allow BAX recruitment to the OMM leading to BAX activation and cell death, conversely, VDAC2 forms a complex with monomeric BAK and thereby restricts BAK oligomerization and activation [80]. Thus, conditional deletion of VDAC in thymocytes results in excessive cell death and enhanced sensitivity to various apoptotic stimuli indicating an essential role for the VDAC2-BAK complex in the regulation of thymocytes survival [96]. In line with this, overexpression of VDAC2 selectively inhibits BAK activation and subsequently mitochondrial apoptosis [13]. In contrast, genetic deletion of VDAC2 inhibits BAX function but not BAK, indicating that efficient BAX-mediated apoptosis is dependent on VDAC2 [14]. These findings reveal a striking differential regulation of BAX and BAK as a result of their interaction with VDAC2.

BH3-only proteins induce their pro-apoptotic functions by two mechanisms: "activators" like tBID and BIM directly activate the proapoptotic effectors BAX and BAK to induce MOMP. However, these activators are normally sequestered by the anti-apoptotic BCL-2 proteins. Therefore, an alternative mechanism is mediated by BH3-only "sensitisers" that sensitise cells to MOMP by the inhibition of the anti-apoptotic BCL-2 repertoire leading to their neutralization. These sensitizers such as BAD and BIK cannot directly activate BAX and BAK, instead they bind anti-apoptotic BCL-2 proteins and can effectively compete with activators for binding BCL-2, for example, displacing BID from BCL-2 enables it to induce cytochrome *c* release and subsequently MOMP [15, 67, 76].

Non-lethal Engagement of Apoptotic Signalling

Until recently, mitochondrial apoptotic signalling was considered a binary event, whereby apoptotic stimuli induce complete widespread and irreversible MOMP in all cells that commit cells to death. However, accumulating evidence demonstrates that under sub-lethal stress, MOMP can occur in the absence of cell death [3, 8, 47, 52]. Remarkably, cells can even survive transient lethal apoptosis after withdrawal of the apoptotic trigger, a process called anastasis [115, 117]. These cells acquire drug resistance and enhanced tumorigenic characteristics [38].

Sub-lethal MOMP can occur in two forms defined by the extent of MOMP: incomplete MOMP and minority MOMP (miMOMP). Incomplete MOMP occurs when most, but not all, mitochondria undergo MOMP following apoptotic stimulus that is promoted by caspase inhibition and hence lacks the distinct apoptotic features induced by caspases [60]. In contrast, miMOMP occurs when the apoptotic stimulus induces minority of mitochondria to undergo MOMP, leading to the release of small quantities of cytochrome *c* that does not cause cell death and instead lead to the activation of caspases and caspase-activated DNase (CAD), resulting in DNA damage and genomic instability [47] (Fig. 5). Sub-lethal apoptotic signalling emerged not only as a resistance mechanism to therapy but counterintuitively as an oncogenic signal that induces DNA damage and transformation leading to aggressive disease.

Similar to miMOMP, non-lethal caspase activation can lead to CAD-dependent genomic instability and mutagenesis [79]. Nonetheless, what factors dictate cellular fate following an apoptotic stimulus is not fully understood. Thus far, it is well known that apoptotic priming which is the state of the apoptosis pathway in a cell before encountering an apoptotic stress can impact cell fate. Apoptotic priming is regulated by the BCL-2 family proteins in which the collective expression of the pro-apoptotic and the anti-apoptotic proteins dictate the propensity of cells to undergo MOMP following apoptotic stimuli. For instance, cells that are dependent

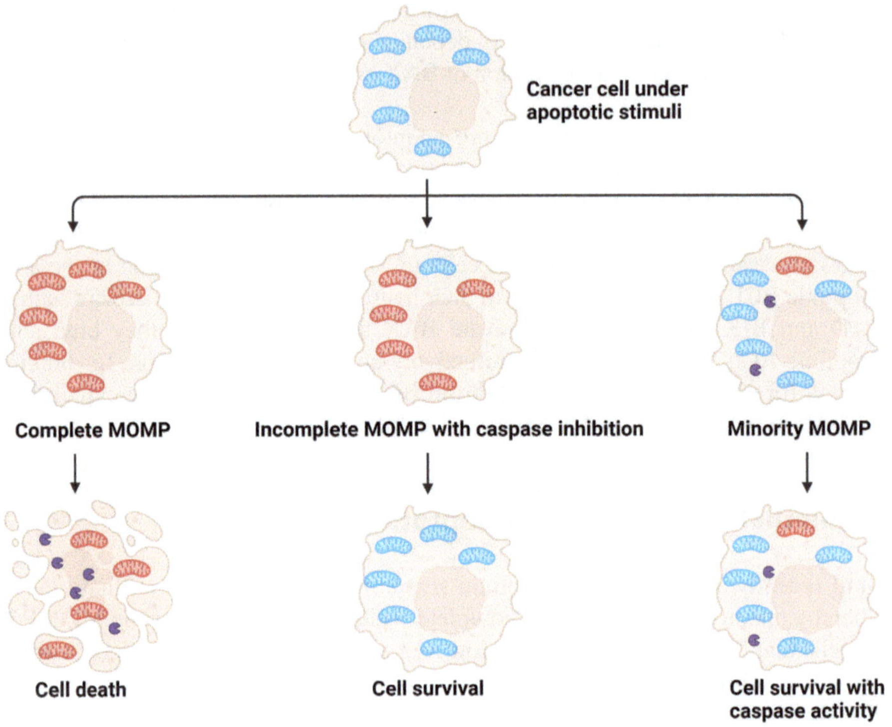

Fig. 5 Non-lethal engagement of apoptotic signalling. Apoptotic stimuli do not always lead to apoptosis. Sub-lethal MOMP can occur in two forms: incomplete MOMP and minority MOMP. When all mitochondria in the cell undergo widespread and complete MOMP, cells will die. However, under incomplete MOMP with caspase inhibition, most, but not all, mitochondria undergo MOMP, leading to cell survival. In contrast, minority MOMP occurs when limited number of mitochondria undergo MOMP, leading to caspase activity and cell survival

on anti-apoptotic BCL-2 proteins for survival are in a "primed for death" state and can readily undergo MOMP upon apoptotic stress in contrast to "unprimed" cells [10]. Mitochondrial dynamics can also regulate MOMP by regulating BCL-2 protein distribution across the mitochondrial network. Whereas mitochondrial fusion promotes complete MOMP, mitochondrial fission promotes sub-lethal MOMP [8].

Therapeutic Targeting of Apoptosis

Evasion of apoptosis is a hallmark of cancer contributing to tumour growth and resistance to therapy. Hence, the disturbance of the physiological balance between pro-survival and pro-apoptotic BCL-2 members in favour of cell survival can facilitate malignant cell growth and induce cancer [43]. This is evident in the upregulation of pro-survival BCL-2 proteins in different cancer types including solid and

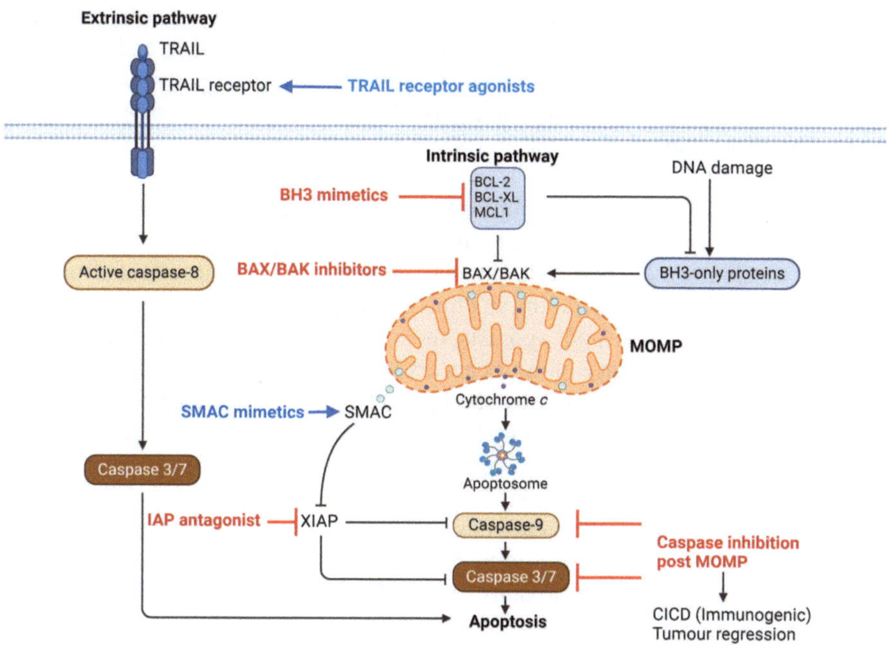

Fig. 6 Therapeutic targeting of apoptosis. Several strategies can be used to therapeutically target apoptosis including the use of BH3-mimetics that neutralize pro-survival BCL2 proteins, agonists for TRAIL receptor, BAK or BAK inhibitors, or small molecules SMAC mimetic/IAP antagonists. Thus far, the only clinically approved pro-apoptotic drug is the BH3-mimetic that selectively targets BCL2, venetoclax. The other drug classes failed in clinical trials due to lack of efficacy or emergence of toxicity. Apoptosis can also be targeted by caspase-inhibitors post MOMP to induce caspase-independent cell death (CICD), which is an immunogenic type of cell death

blood cancers. Accumulating evidence has shown the importance of BCL-2 proteins in cancer development making it an attractive drug target for cancer treatment [7].

Molecular analysis identified the t(14;18) translocation of BCL-2 as a common feature of follicular lymphoma and that increased resistance to apoptosis, attributed to the upregulation of BCL-2, is a key oncogenic mechanism in haematological malignancies [121, 128]. BH3-mimetics, a class of drugs that mimic BH3-only protein and potently inhibit BCL-2, have emerged as a novel strategy for cancer therapy (Fig. 6). The BCL-2 selective inhibitor venetoclax (also known as ABT-199) was the first to receive approval from the US Food and Drug Administration (FDA) for the treatment of chronic lymphocytic leukaemia (CLL) and other B cell malignancies [92, 98]. Clinical trials using venetoclax alone or in combination with other anticancer drugs indicated efficacy and improved response in patients with CLL and acute myeloid leukaemia (AML) with a complete remission in a subset of patients [31, 50, 53, 61, 98, 99].

Thus far, drugs that target MCL1 or BCL-XL are lacking in the clinic despite promising therapeutic potential from preclinical models of blood and solid cancers due to safety concerns such as cardiotoxicity for MCL1 and thrombocytopenia resulting from on-target toxicity to platelets for BCL-XL. BH3 mimetics can be used as monotherapy or in combination to achieve better response and counteract the emergence of resistance. Nonetheless, reasonable balance between safety and tolerability must be carefully considered. BH3 mimetics have been shown to synergize with other anticancer agents, specifically the drugs that increase the levels of BH3-only proteins [27].

The fact that TRAIL induces apoptosis selectively in cancer cells, but not normal cells, prompted the development of TRAIL receptors agonists (TRAs) as cancer therapeutics (Fig. 6). However, despite encouraging anti-tumour effects from preclinical models, TRAs were failing in clinical trials due to variable reasons such as the use of TRAs that were poor agonists, the toxicity of certain TRAs due to immunogenicity and the resistance of most cancers to monotherapy [81]. Cancer cells can resist therapy-induced apoptosis by the upregulation of IAPs, therefore, targeting IAPs represents an attractive strategy for cancer treatment. Small molecule SMAC-mimetic IAPs antagonists have shown promising anti-tumour activity in preclinical models, especially in combination with other anticancer therapies (Fig. 6). IAP antagonists are currently in clinical trials for hematologic and solid cancers either as single agents or in combination with other cancer therapeutics with minimal to moderate efficacy and overt toxicity that led to discontinuation of some trials [16, 33].

Apoptosis can be blocked upstream of MOMP by targeting the effector proteins BAX and BAK (Fig. 6). Modulation of BAX activity represents a promising therapeutic target due to its critical role in apoptosis and disease. Pharmacological inhibition of BAX using peptides, antibodies, and small molecule inhibitors has been shown to reduce executioner caspase activity and enhance cell survival. Furthermore, inhibition of neuronal and cardiomyocyte cell death by BAX inhibitors highlights the therapeutic potential of BAX inhibition in neurodegenerative and cardiovascular diseases. Nonetheless, further research is needed to translate BAX inhibitors to the clinic taking into account which BAX conformation is the most efficient to target with careful consideration to safety [46, 48, 111].

As discussed above, VDAC2 inhibits BAK-mediated apoptosis, hence targeting VDAC2-BAK interaction can effectively inhibit apoptosis [131]. WEHI-9625 is a small molecule selective BAK inhibitor that binds VDAC2 and promotes its ability to block BAK-driven apoptosis. BAK inhibition confers complete and long-term protection from cell death stimulation unlike caspase inhibitors that merely delay apoptosis without preventing it as caspases act downstream of MOMP and cells with permeabilized mitochondria cannot function metabolically [118].

Thus far, caspase inhibitors are failing in clinical trials due to inadequate efficacy, the lack of specificity, and toxicity [25]. Apoptotic caspases actively suppress inflammation in dying cells by inactivating proinflammatory signalling pathways such as cyclic GMP-AMP synthase (cGAS)- stimulator of interferon genes (STING) and NF-κB pathways, alongside shutting down transcription and translation [37, 88, 102, 129]. Consistent with this, caspase inhibition post MOMP induces a form of cell death that is called caspase-independent cell death (CICD) which has been shown to induce anti-tumour immunity and tumour regression [37, 42] (Fig. 6). Moreover, radiation synergizes with emricasan, a pan caspase inhibitor, and immune checkpoint blockade to induce efficient tumour control [42].

Evolutionary Conservation of Apoptosis

Apoptosis is a highly conserved process, demonstrated by its occurrence not only in higher eukaryotes but also in simpler organisms such as the nematode *Caenorhabditis elegans*, with fewer differences in *Drosophila melanogaster* [32]. Studies on caspase substrates revealed a conservation of substrates in these species with differences in cleavage sites [21]. Key components of the apoptotic cascade and their functions were identified by genetic studies in *C. elegans* in which programmed cell death is mediated by cell death abnormal (CED)-3, CED-4, and egg-laying defective-1 (EGL-1) and antagonized by CED-9. The similarity between CED-3 and human IL-1β converting enzyme (caspase-1) identified CED-3 as a protease and prompted the realization that apoptosis is driven by proteolysis. Subsequent studies revealed the biochemical activity of these genes and the close homology between the *C. elegans* and human apoptotic machinery and the identification of CED-9 as BCL-2 homolog, EGL-1 as BH3-only proteins homolog, and CED-4 as APAF-1 homolog [17, 44, 68, 134] .

In *C. elegans*, CED-9 enhances cells survival by directly binding to CED-4, therefore, preventing CED-4 from activating the caspase homologue CED-3. Conversely, apoptosis stimulus upregulates the pro-apoptotic EGL-1 which then binds to CED-9 and disrupts the CED-4-CED-9 interaction, displacing CED-4 and thereby making it free to activate CED-3. Whereas, in *Drosophila*, apoptosis is suppressed by the ubiquitin ligase death-associated IAP1 (DIAP1) that targets the initiator caspase Dronc for proteasomal degradation. Apoptosis stimulation induces the upregulation of the DIAP1 inhibitors Reaper, Hid, and Grim. These inhibitors bind DIAP1 freeing Dronc and thus induces its activation. Dronc then binds to death-associated APAF-1-related killer (DARK). This complex then activates the executioner caspases death-related ICE-like caspase (DrICE) and death caspase-1 (Dcp-1) to induce cell death [86].

A major difference between mitochondrial apoptosis in mammals as compared to *C. elegans* and *Drosophila* is the requirement of MOMP and cytochrome *c* release for the apoptosome formation. Whereas MOMP and cytochrome *c* is essential for the formation of the apoptosome in mammalian cells, cytochrome *c* is not required in *C. elegans* and *Drosophila*. Furthermore, as discussed above, the regulation of mitochondrial apoptosis is highly complex in mammals. Intriguingly, extrinsic apoptosis is absent in *C. elegans* and *Drosophila* [28, 62, 68].

Apoptosis in Health and Disease

Apoptosis plays important roles in tissue homeostasis, development, and immunity. It induces a balance between cell proliferation and the removal of damaged, mutated, or infected cells. Therefore, it is essential that apoptosis is tightly regulated, as failure to induce apoptosis can lead to autoimmunity and cancer, whereas increased apoptosis levels can lead to neurodegenerative diseases [32, 110, 120].

Apoptosis and Tissue Homeostasis

During development, apoptosis is essential for maintaining tissue homeostasis by counteracting the high proliferation rate associated with development that generates superfluous or damaged cells. Apoptosis mediates unwanted cells removal and tissue sculpting [110]. Corroborating this, mice that lack BAX and BAK display multiple developmental defects including accumulation of excess cells within the central nervous and haematopoietic systems. These double knockout mice die perinatally with less than 10% surviving to adulthood [70]. This phenotype is further exacerbated by the loss of BOK [56]. Furthermore, deficiency of caspase-9 or APAF-1 results in perinatal cell death of the majority of mice associated with brain pathology manifested by a markedly enlarged cerebrum as a result of reduced apoptosis during brain development, while the few surviving APAF-1-deficient male mice are infertile [45, 63]. Highlighting the importance of the caspase cascade for cell death and confirming the critical role of caspase-9 as an upstream activator of the caspase cascade, caspase-9-deficient thymocytes fail to cleave caspase-3 and are resistant to mitochondrial apoptotic stimuli [63]. Importantly, cells that lack caspases or APAF-1 will eventually die post MOMP via CICD. Unlike MOMP, caspase are dispensable for cell death and appear to function mainly to accelerate apoptosis and that renders apoptosis immunologically silent [37, 88, 102, 129]. The fact that, in exceptional cases, few mice can survive the loss of key proteins of the intrinsic apoptotic pathway indicates that, although apoptosis is important for development, it is not essential for life, implying that other pathways might compensate for the loss of these proteins [45, 56, 63, 70].

The developing brain is hypersensitive to apoptotic stimuli and thus classified as "primed for apoptosis" due to the high expression of the pro-apoptotic proteins BIM and BID [105]. Therefore, the survival of developing brain cells depends on the balance between these proteins and the pro-survival BCL-XL. In line with this, BCL-XL knockout mice suffer from embryonic lethality due to massive cell death of neuronal cells in addition to immature haematopoietic cells [82]. Apoptosis is also critical for maintenance of homeostasis by removing cells with a high turnover rate, such as haematopoietic cells and intestinal enterocytes, and damaged, infected, or autoreactive cells [55]. Importantly, sensitivity to apoptosis changes during regeneration. For example, intestinal enterocytes and liver hepatocytes are more susceptible to apoptosis when the tissue is damaged and requires repair [110].

Apoptosis in Disease

As discussed previously, apoptosis is essential for the removal of damaged and transformed cells. Therefore, defective apoptosis is associated with tumour development and resistance to cytotoxic therapies. This is demonstrated by the high expression of the pro-survival proteins; BCL2, BCL2A1, and MCL1 in cancer. Furthermore, the tumour suppressor gene p53 that detects DNA damage and subsequently induces apoptosis via the activation of PUMA and NOXA is associated with loss of function mutations in cancer. Apoptosis also curtails pathogen replication by killing infected cells. For instance, the elimination of *Mycobacterium tuberculosis* is facilitated by apoptosis of infected neutrophils and macrophages that phagocytose the bacteria. Hence, many viruses evade apoptosis to sustain replication inside host cells by upregulating BCL-2 homologs [55, 113, 114].

Importantly, one mechanism by which natural killer cells and cytotoxic T cells can kill infected or transformed cells is by releasing perforin and granzyme B. These pore-forming proteins subsequently form pores in the plasma membrane of target cells by activating caspases-3 and -7 or BID. This inspired the development of immunotherapies that target cancer cells such as immune checkpoint inhibitors that unleash the cytotoxicity of T cells or the infusion of chimeric antigen receptor (CAR) T cells which are expanded T cells from the patient engineered to target cancer cells specifically [55, 113].

Insufficient levels of apoptosis is also implicated in autoimmune diseases. For instance, defects in extrinsic apoptosis caused by mutations in FAS gene predispose humans to autoimmune lympho-proliferative syndrome, and high expression of BCL2 is observed in patients with systemic lupus erythematosus, an autoimmune disease characterized by the expression of autoantibodies to double-stranded DNA that causes inflammation and organ damage. Additionally, chronic inflammation of the intestinal epithelium results in excessive cell death and causes autoimmune diseases such as ulcerative colitis and Crohn's disease [18, 110].

Apoptosis is associated with neurodegenerative diseases such as Alzheimer's disease, amyotrophic lateral sclerosis, Parkinson's disease, and Huntington's disease due to excessive neurons death and cardiovascular diseases including heart failure and atherosclerosis due to cardiomyocytes and endothelial cells apoptosis, respectively. Conversely, defective cell death of neuronal cells leads to the development of brain tumours further highlighting the importance of the balance between cell death and survival for tissue homeostasis [83, 87].

Summary

As we have discussed, apoptosis represents a key, evolutionarily conserved, type of regulated cell death. Our understanding of apoptosis is such that we now have a clinically approved drug class called BH3-mimetics that stimulate mitochondrial apoptosis and are effective in the treatment of different leukaemias. Looking ahead, new therapeutic approaches may target cancer apoptosis to make it more immunogenic. On the flip-side, while potent inhibitors of mitochondrial apoptosis are currently lacking, one could envision potential utility in neurodegenerative disease. Finally, there is an emerging paradigm that sub-lethal apoptotic signalling can occur with various biological effects, impacting innate immunity, aging, and cancer. As with apoptosis, understanding the contribution of sub-lethal apoptotic signalling to health and disease may present novel therapeutic opportunities.

References

1. Ahmed, A., & Tait, S. W. G. (2020). Targeting immunogenic cell death in cancer. *Molecular Oncology, 14*, 2994–3006.
2. Annibaldi, A., & Meier, P. (2018). Checkpoints in Tnf-induced cell death: Implications in inflammation and cancer. *Trends in Molecular Medicine, 24*, 49–65.
3. Berthenet, K., Castillo Ferrer, C., Fanfone, D., Popgeorgiev, N., Neves, D., Bertolino, P., Gibert, B., Hernandez-Vargas, H., & Ichim, G. (2020). Failed apoptosis enhances melanoma cancer cell aggressiveness. *Cell Reports, 31*, 107731.
4. Boada-Romero, E., Martinez, J., Heckmann, B. L., & Green, D. R. (2020). The clearance of dead cells by Efferocytosis. *Nature Reviews. Molecular Cell Biology, 21*, 398–414.
5. Bock, F. J., & Tait, S. W. G. (2020). Mitochondria as multifaceted regulators of cell death. *Nature Reviews. Molecular Cell Biology, 21*, 85–100.
6. Broz, P., Pelegrin, P., & Shao, F. (2020). The Gasdermins, a protein family executing cell death and inflammation. *Nature Reviews. Immunology, 20*, 143–157.
7. Campbell, K. J., & Tait, S. W. G. (2018). Targeting Bcl-2 regulated apoptosis in cancer. *Open Biology, 8*.
8. Cao, K., Riley, J. S., Heilig, R., Montes-Gómez, A. E., Vringer, E., Berthenet, K., Cloix, C., Elmasry, Y., Spiller, D. G., Ichim, G., Campbell, K. J., Gilmore, A. P., & Tait, S. W. G. (2022). Mitochondrial dynamics regulate genome stability via control of caspase-dependent dna damage. *Development Cell, 57*, 1211–1225.E6.
9. Carpio, M. A., Michaud, M., Zhou, W., Fisher, J. K., Walensky, L. D., & Katz, S. G. (2015). Bcl-2 family member Bok promotes apoptosis in response to endoplasmic reticulum stress. *Proceedings of the National Academy of Sciences, 112*, 7201–7206.
10. Certo, M., Del Gaizo Moore, V., Nishino, M., Wei, G., Korsmeyer, S., Armstrong, S. A., & Letai, A. (2006). Mitochondria primed by death signals determine cellular addiction to Antiapoptotic Bcl-2 family members. *Cancer Cell, 9*, 351–365.
11. Chen, L., Willis, S. N., Wei, A., Smith, B. J., Fletcher, J. I., Hinds, M. G., Colman, P. M., Day, C. L., Adams, J. M., & Huang, D. C. (2005). Differential targeting of Prosurvival Bcl-2 proteins by their Bh3-only ligands allows complementary apoptotic function. *Molecular Cell, 17*, 393–403.
12. Cheng, E. H., Wei, M. C., Weiler, S., Flavell, R. A., Mak, T. W., Lindsten, T., & Korsmeyer, S. J. (2001). Bcl-2, Bcl-X(L) sequester Bh3 domain-only molecules preventing Bax- and Bak-mediated mitochondrial apoptosis. *Molecular Cell, 8*, 705–711.
13. Cheng, E. H., Sheiko, T. V., Fisher, J. K., Craigen, W. J., & Korsmeyer, S. J. (2003). Vdac2 inhibits Bak activation and mitochondrial apoptosis. *Science, 301*, 513–517.
14. Chin, H. S., Li, M. X., Tan, I. K. L., Ninnis, R. L., Reljic, B., Scicluna, K., Dagley, L. F., Sandow, J. J., Kelly, G. L., Samson, A. L., Chappaz, S., Khaw, S. L., Chang, C., Morokoff, A., Brinkmann, K., Webb, A., Hockings, C., Hall, C. M., Kueh, A. J., Ryan, M. T., Kluck, R. M., Bouillet, P., Herold, M. J., Gray, D. H. D., Huang, D. C. S., Van Delft, M. F., & Dewson, G. (2018). Vdac2 enables Bax to mediate apoptosis and limit tumor development. *Nature Communications, 9*, 4976.

15. Chipuk, J. E., Fisher, J. C., Dillon, C. P., Kriwacki, R. W., Kuwana, T., & Green, D. R. (2008). Mechanism of apoptosis induction by inhibition of the anti-apoptotic Bcl-2 proteins. *Proceedings of the National Academy of Sciences, 105*, 20327–20332.
16. Cong, H., Xu, L., Wu, Y., Qu, Z., Bian, T., Zhang, W., Xing, C., & Zhuang, C. (2019). Inhibitor of apoptosis protein (Iap) antagonists in anticancer agent discovery: Current status and perspectives. *Journal of Medicinal Chemistry, 62*, 5750–5772.
17. Conradt, B., & Horvitz, H. R. (1998). The C. Elegans protein Egl-1 is required for programmed cell death and interacts with the Bcl-2–like protein Ced-9. *Cell, 93*, 519–529.
18. Consonni, F., Gambineri, E., & Favre, C. (2022). Alps, Fas, and beyond: From inborn errors of immunity to acquired Immunodeficiencies. *Annals of Hematology, 101*, 469–484.
19. Cosentino, K., Hertlein, V., Jenner, A., Dellmann, T., Gojkovic, M., Pena-Blanco, A., Dadsena, S., Wajngarten, N., Danial, J. S. H., Thevathasan, J. V., Mund, M., Ries, J., & Garcia-Saez, A. J. (2022). The interplay between Bax and Bak tunes apoptotic pore growth to control mitochondrial-Dna-mediated inflammation. *Molecular Cell, 82*, 933–949.E9.
20. Cowan, A. D., Smith, N. A., Sandow, J. J., Kapp, E. A., Rustam, Y. H., Murphy, J. M., Brouwer, J. M., Bernardini, J. P., Roy, M. J., Wardak, A. Z., Tan, I. K., Webb, A. I., Gulbis, J. M., Smith, B. J., Reid, G. E., Dewson, G., Colman, P. M., & Czabotar, P. E. (2020). Bak core dimers bind lipids and can be bridged by them. *Nature Structural & Molecular Biology, 27*, 1024–1031.
21. Crawford, E. D., Seaman, J. E., Barber, A. E., David, D. C., Babbitt, P. C., Burlingame, A. L., & Wells, J. A. (2012). Conservation of caspase substrates across metazoans suggests hierarchical importance of Signaling pathways over specific targets and cleavage site motifs in apoptosis. *Cell Death & Differentiation, 19*, 2040–2048.
22. Czabotar, P. E., & Garcia-Saez, A. J. (2023). Mechanisms of Bcl-2 family proteins in mitochondrial apoptosis. *Nature Reviews. Molecular Cell Biology, 24*, 732–748.
23. David, L., Borges, J. P., Hollingsworth, L. R., Volchuk, A., Jansen, I., Garlick, E., Steinberg, B. E., & Wu, H. (2024). Ninj1 mediates plasma membrane rupture by cutting and releasing membrane disks. *Cell, 187*, 2224–2235.E16.
24. Degen, M., Santos, J. C., Pluhackova, K., Cebrero, G., Ramos, S., Jankevicius, G., Hartenian, E., Guillerm, U., Mari, S. A., Kohl, B., Müller, D. J., Schanda, P., Maier, T., Perez, C., Sieben, C., Broz, P., & Hiller, S. (2023). Structural basis of Ninj1-mediated plasma membrane rupture in cell death. *Nature, 618*, 1065–1071.
25. Dhani, S., Zhao, Y., & Zhivotovsky, B. (2021). A long way to go: Caspase inhibitors in clinical use. *Cell Death and Disease, 12*.
26. Dickens, L. S., Powley, I. R., Hughes, M. A., & Macfarlane, M. (2012). The 'complexities' of life and death: Death receptor signalling platforms. *Experimental Cell Research, 318*, 1269–1277.
27. Diepstraten, S. T., Anderson, M. A., Czabotar, P. E., Lessene, G., Strasser, A., & Kelly, G. L. (2022). The manipulation of apoptosis for cancer therapy using Bh3-mimetic drugs. *Nature Reviews Cancer, 22*, 45–64.
28. Dorstyn, L., Akey, C. W., & Kumar, S. (2018). New insights into Apoptosome structure and function. *Cell Death & Differentiation, 25*, 1194–1208.
29. Du, C., Fang, M., Li, Y., Li, L., & Wang, X. (2000). Smac, a mitochondrial protein that promotes cytochrome C-dependent caspase activation by eliminating Iap inhibition. *Cell, 102*, 33–42.
30. Edlich, F., Banerjee, S., Suzuki, M., Cleland, M. M., Arnoult, D., Wang, C., Neutzner, A., Tjandra, N., & Youle, R. J. (2011). Bcl-X(L) Retrotranslocates Bax from the mitochondria into the cytosol. *Cell, 145*, 104–116.
31. Eichhorst, B., Niemann, C. U., Kater, A. P., Fürstenau, M., Von Tresckow, J., Zhang, C., Robrecht, S., Gregor, M., Juliusson, G., Thornton, P., Staber, P. B., Tadmor, T., Lindström, V., Da Cunha-Bang, C., Schneider, C., Poulsen, C. B., Illmer, T., Schöttker, B., Nösslinger, T., Janssens, A., Christiansen, I., Baumann, M., Frederiksen, H., Van Der Klift, M., Jäger, U., Leys, M. B. L., Hoogendoorn, M., Lotfi, K., Hebart, H., Gaska, T., Koene, H., Enggaard, L., Goede, J., Regelink, J. C., Widmer, A., Simon, F., De Silva, N., Fink, A. M., Bahlo,

J., Fischer, K., Wendtner, C. M., Kreuzer, K. A., Ritgen, M., Brüggemann, M., Tausch, E., Levin, M. D., Van Oers, M., Geisler, C., Stilgenbauer, S., & Hallek, M. (2023). First-line Venetoclax combinations in chronic lymphocytic Leukemia. *The New England Journal of Medicine, 388*, 1739–1754.
32. Estaquier, J., Vallette, F., Vayssiere, J. L., & Mignotte, B. (2012). The mitochondrial pathways of apoptosis. *Advances in Experimental Medicine and Biology, 942*, 157–183.
33. Ferris, R. L., Harrington, K., Schoenfeld, J. D., Tahara, M., Esdar, C., Salmio, S., Schroeder, A., & Bourhis, J. (2023). Inhibiting the inhibitors: Development of the Iap inhibitor Xevinapant for the treatment of locally advanced squamous cell carcinoma of the head and neck. *Cancer Treatment Reviews, 113*, 102492.
34. Flores-Romero, H., Hohorst, L., John, M., Albert, M. C., King, L. E., Beckmann, L., Szabo, T., Hertlein, V., Luo, X., Villunger, A., Frenzel, L. P., Kashkar, H., & Garcia-Saez, A. J. (2022). Bcl-2-family protein Tbid can act as a Bax-like effector of apoptosis. *The EMBO Journal, 41*, E108690.
35. Galluzzi, L., Vitale, I., Aaronson, S. A., Abrams, J. M., Adam, D., Agostinis, P., Alnemri, E. S., Altucci, L., Amelio, I., Andrews, D. W., Annicchiarico-Petruzzelli, M., Antonov, A. V., Arama, E., Baehrecke, E. H., Barlev, N. A., Bazan, N. G., Bernassola, F., Bertrand, M. J. M., Bianchi, K., Blagosklonny, M. V., Blomgren, K., Borner, C., Boya, P., Brenner, C., Campanella, M., Candi, E., Carmona-Gutierrez, D., Cecconi, F., Chan, F. K., Chandel, N. S., Cheng, E. H., Chipuk, J. E., Cidlowski, J. A., Ciechanover, A., Cohen, G. M., Conrad, M., Cubillos-Ruiz, J. R., Czabotar, P. E., D'angiolella, V., Dawson, T. M., Dawson, V. L., De Laurenzi, V., De Maria, R., Debatin, K. M., Deberardinis, R. J., Deshmukh, M., Di Daniele, N., Di Virgilio, F., Dixit, V. M., Dixon, S. J., Duckett, C. S., Dynlacht, B. D., El-Deiry, W. S., Elrod, J. W., Fimia, G. M., Fulda, S., Garcia-Saez, A. J., Garg, A. D., Garrido, C., Gavathiotis, E., Golstein, P., Gottlieb, E., Green, D. R., Greene, L. A., Gronemeyer, H., Gross, A., Hajnoczky, G., Hardwick, J. M., Harris, I. S., Hengartner, M. O., Hetz, C., Ichijo, H., Jaattela, M., Joseph, B., Jost, P. J., Juin, P. P., Kaiser, W. J., Karin, M., Kaufmann, T., Kepp, O., Kimchi, A., Kitsis, R. N., Klionsky, D. J., Knight, R. A., Kumar, S., Lee, S. W., Lemasters, J. J., Levine, B., Linkermann, A., Lipton, S. A., Lockshin, R. A., Lopez-Otin, C., Lowe, S. W., Luedde, T., Lugli, E., Macfarlane, M., Madeo, F., Malewicz, M., Malorni, W., Manic, G., et al. (2018). Molecular mechanisms of cell death: Recommendations of the nomenclature committee on cell death 2018. *Cell Death and Differentiation, 25*, 486–541.
36. Gavathiotis, E., Reyna, D. E., Davis, M. L., Bird, G. H., & Walensky, L. D. (2010). Bh3-triggered structural reorganization drives the activation of Proapoptotic Bax. *Molecular Cell, 40*, 481–492.
37. Giampazolias, E., Zunino, B., Dhayade, S., Bock, F., Cloix, C., Cao, K., Roca, A., Lopez, J., Ichim, G., Proics, E., Rubio-Patino, C., Fort, L., Yatim, N., Woodham, E., Orozco, S., Taraborrelli, L., Peltzer, N., Lecis, D., Machesky, L., Walczak, H., Albert, M. L., Milling, S., Oberst, A., Ricci, J. E., Ryan, K. M., Blyth, K., & Tait, S. W. G. (2017). Mitochondrial permeabilization engages Nf-Kappab-dependent anti-tumour activity under caspase deficiency. *Nature Cell Biology, 19*, 1116–1129.
38. Glover, H. L., Schreiner, A., Dewson, G., & Tait, S. W. G. (2024). Mitochondria and cell death. *Nature Cell Biology*.
39. Goldstein, J. C., Waterhouse, N. J., Juin, P., Evan, G. I., & Green, D. R. (2000). The coordinate release of cytochrome C during apoptosis is rapid, complete and kinetically invariant. *Nature Cell Biology, 2*, 156–162.
40. Grootjans, S., Vanden Berghe, T., & Vandenabeele, P. (2017). Initiation and execution mechanisms of necroptosis: An overview. *Cell Death and Differentiation, 24*, 1184–1195.
41. Große, L., Wurm, C. A., Brüser, C., Neumann, D., Jans, D. C., & Jakobs, S. (2016). Bax assembles into large ring-like structures remodeling the mitochondrial outer membrane in apoptosis. *The EMBO Journal, 35*, 402–413.
42. Han, C., Liu, Z., Zhang, Y., Shen, A., Dong, C., Zhang, A., Moore, C., Ren, Z., Lu, C., Cao, X., Zhang, C. L., Qiao, J., & Fu, Y. X. (2020). Tumor cells suppress radiation-induced immunity by hijacking caspase 9 signaling. *Nature Immunology, 21*, 546–554.

43. Hanahan, D. (2022). Hallmarks of cancer: New dimensions. *Cancer Discovery, 12*, 31–46.
44. Hengartner, M. O., Ellis, R., & Horvitz, H. R. (1992). Caenorhabditis Elegans gene Ced-9 protects cells from programmed cell death. *Nature, 356*, 494–499.
45. Honarpour, N., Du, C., Richardson, J. A., Hammer, R. E., Wang, X., & Herz, J. (2000). Adult Apaf-1-deficient mice exhibit male infertility. *Developmental Biology, 218*, 248–258.
46. Hui, K. K., Dojo Soeandy, C., Chang, S., Vizeacoumar, F. S., Sun, T., Datti, A., & Henderson, J. T. (2019). Cell-based high-throughput screen for small molecule inhibitors of Bax translocation. *Journal of Cellular and Molecular Medicine, 23*, 1784–1797.
47. Ichim, G., Lopez, J., Ahmed, S. U., Muthalagu, N., Giampazolias, E., Delgado, M. E., Haller, M., Riley, J. S., Mason, S. M., Athineos, D., Parsons, M. J., Van De Kooij, B., Bouchier-Hayes, L., Chalmers, A. J., Rooswinkel, R. W., Oberst, A., Blyth, K., Rehm, M., Murphy, D. J., & Tait, S. W. G. (2015). Limited mitochondrial permeabilization causes Dna damage and genomic instability in the absence of cell death. *Molecular Cell, 57*, 860–872.
48. Jensen, K., Wuwong, D. J., Wong, S., Matsuyama, M., & Matsuyama, S. (2019). Pharmacological inhibition of Bax-induced cell death: Bax-inhibiting peptides and small compounds inhibiting Bax. *Experimental Biology and Medicine (Maywood, N.J.), 244*, 621–629.
49. Julien, O., & Wells, J. A. (2017). Caspases and their substrates. *Cell Death and Differentiation, 24*, 1380–1389.
50. Kadia, T. M., Reville, P. K., Borthakur, G., Yilmaz, M., Kornblau, S., Alvarado, Y., Dinardo, C. D., Daver, N., Jain, N., Pemmaraju, N., Short, N., Wang, S. A., Tidwell, R. S. S., Islam, R., Konopleva, M., Garcia-Manero, G., Ravandi, F., & Kantarjian, H. M. (2021). Venetoclax plus intensive chemotherapy with Cladribine, Idarubicin, and Cytarabine in patients with newly diagnosed acute myeloid leukaemia or high-risk myelodysplastic syndrome: A cohort from a single-centre, single-arm, phase 2 trial. *Lancet Haematol, 8*, E552–E561.
51. Kale, J., Osterlund, E. J., & Andrews, D. W. (2018). Bcl-2 family proteins: Changing partners in the dance towards death. *Cell Death & Differentiation, 25*, 65–80.
52. Kalkavan, H., Chen, M. J., Crawford, J. C., Quarato, G., Fitzgerald, P., Tait, S. W. G., Goding, C. R., & Green, D. R. (2022). Sublethal cytochrome C release generates drug-tolerant Persister cells. *Cell, 185*, 3356–3374.E22.
53. Kater, A. P., Wu, J. Q., Kipps, T., Eichhorst, B., Hillmen, P., D'rozario, J., Assouline, S., Owen, C., Robak, T., De La Serna, J., Jaeger, U., Cartron, G., Montillo, M., Dubois, J., Eldering, E., Mellink, C., Van Der Kevie-Kersemaekers, A. M., Kim, S. Y., Chyla, B., Punnoose, E., Bolen, C. R., Assaf, Z. J., Jiang, Y., Wang, J., Lefebure, M., Boyer, M., Humphrey, K., & Seymour, J. F. (2020). Venetoclax plus rituximab in relapsed chronic lymphocytic Leukemia: 4-year results and evaluation of impact of genomic complexity and gene mutations from the Murano phase iii study. *Journal of Clinical Oncology, 38*, 4042–4054.
54. Kayagaki, N., Kornfeld, O. S., Lee, B. L., Stowe, I. B., O'rourke, K., Li, Q., Sandoval, W., Yan, D., Kang, J., Xu, M., Zhang, J., Lee, W. P., Mckenzie, B. S., Ulas, G., Payandeh, J., Roose-Girma, M., Modrusan, Z., Reja, R., Sagolla, M., Webster, J. D., Cho, V., Andrews, T. D., Morris, L. X., Miosge, L. A., Goodnow, C. C., Bertram, E. M., & Dixit, V. M. (2021). Ninj1 mediates plasma membrane rupture during lytic cell death. *Nature, 591*, 131–136.
55. Kayagaki, N., Webster, J. D., & Newton, K. (2024). Control of cell death in health and disease. *Annual Review of Pathology, 19*, 157–180.
56. Ke, F. F. S., Vanyai, H. K., Cowan, A. D., Delbridge, A. R. D., Whitehead, L., Grabow, S., Czabotar, P. E., Voss, A. K., & Strasser, A. (2018). Embryogenesis and adult life in the absence of intrinsic apoptosis effectors Bax, Bak, and Bok. *Cell, 173*(1217–1230), E17.
57. Ke, F. S., Holloway, S., Uren, R. T., Wong, A. W., Little, M. H., Kluck, R. M., Voss, A. K., & Strasser, A. (2022). The Bcl-2 family member bid plays a role during embryonic development in addition to its Bh3-only protein function by acting in parallel to Bax, Bak and Bok. *The EMBO Journal, 41*, E110300.
58. Kelliher, M. A., Grimm, S., Ishida, Y., Kuo, F., Stanger, B. Z., & Leder, P. (1998). The death domain kinase rip mediates the Tnf-induced Nf-Kappab signal. *Immunity, 8*, 297–303.

59. Kerr, J. F., Wyllie, A. H., & Currie, A. R. (1972). Apoptosis: A basic biological phenomenon with wide-ranging implications in tissue kinetics. *British Journal of Cancer, 26*, 239–257.
60. Killarney, S. T., Tait, S. W. G., Green, D. R., & Wood, K. C. (2024). Sublethal engagement of apoptotic pathways in residual cancer. *Trends in Cell Biology, 34*, 225–238.
61. Konopleva, M., Pollyea, D. A., Potluri, J., Chyla, B., Hogdal, L., Busman, T., Mckeegan, E., Salem, A. H., Zhu, M., Ricker, J. L., Blum, W., Dinardo, C. D., Kadia, T., Dunbar, M., Kirby, R., Falotico, N., Leverson, J., Humerickhouse, R., Mabry, M., Stone, R., Kantarjian, H., & Letai, A. (2016). Efficacy and biological correlates of response in a phase ii study of Venetoclax monotherapy in patients with acute Myelogenous Leukemia. *Cancer Discovery, 6*, 1106–1117.
62. Krasovec, G., Horkan, H. R., Quéinnec, É., & Chambon, J.-P. (2023). Intrinsic apoptosis is evolutionarily divergent among metazoans. *Evolution Letters, 8*, 267–282.
63. Kuida, K., Haydar, T. F., Kuan, C. Y., Gu, Y., Taya, C., Karasuyama, H., Su, M. S., Rakic, P., & Flavell, R. A. (1998). Reduced apoptosis and cytochrome C-mediated caspase activation in mice lacking caspase 9. *Cell, 94*, 325–337.
64. Kuwana, T., Bouchier-Hayes, L., Chipuk, J. E., Bonzon, C., Sullivan, B. A., Green, D. R., & Newmeyer, D. D. (2005). Bh3 domains of Bh3-only proteins differentially regulate Bax-mediated mitochondrial membrane permeabilization both directly and indirectly. *Molecular Cell, 17*, 525–535.
65. Lai, Y.-C., Li, C.-C., Sung, T.-C., Chang, C.-W., Lan, Y.-J., & Chiang, Y.-W. (2019). The role of Cardiolipin in promoting the membrane pore-forming activity of Bax oligomers. *Biochimica Et Biophysica Acta (BBA) – Biomembranes, 1861*, 268–280.
66. Lavrik, I. N., Mock, T., Golks, A., Hoffmann, J. C., Baumann, S., & Krammer, P. H. (2008). Cd95 stimulation results in the formation of a novel death effector domain protein-containing complex. *The Journal of Biological Chemistry, 283*, 26401–26408.
67. Letai, A., Bassik, M. C., Walensky, L. D., Sorcinelli, M. D., Weiler, S., & Korsmeyer, S. J. (2002). Distinct Bh3 domains either sensitize or activate mitochondrial apoptosis, serving as prototype cancer therapeutics. *Cancer Cell, 2*, 183–192.
68. Lettre, G., & Hengartner, M. O. (2006). Developmental apoptosis in C. Elegans: A complex Cednario. *Nature Reviews Molecular Cell Biology, 7*, 97–108.
69. Li, H., Zhu, H., Xu, C.-J., & Yuan, J. (1998). Cleavage of bid by caspase 8 mediates the mitochondrial damage in the Fas pathway of apoptosis. *Cell, 94*, 491–501.
70. Lindsten, T., Ross, A. J., King, A., Zong, W. X., Rathmell, J. C., Shiels, H. A., Ulrich, E., Waymire, K. G., Mahar, P., Frauwirth, K., Chen, Y., Wei, M., Eng, V. M., Adelman, D. M., Simon, M. C., Ma, A., Golden, J. A., Evan, G., Korsmeyer, S. J., Macgregor, G. R., & Thompson, C. B. (2000). The combined functions of Proapoptotic Bcl-2 family members Bak and Bax are essential for Normal development of multiple tissues. *Molecular Cell, 6*, 1389–1399.
71. Liu, X., Zhang, Z., Ruan, J., Pan, Y., Magupalli, V. G., Wu, H., & Lieberman, J. (2016). Inflammasome-activated Gasdermin D causes pyroptosis by forming membrane pores. *Nature, 535*, 153–158.
72. Llambi, F., Wang, Y. M., Victor, B., Yang, M., Schneider, D. M., Gingras, S., Parsons, M. J., Zheng, J. H., Brown, S. A., Pelletier, S., Moldoveanu, T., Chen, T., & Green, D. R. (2016). Bok is a non-canonical Bcl-2 family effector of apoptosis regulated by Er-associated degradation. *Cell, 165*, 421–433.
73. Luo, X., Budihardjo, I., Zou, H., Slaughter, C., & Wang, X. (1998). Bid, a Bcl2 interacting protein, mediates cytochrome C release from mitochondria in response to activation of cell surface death receptors. *Cell, 94*, 481–490.
74. Mcarthur, K., Whitehead, L. W., Heddleston, J. M., Li, L., Padman, B. S., Oorschot, V., Geoghegan, N. D., Chappaz, S., Davidson, S., San Chin, H., Lane, R. M., Dramicanin, M., Saunders, T. L., Sugiana, C., Lessene, R., Osellame, L. D., Chew, T. L., Dewson, G., Lazarou, M., Ramm, G., Lessene, G., Ryan, M. T., Rogers, K. L., Van Delft, M. F., & Kile, B. T. (2018). Bak/Bax macropores facilitate mitochondrial herniation and Mtdna efflux during apoptosis. *Science, 359*.

75. Mcilwain, D. R., Berger, T., & Mak, T. W. (2013). Caspase functions in cell death and disease. *Cold Spring Harbor Perspectives in Biology, 5*, A008656.
76. Mérino, D., Giam, M., Hughes, P. D., Siggs, O. M., Heger, K., O'reilly, L. A., Adams, J. M., Strasser, A., Lee, E. F., Fairlie, W. D., & Bouillet, P. (2009). The role of Bh3-only protein Bim extends beyond inhibiting Bcl-2-like Prosurvival proteins. *The Journal of Cell Biology, 186*, 355–362.
77. Micheau, O., & Tschopp, J. (2003). Induction of Tnf receptor I-mediated apoptosis via two sequential Signaling complexes. *Cell, 114*, 181–190.
78. Micheau, O., Lens, S., Gaide, O., Alevizopoulos, K., & Tschopp, J. (2001). Nf-Kappab signals induce the expression of C-Flip. *Molecular and Cellular Biology, 21*, 5299–5305.
79. Miles, M. A., & Hawkins, C. J. (2017). Executioner caspases and cad are essential for mutagenesis induced by trail or vincristine. *Cell Death & Disease, 8*, E3062–E3062.
80. Moldoveanu, T., & Czabotar, P. E. (2020). *Bax, Bak, and Bok: A coming of age for the Bcl-2 family effector proteins* (p. 12). Cold Spring Harbor Perspectives in Biology.
81. Montinaro, A., & Walczak, H. (2023). Harnessing Trail-induced cell death for cancer therapy: A long walk with thrilling discoveries. *Cell Death & Differentiation, 30*, 237–249.
82. Motoyama, N., Wang, F., Roth, K. A., Sawa, H., Nakayama, K., Nakayama, K., Negishi, I., Senju, S., Zhang, Q., Fujii, S., et al. (1995). Massive cell death of immature hematopoietic cells and neurons in Bcl-X-deficient mice. *Science, 267*, 1506–1510.
83. Moujalled, D., Strasser, A., & Liddell, J. R. (2021). Molecular mechanisms of cell death in neurological diseases. *Cell Death and Differentiation, 28*, 2029–2044.
84. Murphy, K. M., Ranganathan, V., Farnsworth, M. L., Kavallaris, M., & Lock, R. B. (2000). Bcl-2 inhibits Bax translocation from cytosol to mitochondria during drug-induced apoptosis of human tumor cells. *Cell Death & Differentiation, 7*, 102–111.
85. Nakajima, A., Komazawa-Sakon, S., Takekawa, M., Sasazuki, T., Yeh, W. C., Yagita, H., Okumura, K., & Nakano, H. (2006). An Antiapoptotic protein, C-Flipl, directly binds to Mkk7 and inhibits the Jnk pathway. *The EMBO Journal, 25*, 5549–5559.
86. Newton, K., Strasser, A., Kayagaki, N., & Dixit, V. M. (2024). Cell death. *Cell, 187*, 235–256.
87. Nguyen, T. T., Wei, S., Nguyen, T. H., Jo, Y., Zhang, Y., Park, W., Gariani, K., Oh, C. M., Kim, H. H., Ha, K. T., Park, K. S., Park, R., Lee, I. K., Shong, M., Houtkooper, R. H., & Ryu, D. (2023). Mitochondria-associated programmed cell death as a therapeutic target for age-related disease. *Experimental & Molecular Medicine, 55*, 1595–1619.
88. Ning, X., Wang, Y., Jing, M., Sha, M., Lv, M., Gao, P., Zhang, R., Huang, X., Feng, J. M., & Jiang, Z. (2019). Apoptotic caspases suppress type I interferon production via the cleavage of Cgas, Mavs, and Irf3. *Molecular Cell, 74*, 19–31.E7.
89. O'Neill, K. L., Huang, K., Zhang, J., Chen, Y., & Luo, X. (2016). Inactivation of Prosurvival Bcl-2 proteins activates Bax/Bak through the outer mitochondrial membrane. *Genes & Development, 30*, 973–988.
90. Oltvai, Z. N., Milliman, C. L., & Korsmeyer, S. J. (1993). Bcl-2 heterodimerizes in vivo with a conserved homolog, Bax, that accelerates programmed cell death. *Cell, 74*, 609–619.
91. Orning, P., Weng, D., Starheim, K., Ratner, D., Best, Z., Lee, B., Brooks, A., Xia, S., Wu, H., Kelliher, M. A., Berger, S. B., Gough, P. J., Bertin, J., Proulx, M. M., Goguen, J. D., Kayagaki, N., Fitzgerald, K. A., & Lien, E. (2018). Pathogen blockade of Tak1 triggers Caspase-8-dependent cleavage of Gasdermin D and cell death. *Science, 362*, 1064–1069.
92. Perini, G. F., Ribeiro, G. N., Pinto Neto, J. V., Campos, L. T., & Hamerschlak, N. (2018). Bcl-2 as therapeutic target for hematological malignancies. *Journal of Hematology & Oncology, 11*, 65.
93. Piao, X., Komazawa-Sakon, S., Nishina, T., Koike, M., Piao, J. H., Ehlken, H., Kurihara, H., Hara, M., Van Rooijen, N., Schutz, G., Ohmuraya, M., Uchiyama, Y., Yagita, H., Okumura, K., He, Y. W., & Nakano, H. (2012). C-Flip maintains tissue homeostasis by preventing apoptosis and programmed necrosis. *Science Signaling, 5*, Ra93.
94. Pourmal, S., Truong, M. E., Johnson, M. C., Yang, Y., Zhou, L., Alegre, K., Stowe, I. B., Gupta, S., Chen, P. A., Zhang, Y., Rohou, A., Newton, K., Kayagaki, N., Dixit, V. M., &

Deshpande, I. (2025). Autoinhibition of dimeric Ninj1 prevents plasma membrane rupture. *Nature, 637*, 446–452.

95. Ramos, S., Hartenian, E., Santos, J. C., Walch, P., & Broz, P. (2024). Ninj1 induces plasma membrane rupture and release of damage-associated molecular pattern molecules during Ferroptosis. *The EMBO Journal, 43*, 1164–1186.
96. Ren, D., Kim, H., Tu, H. C., Westergard, T. D., Fisher, J. K., Rubens, J. A., Korsmeyer, S. J., Hsieh, J. J., & Cheng, E. H. (2009). The Vdac2-Bak rheostat controls Thymocyte survival. *Science Signaling, 2*, Ra48.
97. Riley, J. S., Quarato, G., Cloix, C., Lopez, J., O'Prey, J., Pearson, M., Chapman, J., Sesaki, H., Carlin, L. M., Passos, J. F., Wheeler, A. P., Oberst, A., Ryan, K. M., & Tait, S. W. (2018). Mitochondrial inner membrane permeabilisation enables Mtdna release during apoptosis. *The EMBO Journal, 37*, E99238.
98. Roberts, A. W., & Huang, D. (2017). Targeting Bcl2 with Bh3 Mimetics: Basic science and clinical application of Venetoclax in chronic lymphocytic Leukemia and related B cell malignancies. *Clinical Pharmacology and Therapeutics, 101*, 89–98.
99. Roberts, A. W., Davids, M. S., Pagel, J. M., Kahl, B. S., Puvvada, S. D., Gerecitano, J. F., Kipps, T. J., Anderson, M. A., Brown, J. R., Gressick, L., Wong, S., Dunbar, M., Zhu, M., Desai, M. B., Cerri, E., Heitner Enschede, S., Humerickhouse, R. A., Wierda, W. G., & Seymour, J. F. (2016). Targeting Bcl2 with Venetoclax in relapsed chronic lymphocytic Leukemia. *The New England Journal of Medicine, 374*, 311–322.
100. Rogers, C., Fernandes-Alnemri, T., Mayes, L., Alnemri, D., Cingolani, G., & Alnemri, E. S. (2017). Cleavage of Dfna5 by Caspase-3 during apoptosis mediates progression to secondary necrotic/Pyroptotic cell death. *Nature Communications, 8*, 14128.
101. Rogers, C., Erkes, D. A., Nardone, A., Aplin, A. E., Fernandes-Alnemri, T., & Alnemri, E. S. (2019). Gasdermin pores Permeabilize mitochondria to augment Caspase-3 activation during apoptosis and Inflammasome activation. *Nature Communications, 10*, 1689.
102. Rongvaux, A., Jackson, R., Harman, C. C., Li, T., West, A. P., De Zoete, M. R., Wu, Y., Yordy, B., Lakhani, S. A., Kuan, C. Y., Taniguchi, T., Shadel, G. S., Chen, Z. J., Iwasaki, A., & Flavell, R. A. (2014). Apoptotic caspases prevent the induction of type I interferons by mitochondrial DNA. *Cell, 159*, 1563–1577.
103. Safa, A. R. (2012). C-Flip, a master anti-apoptotic regulator. *Experimental Oncology, 34*, 176–184.
104. Salvador-Gallego, R., Mund, M., Cosentino, K., Schneider, J., Unsay, J., Schraermeyer, U., Engelhardt, J., Ries, J., & García-Sáez, A. J. (2016). Bax assembly into rings and arcs in apoptotic mitochondria is linked to membrane pores. *The EMBO Journal, 35*, 389–401.
105. Sarosiek, K. A., Fraser, C., Muthalagu, N., Bhola, P. D., Chang, W., Mcbrayer, S. K., Cantlon, A., Fisch, S., Golomb-Mello, G., Ryan, J. A., Deng, J., Jian, B., Corbett, C., Goldenberg, M., Madsen, J. R., Liao, R., Walsh, D., Sedivy, J., Murphy, D. J., Carrasco, D. R., Robinson, S., Moslehi, J., & Letai, A. (2017). Developmental regulation of mitochondrial apoptosis by C-Myc governs age- and tissue-specific sensitivity to cancer therapeutics. *Cancer Cell, 31*, 142–156.
106. Schulman, J. J., Szczesniak, L. M., Bunker, E. N., Nelson, H. A., Roe, M. W., Wagner, L. E., 2nd, Yule, D. I., & Wojcikiewicz, R. J. H. (2019). Bok regulates mitochondrial fusion and morphology. *Cell Death and Differentiation, 26*, 2682–2694.
107. Shamas-Din, A., Kale, J., Leber, B., & Andrews, D. W. (2013). Mechanisms of action of Bcl-2 family proteins. *Cold Spring Harbor Perspectives in Biology, 5*, A008714.
108. Shi, J., Zhao, Y., Wang, K., Shi, X., Wang, Y., Huang, H., Zhuang, Y., Cai, T., Wang, F., & Shao, F. (2015). Cleavage of Gsdmd by inflammatory caspases determines Pyroptotic cell death. *Nature, 526*, 660–665.
109. Simmons, M. J., Fan, G., Zong, W. X., Degenhardt, K., White, E., & Gélinas, C. (2008). Bfl-1/A1 functions, similar to mcl-1, as a selective Tbid and Bak antagonist. *Oncogene, 27*, 1421–1428.

110. Singh, R., Letai, A., & Sarosiek, K. (2019). Regulation of apoptosis in health and disease: The balancing act of Bcl-2 family proteins. *Nature Reviews. Molecular Cell Biology, 20*, 175–193.
111. Spitz, A. Z., & Gavathiotis, E. (2022). Physiological and pharmacological modulation of Bax. *Trends in Pharmacological Sciences, 43*, 206–220.
112. Srivastava, R., Cao, Z., Nedeva, C., Naim, S., Bachmann, D., Rabachini, T., Gangoda, L., Shahi, S., Glab, J., Menassa, J., Osellame, L., Nelson, T., Fernandez-Marrero, Y., Brown, F., Wei, A., Ke, F., O'Reilly, L., Doerflinger, M., Allison, C., Kueh, A., Ramsay, R., Smith, B. J., Mathivanan, S., Kaufmann, T., & Puthalakath, H. (2019). Bcl-2 family protein Bok is a positive regulator of uridine metabolism in mammals. *Proceedings of the National Academy of Sciences of the United States of America, 116*, 15469–15474.
113. Strasser, A., & Vaux, D. L. (2020). Cell death in the origin and treatment of cancer. *Molecular Cell, 78*, 1045–1054.
114. Stutz, M. D., Allison, C. C., Ojaimi, S., Preston, S. P., Doerflinger, M., Arandjelovic, P., Whitehead, L., Bader, S. M., Batey, D., Asselin-Labat, M. L., Herold, M. J., Strasser, A., West, N. P., & Pellegrini, M. (2021). Macrophage and neutrophil death programs differentially confer resistance to tuberculosis. *Immunity, 54*(1758–1771), E7.
115. Sun, G., Guzman, E., Balasanyan, V., Conner, C. M., Wong, K., Zhou, H. R., Kosik, K. S., & Montell, D. J. (2017). A molecular signature for Anastasis, recovery from the brink of apoptotic cell death. *The Journal of Cell Biology, 216*, 3355–3368.
116. Suzuki, M., Youle, R. J., & Tjandra, N. (2000). Structure of Bax: Coregulation of dimer formation and intracellular localization. *Cell, 103*, 645–654.
117. Tang, H. L., Tang, H. M., Mak, K. H., Hu, S., Wang, S. S., Wong, K. M., Wong, C. S., Wu, H. Y., Law, H. T., Liu, K., Talbot, C. C., Jr., Lau, W. K., Montell, D. J., & Fung, M. C. (2012). Cell survival, Dna damage, and oncogenic transformation after a transient and reversible apoptotic response. *Molecular Biology of the Cell, 23*, 2240–2252.
118. Van Delft, M. F., Chappaz, S., Khakham, Y., Bui, C. T., Debrincat, M. A., Lowes, K. N., Brouwer, J. M., Grohmann, C., Sharp, P. P., Dagley, L. F., Li, L., Mcarthur, K., Luo, M. X., Chin, H. S., Fairlie, W. D., Lee, E. F., Segal, D., Duflocq, S., Lessene, R., Bernard, S., Peilleron, L., Nguyen, T., Miles, C., Wan, S. S., Lane, R. M., Wardak, A., Lackovic, K., Colman, P. M., Sandow, J. J., Webb, A. I., Czabotar, P. E., Dewson, G., Watson, K. G., Huang, D. C. S., Lessene, G., & Kile, B. T. (2019). A small molecule interacts with Vdac2 to block mouse Bak-driven apoptosis. *Nature Chemical Biology, 15*, 1057–1066.
119. Van Opdenbosch, N., & Lamkanfi, M. (2019). Caspases in cell death, inflammation, and disease. *Immunity, 50*, 1352–1364.
120. Vaux, D. L., & Korsmeyer, S. J. (1999). Cell death in development. *Cell, 96*, 245–254.
121. Vaux, D. L., Cory, S., & Adams, J. M. (1988). Bcl-2 gene promotes Haemopoietic cell survival and cooperates with C-Myc to immortalize pre-B cells. *Nature, 335*, 440–442.
122. Verhagen, A. M., Ekert, P. G., Pakusch, M., Silke, J., Connolly, L. M., Reid, G. E., Moritz, R. L., Simpson, R. J., & Vaux, D. L. (2000). Identification of Diablo, a mammalian protein that promotes apoptosis by binding to and antagonizing Iap proteins. *Cell, 102*, 43–53.
123. Wajant, H., Pfizenmaier, K., & Scheurich, P. (2003). Tumor necrosis factor signaling. *Cell Death and Differentiation, 10*, 45–65.
124. Wang, Y., Gao, W., Shi, X., Ding, J., Liu, W., He, H., Wang, K., & Shao, F. (2017). Chemotherapy drugs induce Pyroptosis through Caspase-3 cleavage of a Gasdermin. *Nature, 547*, 99–103.
125. Wang, Q., Wang, Y., Ding, J., Wang, C., Zhou, X., Gao, W., Huang, H., Shao, F., & Liu, Z. (2020). A bioorthogonal system reveals antitumour immune function of pyroptosis. *Nature, 579*, 421–426.
126. Wei, M. C., Zong, W. X., Cheng, E. H. Y., Lindsten, T., Panoutsakopoulou, V., Ross, A. J., Roth, K. A., Macgregor, G. R., Thompson, C. B., & Korsmeyer, S. J. (2001). Proapoptotic Bax and Bak: A requisite gateway to mitochondrial dysfunction and death. *Science, 292*, 727–730.

127. Weinlich, R., & Green, D. R. (2014). The two faces of receptor interacting protein Kinase-1. *Molecular Cell, 56*, 469–480.
128. Weiss, L. M., Warnke, R. A., Sklar, J., & Cleary, M. L. (1987). Molecular analysis of the T(14;18) chromosomal translocation in malignant lymphomas. *The New England Journal of Medicine, 317*, 1185–1189.
129. White, M. J., Mcarthur, K., Metcalf, D., Lane, R. M., Cambier, J. C., Herold, M. J., Van Delft, M. F., Bedoui, S., Lessene, G., Ritchie, M. E., Huang, D. C., & Kile, B. T. (2014). Apoptotic caspases suppress Mtdna-induced sting-mediated type I Ifn production. *Cell, 159*, 1549–1562.
130. Willis, S. N., Chen, L., Dewson, G., Wei, A., Naik, E., Fletcher, J. I., Adams, J. M., & Huang, D. C. (2005). Proapoptotic Bak is sequestered by mcl-1 and Bcl-xl, but not Bcl-2, until displaced by Bh3-only proteins. *Genes & Development, 19*, 1294–1305.
131. Yuan, Z., Dewson, G., Czabotar, P. E., & Birkinshaw, R. W. (2021). Vdac2 and the Bcl-2 family of proteins. *Biochemical Society Transactions, 49*, 2787–2795.
132. Zamaraev, A. V., Kopeina, G. S., Zhivotovsky, B., & Lavrik, I. N. (2015). Cell death controlling complexes and their potential therapeutic role. *Cellular and Molecular Life Sciences, 72*, 505–517.
133. Zhang, Z., Zhang, Y., Xia, S., Kong, Q., Li, S., Liu, X., Junqueira, C., Meza-Sosa, K. F., Mok, T. M. Y., Ansara, J., Sengupta, S., Yao, Y., Wu, H., & Lieberman, J. (2020). Gasdermin E suppresses tumour growth by activating anti-tumour immunity. *Nature, 579*, 415–420.
134. Zou, H., Henzel, W. J., Liu, X., Lutschg, A., & Wang, X. (1997). Apaf-1, a human protein homologous to C. Elegans Ced-4, participates in cytochrome C-dependent activation of Caspase-3. *Cell, 90*, 405–413.

Efferocytosis in Homeostasis

Christian Maueröder, Caitlin M. Gillis,
and Kodi S. Ravichandran

Abstract

Apoptotic cells are removed by professional and nonprofessional phagocytes via the process of "efferocytosis." Efferocytosis is highly efficient such that accumulation of dead cells is rarely seen in healthy tissues, even though our bodies remove 200–400 billion apoptotic cells daily. Besides clearing apoptotic cells, efferocytosis also has many beneficial roles, including secretion of anti-inflammatory mediators and promotion of local tissue repair. Thus, impairments in efferocytosis interfere with normal function in tissues. In this review, we will focus on how efferocytosis supports the return to homeostasis following inflammation and injury. Further, we highlight maladaptive responses when efferocytosis is disrupted, contributing to autoimmunity and fibrosis.

Keywords

Apoptosis · Efferocytosis · Homeostasis · Dead cell clearance in the testes · Mammary involution · SLE · Immune tolerance · Asthma · Lung inflammation · Fibrosis · Tissue repair

C. Maueröder · C. M. Gillis
VIB-UGent Center for Inflammation Research, Ghent University, Ghent, Belgium

Department of Biomedical Molecular Biology, Ghent University, Ghent, Belgium
e-mail: christian.maueroder@irc.vib-ugent.be; caitlin.gillis@irc.vib-ugent.be

K. S. Ravichandran (✉)
VIB-UGent Center for Inflammation Research, Ghent University, Ghent, Belgium

Department of Biomedical Molecular Biology, Ghent University, Ghent, Belgium

Division of Immunobiology, Department of Pathology and Immunology, Washington University School of Medicine, St. Louis, MO, USA
e-mail: kodi@wustl.edu

C. D. Gregory et al. (eds.), *Apoptosis and Other forms of Cell Death in the Regulation of Inflammation in Health and Disease*, Advances in Experimental Medicine and Biology 1481, https://doi.org/10.1007/978-3-031-92785-0_2

Abbreviations

MDM	monocyte-derived macrophages
BMDM	bone-marrow-derived macrophages
mTEC	medullary thymic epithelial cells
OVA	ovalbumin
SLE	systemic lupus erythematosus
CHEF	chimeric efferocytosis receptor
T_{reg}	T-regulatory lymphocyte
TLR	Toll-like receptor
ANA	autoantibodies against nuclear antigens
COPD	chronic obstructive pulmonary disease
HDM	house dust mite
BM-MSC	bone-marrow-derived mesenchyma
LE	lupus erythematosus

Efferocytosis, which is the clearance of apoptotic cells by professional and non-professional phagocytes, is a basic feature of any tissue. Even in the absence of an inflammatory trigger, 200-400 billion apoptotic cells are turned over on average in a healthy human adult [1–4]. The role of apoptotic cell clearance in homeostasis can roughly be divided into three different layers: First, prompt removal of apoptotic cells prevents these dying cells from lingering around in the tissue. Remnants of dead cells are likely to interfere with homeostatic function, either simply by occupying space that is needed for new cells or through the release of danger-associated molecular patterns that stimulate inflammatory responses. Second, efferocytosis is also involved in more qualitative aspects, i.e., how the "apoptotic signal" is communicated within a tissue. Both the apoptotic cells themselves and the phagocytes, after they engulf them, release anti-inflammatory mediators, including metabolites that promote apoptosis-induced compensatory proliferation, regeneration, or the polarization of macrophages, and thereby impart an anti-inflammatory tone within the tissue [5–8]; in fact, one could even speculate that efferocytosis is part of the core signature of tissue-resident macrophages [9]. Third, there are systemic responses that go beyond the tissue: for example, apoptotic cells are a source of autoantigens and the presentation of these through tolerogenic dendritic cells can contribute to peripheral tolerance [10]. Similarly, apoptotic cell-derived vesicles may also carry messages outside of the tissue [11].

Efferocytosis in Tissue Homeostasis

Efferocytosis plays a major role in maintaining tissue integrity, both under steady state and during remodeling. Similarly, efferocytosis is implicated in embryogenesis; however, most studies have rather looked at how cell death shapes the embryo instead of how these dying cells are removed. Prominent examples for this are neuronal cell death in the developing fly or the removal of the interdigital space during shaping of the fingers. Only recent work focused on the clearance aspect of these processes [12–14]. Therefore, in this review, we will mainly discuss the role of efferocytosis in adult tissues, both under homeostasis and during physiological remodeling. Efferocytosis comes with remarkable efficiency, and most studies have focused on tissues with a high homeostatic turnover of cells and limited capacity for luminal shedding. Such tissues are the germinal center, the thymus, or the testis. Here, the rate of cell death is rather close to the phagocytic capacity of the tissue, so even minor disturbances are expected to tip the balance.

Efferocytosis in Immunological Compartments

During development, B and T cells go through several checkpoints to prevent nonfunctional or autoreactive cells from being released into the periphery. During this process, a high rate of cell death is expected to occur in the bone marrow and in the thymus [15, 16]. Few studies exist that focus on the clearance of apoptotic cells in the bone marrow, in part because of the challenges in isolating/imaging the "differentiated macrophages" that are adherent in the bone marrow (these should be distinguished from the ex vivo differentiation of progenitors from the bone marrow, the so-called bone-marrow-derived macrophages). The thymus also has a remarkable capacity to silently dispose of apoptotic cells [17]. Here, new T cells are constantly produced, yet only few are allowed to exit the thymus, with large numbers undergoing apoptosis. The macrophages in this compartment have a unique metabolic adaption and by deploying the pentose phosphate pathway, they cope with efferocytosis-associated ROS-production and maintain a high efferocytic capacity [18]. If the rate of apoptosis is further increased, for example, by administration of dexamethasone that causes death of immature thymocytes, thymic macrophages engulf prolifically and can undergo proliferation to manage the increased apoptotic load [19]. High rates of apoptosis bring the necessity of consecutive uptake of several apoptotic thymocytes by individual macrophages. Thereby, engulfment of multiple corpses has been linked to mitochondrial changes, such as mitochondrial fission, which mediates cytosolic calcium increase necessary for phagosome formation around secondary corpses, as well as the mitochondrial uncoupling protein UCP2 [20, 21].

B cells home to areas referred to as follicles in the cortical region of lymph nodes. In these regions, resident macrophages are also present, and some show vacuolar content, suggesting a baseline level of apoptosis and efferocytosis [22]. After immunization, germinal centers form, and the rate of apoptosis increases due to many B cells being eliminated during affinity maturation. Macrophages that locate to the germinal center become stationary and start to engulf apoptotic B cells. Interestingly, macrophages do not migrate toward apoptotic cells, but the movement is predominantly coming from the apoptotic cells, which is in line with data from the developing brain in zebrafish, where the movement of the apoptotic cells toward the macrophages, rather than the movement of the macrophage, mediates their clearance [23]. Intravital imaging of follicles further suggests that tingible body macrophages do not engulf whole apoptotic cells, but rather take up small pieces of dying cells. It is estimated that under these conditions, every macrophage needs to engulf 30–40 apoptotic cells per hour [22]. Efferocytosis in the follicle is not restriced to macrophages, as also marginal reticular cells can contribute to the removal of apoptotic B cells [24]. The concept that both nonprofessional and professional phagocytes participate in the clearance of apoptotic cells is not restricted to the germinal center. For example, in the lung, engulfing alveolar macrophages secrete IGF-1, which acts on nonprofessional phagocytes to reroute uptake such that the epithelial cells engulf small particles, whereas whole apoptotic cells are engulfed by alveolar macrophages [25].

Homeostatic Efferocytosis in the Testis and Its Implications

It is estimated that during spermatogenesis, approximately 75% of the spermatogonia undergo apoptosis [26]. However, despite the lack of macrophages within the seminiferous tubules, no accumulation of dead cells occurs, which is due to the profound capacity of Sertoli cells to engulf apoptotic cells [27]. Sertoli cells are located on the basal membrane of the seminiferous tubules, where spermatogenesis occurs, and they surround adjacent germ cells. Tight junctions between Sertoli cells also impart the blood-testis barrier that separates the basal compartment from the adluminal compartment [28]. The male spermatogonial stem cells are located basally with their progeny gradually maturing on their journey toward the luminal space [28]. The profound phagocytic capacity of Sertoli cells can especially be seen in seasonal breeders, such as the large hairy armadillo and the Syrian hamster [29, 30], which show pronounced testicular regression through sperm cell apoptosis.

Evidence for Sertoli cell-mediated clearance of apoptotic germ cells has come from several lines of investigation. Mice lacking all three TAM-family receptors TYRO3, AXL, and MERTK render male mice infertile, with their Sertoli cells showing drastically decreased capacity to engulf apoptotic germ cells [31, 32]. Interestingly, the lack of two of these receptors, such as TYRO3 and MERTK, or AXL and MERTK, is sufficient to cause impairments in spermatogenesis [33]. This suggests a dominant role for MERTK, and indeed, single knockout of MERTK causes a progressive decrease in sperm count, which only becomes apparent in older mice [34]. Sertoli cells lacking the TAM receptors also respond to engulfment of apoptotic germ cells with inflammatory cytokine production and recruitment of immune cells [35]. Further, DAMPs released by uncleared apoptotic germ cells can stimulate inflammatory cytokine production by Sertoli cells [36]. Other receptors also contribute to the removal of apoptotic germ cells in the testes. The class B scavenger receptor type I (SRBI) is implicated in efferocytosis by Sertoli cells [37, 38], and injection of inhibitors of SRBI into the seminiferous tubules impairs clearance of apoptotic germ cells [39, 40]. Signaling occurs through a pathway similar to the one deployed by the phosphatidylserine receptor STAB2 and involves GULP,

MAPK, and RAC1 [41]. It is worth speculating whether SRBI also signals through the ELMO1-DOCK180 axis, which is also used by STAB2, as mice lacking ELMO1 show impairments in spermatogenesis as well [42]. Interestingly, the phosphatidylserine receptor BAI1, which signals through ELMO1 and DOCK180, is highly expressed in the testis, further pointing to efferocytosis as an important aspect of Sertoli cell biology. It is unclear at this point whether these different receptors are simply adding robustness to a system requiring fast kinetics or may contribute differently at the signaling stage (to interpret engagement with apoptotic germ cells) or may participate differently in clearing germ cells at different stages of maturation.

Efferocytosis in the testes at the homeostasis level also provides an excellent example of crosstalk within the tissue and how clearing dying cells can impact multiple other aspects of a tissue. Dong and colleagues [43, 44] found that the TXLNA, a member of the syntaxin family, is upregulated in Sertoli cells upon engulfment of apoptotic germ cells and that recovery from testicular heat stress is impaired in mice lacking TXLNA. However, TXLNA did not directly influence the uptake of apoptotic germ cells, but it rather stabilizes ABCA1 and prevents pathological accumulation of lipid droplets in Sertoli cells [43]. This is striking, as mice lacking ABCA1 show accumulation of lipid droplets in Sertoli cells and depletion of the lipid content in Leydig cells, resulting in decreased testosterone production and impaired fertility [45]. Thus, uptake of apoptotic cells could potentially serve as a quality control such that apoptotic germ cell removal and the subsequent response of Sertoli cells functions correctly, cholesterol is provided to Leydig cells to sustain testosterone production and fertility in the long run.

Further, efferocytosis seems to be critical for the generation of a microenvironment that favors spermatogenesis: SMTLN2 in Sertoli cells orchestrates actin assembly by recruitment of the actin nucleator ARP3, with these cytoskeletal rearrangements necessary for both phagocytosis and the maintenance of the blood-testis barrier [46]. Further, uptake with involvement of SMTLN2 favors lactate production in Sertoli cells that serves as an energy source for developing sperm cells. SPTBN2 is a member of the spectrin family of proteins that form a meshwork at the plasma membrane, thereby retaining several proteins such as transporters and receptors. Knockout of SPTBN2 resulted in impaired glycolysis and reduced capacity to engulf apoptotic germ cells [47]. Thus, glycolysis is necessary for efferocytosis by Sertoli cells and is further enhanced during efferocytosis as highlighted by upregulation of the glycolytic enzyme ENO1 [48]. Panneerdoss and colleagues revealed that expression of DOCK180, a guanosine exchange factor that forms a functional complex with ELMO1 (which has been genetically shown to be relevant for Sertoli cell-mediated clearance of apoptotic germ cells), is negatively regulated by *miR-471-5p*, a miRNA, that is induced by low levels of testosterone [49]. Overexpression of *miR-471-5p* in the testis resulted in disruption of LC3-associated phagocytosis and impaired spermatogenesis.

Efferocytosis in the testes may also help coordinate some of the peculiar metabolic adaptations found in this organ. Sertoli cells meet their own energy requirements predominantly through fatty acid metabolism yet show a distinct glycolytic activity to provide lactate to developing sperm cells [50]. As mentioned above, uptake of apoptotic germ cells through mechanisms involving SMTLN2 or STBN2 enforces lactate production [46, 47]. On the contrary, Sertoli cells rely on using fatty acids derived from apoptotic cargo to produce ATP [51]. This moves mitochondria, the site of beta oxidation, into the center of attention, and indeed work by Gong and colleagues highlights the role of mitochondrial fission in the clearance of apoptotic germ cells by Sertoli cells [52]. Here, inactivation of GSK3α during phagocytosis leads to activation of DRP1, which has previously been shown to mediate the mitochondrial changes that allow continued uptake of apoptotic cells by macrophages [20]. This is especially interesting in the context of more recent research, suggesting that fragmented mitochondria are hotspots of beta oxidation [53].

Other findings also point to the fact that it is not just the removal that matters, but also which engulfment pathways are engaged. For example, overexpression of BAI1 in the testis rescued the deficiency for MERTK, yet fails to counteract the degenerative changes in the eye caused by loss of MERTK [54]. Further, Sertoli cells engulfing apoptotic germ cells in the absence of the TAM receptors react similarly to Sertoli cells that have encountered DAMPs derived from secondary necrotic cells. Such changes can lead to various maladaptive changes, reaching from the breakdown of immune homeostasis to the loss of tolerance to sperm-derived antigens and the development of autoimmune orchitis [55]. Although the above discussion largely centered on the receptors on Sertoli cells, the display of "eat-me" signals on apoptotic germ cells is also essential for proper removal. XKR8 is a scramblase involved in caspase-dependent externalization of phosphatidylserine during apoptosis, and genetic depletion of XKR8 renders male mice infertile [56]. Similarly, injection of ANXA5, which masks phosphatidylserine displayed by apoptotic cells, into the seminiferous tube leads to delayed clearance of apoptotic germ cells [57]. There is also an interesting aspect of phosphatidylserine on fully mature live sperm cells, with classical efferocytic receptors on oocytes binding to phosphatidylserine on the sperm and thereby promoting the fusion of sperm and the oocyte [58].

Efferocytosis in the Mammary Gland

Post-lactational involution causes massive apoptosis within the mammary glands. Failure to properly clear these apoptotic cells, such as in mice lacking the engulfment receptor MERTK, shows prolonged presence of milk in the ductal lumen and a higher burden of apoptotic cells. These changes are accompanied by increased proliferation of epithelial cells, more periductal collagen deposition, and increased vascularization of the mammary gland. Strikingly, these pathological changes prevent lactation following a second pregnancy [59]. Several other findings support the involvement of efferocytosis in post-lactational involution. Mice lacking MFGE8, a bridging molecule that links apoptotic cells with their cognate receptor on the phagocyte, display the accumulation of phosphatidylserine exposing large milk fat globules and signs of mastitis. Again, redevelopment of the mammary gland following a secondary pregnancy was impaired [60]. More indirect evidence for the role of apoptotic cell clearance in post-lactational involution exists for plasminogen: plasminogen gets activated by apoptotic cells and induces their phagocytic removal [61]. Interestingly, mice lacking plasminogen show abnormal development of the mammary gland and impaired post-lactational involution, which might be linked to the involvement of plasminogen in apoptotic cell clearance [62]. The remodeling occurring during mammary gland involution could come with a downside, as breast cancer diagnosed 2–5 years postpartum is correlated with poorer prognosis: using spontaneous and allograft models, Stanford and colleagues found that cell death was not restricted to mammary epithelial cells, but also tumor cells, and that MERTK-mediated efferocytosis contributed to establishing a pro-metastatic niche [63].

Efferocytosis in Maintaining Self-Tolerance

Multiple studies suggest that uptake of apoptotic cells polarizes engulfing phagocytes. The overwhelming majority of these studies have been performed on macrophages, as these are considered to be most prolific eaters [64]. Macrophages rapidly remove and degrade apoptotic cells, thereby obtaining a pro-resolving, immunosuppressive phenotype characterized by the secretion of IL-10 and TGF-β [65, 66]. Since cargo processing of macrophages is rather degradative, macrophages are considered to only sporadically present antigen and they contribute to tolerance primarily by antigen removal and generalized immune suppression, either locally or after migration to draining lymph nodes, as has been reported for "satiated" macrophages [64, 67, 68].

In contrast, dendritic cells can also take up apoptotic cells and present apoptotic cell-derived antigen in a tolerogenic manner. Most of these studies have administered apoptotic cells to dendritic cells and only few have assessed sampling of apoptotic cells by dendritic cells during true tissue homeostasis (Fig. 1).

A hallmark study from Liu and colleagues reported that ovalbumin (OVA)-loaded TAP-knockout splenocytes are taken up by CD8-positive dendritic cells, which induce short-lived antigen-specific expansion of OT-I cells followed by long-term tolerance via deletion [69]. In the context of cell death, this study comes with the limitation that cell death was not assessed but assumed to occur due to the osmotic loading process. However, the use of TAP-knockout cells implied that antigen presentation by the administered splenocytes themselves or via trogocytic transfer of MHC class I molecules to recipient dendritic cells is likely not the mechanism for tolerance induction [69].

Desch and colleagues investigated the uptake of instilled CFSE-labeled apoptotic thymocytes by pulmonary dendritic cells [70]. CFSE signal was restricted to the $CD103^+$ migratory dendritic cell compartment, and after engulfment, these cells were predominantly found in the T lymphocyte compartment of the draining lymph nodes. This pattern was not changed when LPS was co-administered. Similar to the study of Liu and colleagues, administration of thymocytes labeled with OVA resulted in antigen-specific CD8-positive T lymphocyte expansion 3 days later [69,

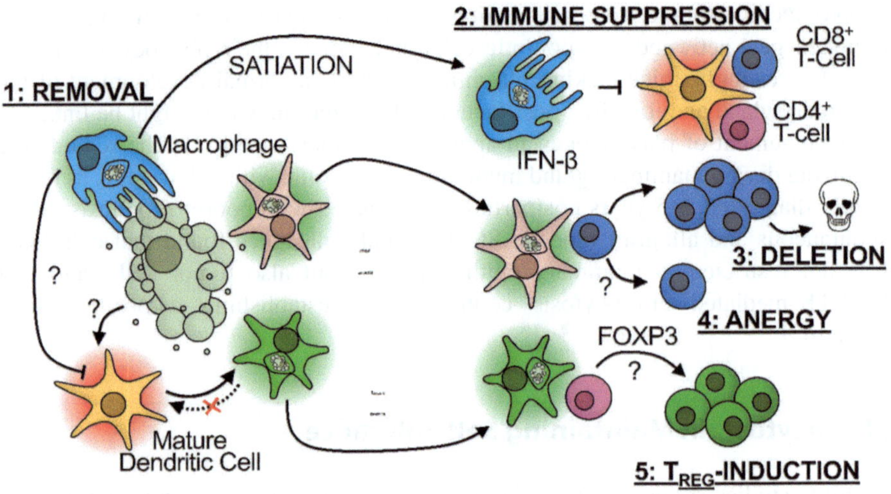

Fig. 1 The role of efferocytosis in maintaining tolerance: (1) By removing and degrading apoptotic cells, presentation by immunogenic cell types is prevented. (2) Localized or systemic immune suppression by factors released from apoptotic cells and phagocytes that have engulfed apoptotic cells. (3) Presentation of antigens by dendritic cells may lead to short-term proliferation and activation-induced cell death. (4) Similarly, presentation of apoptotic-cell-derived antigen might directly induce anergy. (5) Further evidence suggests that the uptake of apoptotic cells by tolerogenic dendritic cells can lead to the induction of T regulatory cells

70]. The authors did not investigate whether this response resulted in tolerance by deletion or prolonged antigen reactivity.

Assessing the functions of efferocytic dendritic cells in situ is technically challenging. Huang and colleagues reported that CD4-negative/OX41-negative dendritic cells of the intestine take up apoptotic cells and migrate to the T lymphocyte area of lymph nodes [71, 72]. Years later, this finding was confirmed using an elegant DTR-system to induce intestinal epithelial cell apoptosis below the threshold that would cause inflammation [10]. In addition to macrophages, migratory $CD103^+$ dendritic cells captured apoptotic cells and trafficked to the mesenchymal lymph node. These dendritic cells showed a transcriptional profile typical for regulatory T lymphocyte (T_{reg}) activation and supported the differentiation of T_{reg} in vitro, however, no antigen-specific responses were evaluated [10]. Recently, Bosteels and colleagues showed that splenic cDC1 engulf apoptotic cells from circulation and undergo a homeostatic maturation program into tolerogenic dendritic cells [73]. Of note, cholesterol uptake was necessary and sufficient to drive homeostatic maturation, and engulfment of apoptotic cells in the absence of LXRα and LXRβ resulted in immunogenic maturation [73]. Since this was also accompanied by decreased survival, it could be related to a maladaptive response toward engulfment, which has also been observed in macrophages and referred to as "unhealthy eating." [74]

Investigating antigen-specific effects of apoptotic cells, Kushwah et al. observed that apoptotic cell uptake by dendritic cells prevented dendritic cell maturation and the administration of dendritic cells having engulfed apoptotic cell can lead to the induction of antigen-specific T_{reg} [75]. However, the apoptotic cells did not serve as the source of the model antigen. Therefore the effects observed could be due to systemic or local immune suppression, which has been observed in several other studies as well.

So far, there is experimental evidence for the induction of tolerance by efferocytic dendritic cells via deletion of autoreactive T lymphocytes [69]. Remarkably, this effect becomes first apparent 10 days after exposure to the apoptotic cells and some studies only investigated short-term T lymphocyte responses, thus seemingly contrasting these finding [70]. Further, it remains unclear whether dendritic cells can also present apoptotic-cell-derived antigens to directly induce T_{reg}, although the DC activation profile suggests this functionality [10]. Another potential mechanism for the induction of tolerance is the localized immune suppression in lymphoid organs: in contrast to macrophages, efferocytic dendritic cells can translocate to the draining lymph nodes and might modulate immune responses directly in the control center of the immune system. A recent study has further shown that apoptotic cell clearance is also involved in central tolerance: medullary thymic epithelial cells (mTEC) express peripheral antigens via action of the transcription factor AIRE. In this manner, autoreactive T cells are eliminated, and T regulatory lymphocytes are selected. Perry and colleagues showed that selection of T regulatory cells significantly relied on the transfer of antigen to CD8-positive dendritic cells. Strikingly, cell surface antigen was transferred in a CD36-dependent manner and relied on apoptosis of mTEC [76]. Such transfer was found in steady state and was necessary for establishing allo-tolerance.

In healthy individuals, apoptotic cells are rapidly removed; however, when efferocytosis is impaired, as in patients with SLE, apoptotic cells may accumulate and progress into secondary necrosis [77]. In such settings, the relative contribution of different cell subsets to dead cell clearance, as well as the functional outcome might change. Indeed, mice deficient for the phosphatidylserine receptor CD300F show an autoimmune phenotype characterized by less efferocytic macrophages and increased uptake of apoptotic cells by dendritic cells with subsequent expansion of T lymphocytes [78].

Rovere and colleagues have shown that increasing the ratio of apoptotic cells to D1 dendritic cells dose-dependently induces secretion of the cytokines IL-1β, IL-10, and TNF-α [79]. Likewise, exposing D1 cells to low numbers of UV-treated cells failed to induce dendritic cell maturation and prevented progression of apoptotic cells into secondary necrosis, whereas D1 cells were not able to properly clear high numbers of cells undergoing apoptosis and induced dendritic cell maturation [80].

Tolerance to Apoptotic Self During Infection

Another scenario that comes with specific challenges toward tolerance is dead cell removal during infection. Here, apoptotic cells and pathogen-infected apoptotic cells are found within the same environment, yet dendritic cells make the right choice and mount the correct immune response. What mechanisms might ensure protective immunity against the pathogen without risking loss of tolerance to self?

Interestingly, several of the deadliest pathogens deploy molecular mimicry and disguise themselves as apoptotic material by expressing phosphatidylserine [81]. Thereby they evade immune responses and ensure non-inflammatory uptake by host cells. In contrast, once a cell is infected, a common strategy of intracellular pathogens is to inhibit apoptosis, suggesting that host cell apoptosis counteracts bacterial survival [82, 83]. For example, the virulence of *Mycobacterium tuberculosis* directly correlates with its ability to prevent host cell apoptosis; apoptotic-cell-derived mycobacterial antigens are presented to CD8-positive T lymphocytes by dendritic cells, and immunization with apoptotic corpses infected with *Mycobacterium tuberculosis* protects animals from developing tuberculosis in a DC-dependent manner [84–87].

A hallmark study from Torchinsky and colleagues highlighted that dendritic cells that had engulfed infected apoptotic cells show a unique phenotype characterized by the secretion of both inflammatory (IL-6, IL-12) and immunomodulatory (TGF-β) mediators [88]. In addition, co-culture of naïve CD4-positive T lymphocytes and dendritic cells fed with either apoptotic cells or infected apoptotic cells induced T_{reg} or T_H17 differentiation, respectively. Strikingly, inhibition of apoptosis during infection with *Citrobacter rodentium* impaired development of a protective T_H17 response [88].

Challenges and Links to Autoimmunity

During infection, dendritic cells find themselves in an inflammatory environment and exposed to pathogens, (infected) apoptotic cells, and bystander host cell death. Influenced by contrasting stimuli, how do phagocytes ensure that they react appropriately to these different targets? In a model of experimental peritonitis, Uderhardt and colleagues have shown that apoptotic cell clearance is limited to resident macrophages and that these cells actively prevent infiltrating monocytes from taking up apoptotic cells [89]. Mechanistically, 12/15-lipoxygenase translocates to the plasma membrane of resident macrophages and catalyzes the oxidation of phosphatidylethanolamine that acts as a sink for MFGE8. As infiltrating monocytes rely on MFGE8 to engulf apoptotic cells, the action of 12/15-lipoxygenase ensures that uptake of apoptotic cells is restricted to resident macrophages and tolerance to self is not broken [89]. In a similar manner, priming of dendritic cells with LPS has been shown to reduce their capacity to engulf apoptotic cells, which could constitute a mechanism to prevent apoptotic cell uptake by dendritic cells with an inflammatory profile [90]. But how do dendritic cells ensure that immune responses against pathogens are mounted but tolerance to self is not broken? This is particularly interesting for infected cells, where self and foreign antigens are taken up together.

Blander and colleagues reported that antigens from apoptotic cells are not presented via MHCII, even if co-administered with bacteria, suggesting that the decision of whether antigen is presented or not is made on the level of individual phagosomes [91]. Indeed, spiking apoptotic cells with LPS, but not exogeneous

LPS administration, lead to the presentation of antigens derived from the apoptotic cargo, supporting a phagosome-autonomous mechanism regulated by Toll-like receptors (TLR). Such a sorting mechanism is likely critical during infection to discriminate between pathogen- and self-derived antigens. However, it remains unclear how autoimmunity is exactly prevented in case of infected cells. It is conceivable that a naïve autoreactive lymphocyte encounters its antigen for the first time during infection and might therefore give rise to an autoimmune response. To prevent such from happening, autoreactive lymphocytes need to either be efficiently depleted prior to infection or tolerance needs to be re-established after infection. In both scenarios, presentation of antigens derived from apoptotic cells by dendritic cells could play a crucial role. Yet, infection has long been discussed as a trigger of autoimmunity [92] and so has the failure to properly clear apoptotic cells [77]. More recently, aberrant T_H17 responses have been identified as a major contributor to several inflammatory disorders and autoimmune diseases [93]. Given the importance of the clearance of infected apoptotic cells for the development of T_H17 responses, it is tempting to speculate that infected apoptotic cells might predispose to autoimmunity (Fig. 2).

The relatively low abundance of autoimmune disease highlights the efficiency to retain tolerance to self. Maintaining tolerance to self during infection is of high importance, yet mounting an immune response against the pathogen might be even more critical for immediate survival. A recent study indicates that efferocytosis mediated by AXL dampens M1 macrophage responses and thereby heart pathology mediated by infection with the parasite *Trypanosoma cruzi* [94]. Similarly, *Encephalitozoon cuniculi* uses apoptotic cells in a "Trojan horse" strategy to downregulate inflammatory responses and allow intracellular proliferation, similar to what has been described for *Streptococcus pneumoniae* and *Leishmania major* [95–97].

SLE: A Prototypic Disease of Clearance Deficiency

Systemic lupus erythematosus (SLE) is an autoimmune disease that usually presents with nonspecific symptoms such as joint pain and fever [98]. The disease can affect numerous organs, the most common being the skin and the kidneys. Although the clinical manifestations vary between individuals, autoantibodies against nuclear antigens (ANA) are shared by virtually all patients [99]. Now we know that the antigens for these autoantibodies are constituents of uncleared apoptotic cells that have progressed into secondary necrosis; and that impaired clearance of apoptotic cells is thought to be a major driver for the etiopathology of SLE [100].

A Historic Perspective on Clearance Deficiency in SLE

In retrospect, several early findings indicated that defective apoptotic clearance is involved in SLE pathogenesis. In the 1940s, lupus erythematosus cells (LE cells) were identified in the bone marrow and blood of patients with SLE [101, 102]

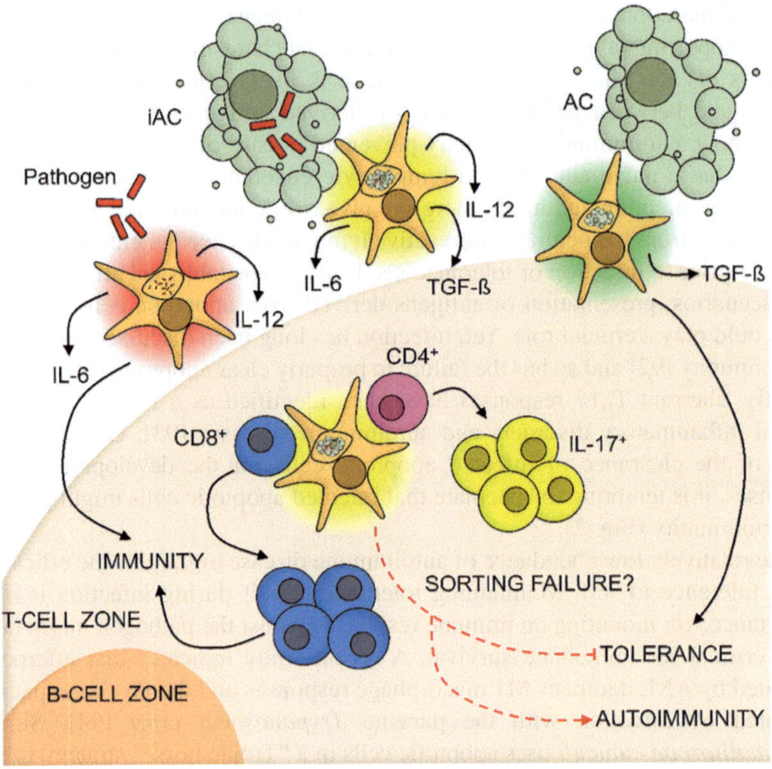

Fig. 2 Dealing with sorting challenges: Intracellular pathogen will cause dendritic cell maturation and presentation of pathogen-derived antigens via MHCI to $CD8^+$ T cells. Similarly, dendritic cells engulfing pathogens will mature and present antigens to $CD4^+$ T cells via MHCII. On the contrary, engulfment of apoptotic cells leads to production of TGF-β, exerting a tolerogenic profile. Interestingly, engulfment of infected apoptotic cells leads to the production of both pro- and anti-inflammatory mediators, which induce T_H17-mediated immunity. Failure to properly sort pathogen-derived and self-antigen might cause autoimmunity, which could explain the link between T_H17 responses and autoimmunity

characterized largely by a lobulated nucleus and a hematoxylin body. Soon after the discovery of LE cells, studies showed that incubation of leukocytes from healthy subjects with serum from patients with SLE can generate LE cells, indicating a role for serum factors [103]. Indeed, these LE cells are granulocytes that have taken up the opsonized remains of apoptotic or secondary necrotic cells via complement- or antibody-mediated phagocytosis. While in a healthy state, one would expect that efficient removal of apoptotic cells would occur before progression into secondary necrosis, in SLE patients this process is impaired.

Similarly, Civatte bodies, that have already been described in 1910, are commonly found in skin biopsies of patients with SLE [104, 105]. They might explain another typical feature of skin sections from SLE patients, i.e., the lupus band, which is an IgG-positive region at the dermo-epidermal junction [105, 106].

Fragmentation of Civatte bodies leads to the accumulation of necrotic debris above the basement membrane [107]. Direct evidence for a clearance deficiency in the skin comes from a study reporting that exposing skin of SLE patients but not healthy subjects to ultraviolet light leads to uncleared apoptotic keratinocytes [108].

Murine Models of Lupus Reflect the Heterogenicity of Human SLE

Some researchers propose that SLE is not a single disease, rather a syndrome with several different etiologies giving rise to similar clinical presentations [109]. It is still debated whether clearance deficiency causes or exacerbates SLE; however, given the disease heterogeneity it is likely that either or both could apply.

This heterogeneity is also reflected in preclinical models of lupus in mice [110]. Female NZBWF1 mice spontaneously develop lupus-like pathology driven by multiple susceptibility loci, including genes that are involved in the clearance of apoptotic cells, such as DEL1, a homologue of MFGE8 [111]. In contrast, the MRL/MpJ-Fas^{lpr} mouse model is predominantly driven by unchecked lymphoproliferation that leads to autoimmune presentations on the MRL background [112]. Accordingly, studies suggest that macrophages from pre-morbid NZBWF1, but not MRL/MpJ-Fas^{lpr} mice, show decreased uptake of apoptotic cells compared to controls, which supports the notion that defective clearance plays a distinct role in disease initiation in the former model, but not the latter [113]. Nonetheless, once autoimmunity is established, both models are characterized by reduced clearance, indicating that impaired efferocytosis does contribute to disease exacerbation [114]. The NZBWF1 model also reflects that similar to what is observed in humans, females are more likely to develop lupus-like disease. Interestingly, transfer of the microbiome of male NZBWF1 into female NZBWF1 mice boosted efferocytic capacity, suggesting that the microbiome may contribute to clearance deficiency [115]. Indeed, recent work by Saavedra et al. links butyrate, produced by the gut microbiome, to the capacity of macrophages to engulf apoptotic cells and treatment of mice with antibiotics exacerbated murine lupus [116].

Cell-Autonomous Factors Contribute to Clearance Deficiency in SLE

In human SLE, several findings highlight an intrinsic efferocytosis deficiency. Monocyte-derived macrophages from SLE patients have an impaired capacity to engulf apoptotic cells, and blood from SLE patients shows an increased percentage of apoptotic cells [117–119]. Mechanistically, monocyte-derived macrophages from SLE patients express lower levels of the bridging molecule MFGE8 and the integrin ITGB5, that together form a functional engulfment receptor [120]. In addition, macrophages from patients with clinically active lupus are marked by downregulation of the bridging molecule GAS6 and the engulfment proteins DOCK1 and ELMO2 [120]. Given that these studies were performed ex vivo in the absence of autologous serum, it is likely that the underlying driver for defective macrophage efferocytosis is genetic or epigenetic. Indeed, genome-wide association studies have linked polymorphisms in the coding region of MFGE8 with SLE [121]. Interestingly, serum concentrations of MFGE8 increase in patients treated with corticosteroids, possibly counteracting the clearance deficiency seen in some patients [122]. Studies in mice have shown that the knockout of a single gene involved in apoptotic cell clearance, such as MERTK, MFGE8, or C1q results in the development of lupus-like disease [123–125]. In humans, analogous to mouse knockout models, a single gene defect can also cause monogenic lupus [126]. For example, more than 90% of patients with C1q deficiency develop SLE [127]. Complement plays a paradoxical role in lupus: on the one hand, it promotes vascular injury, while on the other hand, the deficiency of early complement proteins strongly predisposes to autoimmunity [127]. While the classical pathway is protective in SLE, activation of the alternative pathway may exacerbate disease [128]. This is likely due to the critical role of C1q as an opsonin: both directly enabling apoptotic cell recognition and removal by phagocytes, as well as mediating the removal of immune complexes that would otherwise predispose for glomerulonephritis [125, 129–131]. Dysfunction of the complement regulator Factor H is also found in some patients with SLE [132]. Factor H is internalized by apoptotic cells, thereby stimulating cell surface translocation of iC3b. In this way, apoptotic cells are opsonized in the absence of alternative complement pathway activation [133] and are targeted for removal [133–135].

Another cause of monogenic lupus can be loss of SLC7A7 [136–139], an amino acid transporter with a profound role in metabolism. Interestingly, SLC7A7 is necessary for the continuous clearance of apoptotic cells in zebrafish, once again associating the failure to properly clear apoptotic cells with SLE [140]. Further, the human SLE variant NCF1-R90H results in clearance deficiency in mice [141]. NCF1 is part of the NOX2 complex that mediates phagosomal ROS production and the variant NCF1-R90H has a decreased electrogenic response, leading to lower acidification and defective phagosome maturation. This leads to incorrect processing of the cargo, ultimately causing the upregulation of the M1-like macrophage marker CD40 that activates follicular T lymphocytes [141]. Drugs that impair apoptotic cell clearance, such as quinidine and procainamide, predispose to drug-induced lupus, further linking apoptotic cell clearance and SLE [142].

Clearance Deficiency Can Exacerbate SLE Pathology

As mentioned above, clearance deficiency may also occur secondary to an otherwise instigated break in self-tolerance and/or be exacerbated by the inflammatory changes accompanying SLE (Fig. 3). For example, one major characteristic of patients with SLE is a so-called interferon signature: peripheral blood leukocytes have elevated expression of genes induced by type-I interferons [143]. Type I interferons have been shown to establish mechanoenzyme defects in marginal zone macrophages in mice, thereby allowing follicular entry of autoantigens derived from apoptotic cells and presentation of these to autoreactive B lymphocytes to trigger

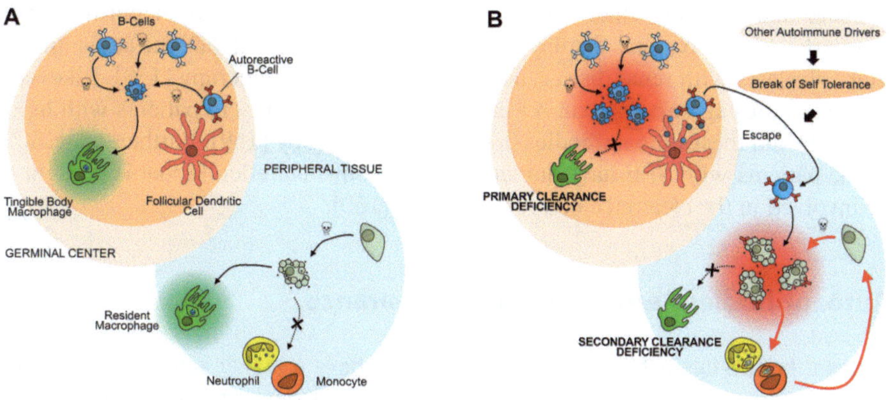

Fig. 3 (**a**) Physiological clearance of apoptotic cells: Within the germinal centers, tingible body macrophages readily engulf B lymphocytes that fail to recognize their cognate antigen following affinity maturation. Follicular dendritic cells do not present any self-antigen. In the periphery, homeostatic apoptosis is taken care of by tissue resident macrophages, thereby preventing the accumulation of apoptotic cells. Only few inflammatory cells are present within the tissue. (**b**) Pathological outcomes of clearance deficiency: In the case of primary clearance deficiency, the tingible body macrophages fail to efficiently clear apoptotic B lymphocytes, resulting in the accumulation of apoptotic cells that transition into secondary necrosis. Secondary necrotic cells disintegrate and release danger-associated molecular patterns that generate an inflammatory microenvironment. Material derived from disintegrating apoptotic cells is captured by follicular dendritic cells that present the autoantigen to autoreactive B lymphocytes, effectively mediating the break of tolerance to apoptotic self. In the periphery, these autoreactive B cells produce autoantibodies that opsonize apoptotic cells, thereby rerouting the uptake of apoptotic material to neutrophils and monocytes. Fc-mediated uptake induces robust production of inflammatory cytokines, which results in further recruitment of inflammatory cells. Inflamed tissue is characterized by a heightened level of apoptosis, and the failure to clear these cells further aggravates inflammation, thereby creating a vicious cycle of cell death, failure to properly clear apoptotic cells, and inflammation. Of note, this scenario is independent of how the break of tolerance to apoptotic self-occurred. Thus, even if there was no primary clearance deficiency, it will result in secondary clearance deficiency because of how autoantibodies prevent clearance of apoptotic cells by tissue-resident macrophages

autoantibody production [144]. Further, the find-me signal LPC is increased in the serum of SLE patients, letting Grossmayer and colleagues hypothesize that such high concentrations could disrupt chemotactic gradients and interfere with migration of phagocytes toward apoptotic cells [145]. Similarly, lupus patients show increased serum levels of cleaved AXL, MER, and TYRO3, effectively decreasing the abundance of these receptors on phagocytes [146–149].

Similarly, autoantibodies against phagocytic receptors can impair proper execution of efferocytosis and autoantibodies against TYRO3 and the scavenger receptor SCARF1 have been reported in patients with SLE [150, 151]. Further, autoantibodies against other pro-efferocytic factors found on the surface of apoptotic cells are common in SLE: these include opsonins such as C1q and CRP [130, 152–154], but also proteins that translocate to the cell surface during apoptosis such as ANXA1 and calreticulin [155, 156]. Autoantibodies against C1q occur in one third of SLE patients and in more than 90% of patients with proliferative nephritis [157]. Similarly, 45% of patients display autoantibodies against CRP and the autoantibody titer correlates with renal disease activity and response to therapy [158]. Likewise, autoantibodies against annexin A1 are common, particularly in patients with lupus nephritis and these autoantibodies decrease upon therapy [159]. Further, up to 80% of the patients with SLE have autoantibodies against calreticulin that is thought to be involved in the clearance of apoptotic cells [160].

Autoantibodies Reroute Dead Cell Clearance

Autoantibodies against secondary necrotic cells have a higher sensitivity, specificity, positive, and negative predictive value than tests detecting antibodies against dsDNA or nucleosomes [161], yet contrasting evidence exists on how autoantibodies against late apoptotic or secondary necrotic cells influence their removal or contribute to clearance deficiency: Reefman and colleagues observed that late apoptotic cells coated with IgG fractions from lupus patients reduce their uptake by monocyte-derived macrophages in medium containing serum from healthy donors [162]. In contrast, Munoz and colleagues reported that granulocytes and monocytes engulf secondary necrotic cells only when added to the anti-coagulated blood of SLE patients and not healthy donors [163]. The phagocytosis index correlated with anti-dsDNA antibody titer and treatment of secondary necrotic cells with DNAse I or removal of immunoglobulins decreased phagocytosis and inflammatory cytokine production. Others observed a dichotomic pattern, with some SLE sera inducing more engulfment and others inducing less engulfment than serum from healthy donors [164]. These findings indicate that dependent on the subgroup of patients, autoantibodies may inhibit the uptake of late apoptotic or secondary necrotic cells but also cause Fcγ-mediated uptake and inflammatory responses in other cells. Immune complexes of autoantibodies bound to secondary necrotic cells could mimic opsonized viral particles, thereby stimulating an interferon response and contribute to the interferon signature commonly seen in SLE patients [165]. Similarly, uptake of secondary necrotic cells by monocytes and neutrophils explains the well-known phenomena of the LE cell.

Clearance deficiency might be particularly problematic in locations with a high cellular turn over, such as the germinal centers. Indeed, germinal center biopsies from patients with SLE show more free apoptotic cells, and their tingible body macrophages are smaller and lower in number [166]. Autoantigens arising from these uncleared cells are hypothesized to be presented to autoreactive B cells via follicular dendritic cells, resulting in the production of autoantibodies [166, 167]. However, autoantibodies might also interfere with the clearance of apoptotic cells in the periphery and therefore allow them to progress into secondary necrosis [168]. They might also opsonize secondary necrotic cells and reroute their clearance to infiltrating inflammatory cells, further aggravating inflammation [163, 169]. Importantly, even if the break in tolerance to apoptotic self was independent of impaired efferocytosis, the action of the autoantibodies and the ongoing systemic inflammation is likely to cause a secondary deficiency in apoptotic clearance [170]. The etiopathogenesis of SLE may be depicted as a vicious, self-aggravating cycle that can be initialized by clearance deficiency or loss of self-tolerance, with the fuel behind this machinery being uncleared apoptotic cells [100].

The Role of Efferocytosis in Lung Disease

Several important studies on the role of efferocytosis in the resolution of inflammation have been conducted in mice. However, in the context of lung injury, evidence exists for both COPD and asthma in man:

Clearance Deficiency in Chronic Obstructive Pulmonary Disease (COPD)

COPD is a progressive lung disease characterized by shortness of breath and periodic exacerbations. Smoking and air pollution are the most common risk factors for developing COPD, with a history of smoking found in more than 70% of patients [171]. Interestingly, proteomic profiling of the bronchoalveolar fluid of COPD patients, as compared to healthy individuals, suggests an impairment of wound healing and efferocytosis, as highlighted by downregulation of CD163, MARCO, and VSIG4 [172]. Further, COPD patients show decreased serum levels of MFGE8, and serum concentration positively correlated with lung function, while cigarette smoke decreases the concentration of MFGE8 in the lung tissue of mice [173, 174]. These findings are supported by functional data showing that isolated alveolar macrophages from COPD patients, either current or previous smokers, showed decreased capacity to engulf apoptotic bronchial epithelial cells as compared to healthy individuals [175, 176]. Further, expression of RUBICON, which is essential for LC3-associated phagocytosis, is downregulated in COPD patients and mice exposed to cigarette smoke [177]. Treatment of alveolar macrophages from healthy individuals with cigarette smoke extract reduced the expression of receptors involved in the uptake of apoptotic cells, such as CD44 [175]. Cigarette smoke exposure in mice leads to oxidative-stress mediated activation of RhoA [178]. Indeed, COPD patients show decreased availability of the antioxidant GSH in the bronchoalveolar fluid, and feeding mice procysteine, a potent antioxidant, during exposure to cigarette smoke, ameliorates the impairment of lung macrophages to engulf apoptotic cells [179]. Similar results have been obtained with carbocysteine [180]. It has further been reported that cigarette smoke extract induces activation of RhoA by induction of ER stress [181]. However, the influence of cigarettes is not limited to classical cigarettes, as the administration of aerosols from E-cigarettes also causes impaired efferocytosis in macrophages [182, 183]. Ceramide synthesis is upregulated by cigarette smoke and administration of ceramide inhibits efferocytosis through accumulation of sphingosine-1-phosphate, that is linked to decreased recruitment of RACc1 to the plasma membrane [184, 185]. Further, SPHK1 and SPHK2, the enzymes catalyzing the production of sphingosine-1-phopshate and their respective receptors S1PR2 and S1PR5, are upregulated in COPD patients [186], with the gene encoding for S1PR5 being less methylated in COPD patients. This suggest an epigenetic dysfunction associated with COPD [187]. Thymoquinone, a bioactive constituent of black seed oil, prevented the changes in both the S1P synthesis and signaling pathway and counteracted the negative effects of cigarette smoke extract on macrophages [188]. Similarly, treatment of alveolar macrophages with macrolides, such as azithromycin or nonantibiotic macrolides, improved the efferocytic function, presumably through their anti-inflammatory function that influences macrophage phenotype [189, 190]. Indeed, a clinical study showed that treatment of COPD patients with low-dose azithromycin increased the phagocytic capacity of alveolar macrophages [191]. A subgroup of patients with COPD shows eosinophilic airway inflammation. Efferocytosis of eosinophils by alveolar macrophages can be indirectly assessed via quantification of eosin staining in macrophages. Indeed, monocyte-derived macrophages (MDM) from patients with a high percentage of eosinophils in the absence of eosin staining in macrophages have a decreased capacity to engulf apoptotic eosinophils [192]. Further, the capacity of MDM to engulf apoptotic eosinophils correlated with the number of disease exacerbations in the last year, suggesting that a systemic clearance deficiency contributes to the etiopathology of COPD [192].

Clearance Deficiency in Asthma

Asthma is an immune-mediated disease of the lung characterized by episodic bronchial spasms. Therefore, asthma and models of allergic airway inflammation can be used to understand the role of apoptotic cell clearance in the resolution of inflammation. Indeed, alveolar macrophages from patients with severe asthma show a decreased capacity to engulf apoptotic cells; however, this could be reversed by ex vivo treatment with dexamethasone, or a corticosteroid boost in vivo [193]. This effect has been well documented, as treatment of macrophages with glucocorticoids leads to RAC1 activation and reroutes efferocytosis to occur through MERTK and the bridging molecule PROS1 [194–197]. The involvement of members of the TAM family in asthma is further supported by the finding that AXL expression is downregulated in asthma and negatively correlates with the accumulation of cellular debris [198]. Administration of steroids might help to overcome clearance deficiency; however, non-eosinophilic asthma, which affects up to 50% of asthma patients, comes without an eosinophilic response and displays relative insensitivity to inhaled corticosteroids [199]. Simpson and colleagues found that the impairment of alveolar macrophages to engulf apoptotic cells is specific for non-eosinophilic asthma [200]. Similarly, neutrophilic asthma, which is a special form of non-eosinophilic asthma characterized by persistent lung neutrophilia, is characterized by a decreased abundance of LGAL3 and lower ratios of LGAL3/LGAL3BP [201]. LGAL3 is an opsonin of apoptotic cells that increases the capacity to engulf apoptotic cells, particularly in patients with neutrophilic asthma [202, 203]. Asthma tends to be more severe in obese patients with decreased sensitivity to corticosteroid treatment. Alveolar macrophages from obese asthma patients show a decreased efferocytic capacity and efferocytosis correlates with inducibility of MKP-1 and GCRα by dexamethasone [204]. The role of efferocytosis in asthma is further supported by murine studies. Administration of apoptotic cells prior to sensitization and challenge with house dust mite (HDM) ameliorates HDM-driven airway inflammation [205]. In contrast, mice with knockout of the gene encoding for MERTK showed delayed resolution of inflammation in OVA-induced allergic airway inflammation [206]. On the contrary, knockout of PGRN, which renders macrophages more efferocytic, ameliorated OVA-induced allergic airway inflammation [207]. Efferocytosis within the alveolar space is not restricted to alveolar macrophages, and deletion of RAC1 in type-2 epithelial cells exacerbated OVA-induced allergic airway inflammation, presumably through upregulation of IL-33 expression [208]. Importantly, in the lung, there seems to be division of labor, as macrophages predominantly engulf whole apoptotic cells and produce IGF1, which redirects engulfment of small particles to type-2 epithelial cells, suggesting that not only the removal of apoptotic cells but also the division of this task between professional and nonprofessional phagocytes is an important aspect of tissue biology [25].

Boosting of Efferocytosis to Promote Resolution of Inflammation

In the previous paragraphs, we have presented that impairments in efferocytosis contribute to COPD and asthma. We have also touched upon evidence that certain drugs may ameliorate disease and promote resolution of inflammation. Here we want to further elaborate on how boosting of engulfment contributes to resolution of lung inflammation:

Statins are well known for their action on Rho GTPases, and indeed, treating macrophages with lovastatin enhances uptake of apoptotic cells [209]. This effect is also observed in humans, as the efferocytic capacity of alveolar macrophages from patients recovering from community acquired pneumonia is positively correlated with statin use [210]. This is interesting, given that alveolar macrophages from statin-treated smokers showed a higher rate of efferocytosis than never smokers that were not treated with statins, suggesting a dominant effect of the action of statins on RHOA. As mentioned previously, pro-resolving lipid mediators play an important role in resolution of inflammation. The aspirin-triggered resolvins 15-Epi-LXA4 and 17-epi-RvD1 restore impaired clearance of pulmonary *Escherichia coli* after challenge with CpG and promote efferocytosis of apoptotic neutrophils [211]. Similarly, aspirin-triggered resolvin D1 mediates the resolution of gram-negative bacterial pneumonia [212]. The beneficial effects of SPMs are not limited to the context of infection, since D-series resolvins cause a functional switch from M1-like to M2-like macrophages, thereby ameliorating ischemic lung injury in a PLD2-S6 kinase-dependent manner [213]. Old mice are more susceptible to this kind of lung injury; however, resolvin D1 is effective in this context as well [214]. Rymut and colleagues report that aging leads to more cleavage of MERTK, which is phenocopied by administration of conditioned medium from senescent cells, and that resolvin D1 prevents the cleavage of MERTK [214]. Interestingly, 25-hydroxycholesterol, which is thought to have senolytic properties, prevents lipid overload in engulfing macrophages during resolution of inflammation in the lung [215]. Resolvin D1 also proves to boost efferocytosis in a model of acute inflammation caused by cigarette smoke, with administration of resolvin D1 following the exposure to cigarette smoke promoting the resolution of inflammation, with a higher percentage of lung macrophages showing co-staining for neutrophils, which is suggestive of efferocytosis [216]. Not only active but also passive smoking leads to impaired efferocytosis, which is rescued by intranasal administration of GM-CSF. Such treatment is even able to boost engulfment in mice exposed to secondhand smoke with an ongoing influenza infection, thereby ameliorating the disease [217]. Ulinastatin, a urinary trypsin inhibitor used to treat pancreatitis and other inflammatory conditions, was further shown to induce MERTK expression and boost the resolution of LPS-induced lung injury [218]. Similarly, the MEK-inhibitor mirdametinib, which is currently undergoing phase 2b trials for plexiform neurofibroma, increased expression of MERTK, in addition to TYRO3 and ABCA1 [219]. In vivo treatment with mirdametinib resulted in increased efferocytic capacity in peritoneal macrophages and amelioration of LPS-induced lung injury [219]. Given the success of CAR-T cell therapy, modification of macrophage function is a logical next step [220]. While a lot of interest resolves around the phagocytosis of tumor cells [221], manipulating the efferocytic function of macrophages could be a prime target for inflammatory disease. Indeed, expression of a chimeric receptor for efferocytosis (CHEF), which consists of an extracellular domain binding phosphatidylserine (derived from either BAI1 or TIM4) and an intracellular domain derived from ELMO1, which mediates the intracellular signaling, boosted efferocytosis in vitro and in vivo and proved effective in several inflammatory diseases [222].

Efferocytosis in Tissue Repair and Remodeling

Although some patient data exists for lung pathologies, such as restoring the function of alveolar macrophages from COPD patients with corticosteroids, most interventional studies on the role of apoptotic cell clearance in the resolution of inflammation have been conducted in mice. However, work from the Gilroy group suggests that similar mechanisms apply in humans. Topically applied cantharidin induces blisters with neutrophilic inflammation that spontaneously resolve within a few days [223]. De Maeyer and colleagues found that compared to young adults, resolution of inflammation is delayed in the elderly and that their efferocytosis is impaired [224]. Blisters from elderly adults show a higher percentage of uncleared apoptotic neutrophils and their macrophages express lower amounts of TIM4. Strikingly, TIM4 expression and the resolution of inflammation could be restored by administration of losmapimod, an inhibitor of the p38 MAPK pathway, which regulates TIM4 expression and was more active in the elderly [224]. In vitro research suggests that senescent cells, through cell–cell contact with macrophages involving the don't eat me signal CD47, negatively affect the capacity of macrophages to engulf apoptotic cells [225]. Hu and colleagues further report that aged macrophages show increased cleavage of MERTK by ADAM17. Such impairment in efferocytosis leads to the stimulation of the STING pathway through DNA from uncleared secondary necrotic cells [226]. These two studies provide alternative mechanisms for how efferocytosis is impaired during aging and support the notion that therapies targeting efferocytosis may help ameliorate some age-related pathologies.

Many important studies on the resolution of inflammation have focused on the sterile inflammation induced by microbial components. However, there is strong evidence that efferocytosis is also involved in wound healing and repair following other insults. For example, efferocytosis is involved in recovery from various injuries, including, but not limited to, dermal wound healing [227–229], ischemic stroke [230–232], myocardial infarction [233–240], and osteoarthritis [241–243], as well as in models of injury of the kidney [244–246], muscle [247–249], intestine [250–252], and liver [253–255]. Further, several lines of evidence suggest that efferocytosis is important for vascular health and an important factor in the prevention of plaque progression in atherosclerosis [256–258].

Fibrosis: Resolution Gone Wrong

Upon engulfment of apoptotic cells, macrophages secrete factors such as TGF-β, which contribute to the resolution of inflammation and wound healing, but also play a major role in the pathogenesis of fibrosis [259]. Fibrosis often occurs in the context of chronic inflammation, where inflammatory, reparative, and angiogenic factors are present in an unbalanced manner [260]. Fibrosis is characterized by excessive extracellular matrix deposition and decreased tissue cellularity and compromises the function of the tissue [261]. It is estimated that up to 45% of deaths in the industrialized world are due to fibrotic diseases [262]. The average life span of idiopathic pulmonary fibrosis is 3–5 years and few effective treatments exist [263]. Excellent reviews have highlighted the dual role of macrophages in promoting and limiting fibrosis [264, 265], and there is good evidence that the same applies for efferocytosis [266, 267]. In some contexts, the clearance of apoptotic cells contributes to fibrosis, whereas in others, it prevents or even helps to reverse fibrosis.

Evidence for Efferocytosis Limiting Fibrosis

As mentioned above, the use of chimeric receptors for efferocytosis that boost efferocytosis in vivo can improve kidney function following reperfusion injury and prevented the development of fibrosis [222]. In the context of the liver, engulfment of apoptotic cholangiocytes induces a fibrolytic phenotype in macrophages and helps to reverse fibrosis following bile-duct ligation [268]. Using an elegant approach that harnesses the differences in the susceptibility of developing fibrosis in different mice strains, An and colleagues showed that one dose of TAA induces comparable damage in FVB and BALB/C mice, yet the necrotic areas are cleared more rapidly in FVB mice [269]. This correlated with higher co-localization of apoptotic cells with macrophages in FVB mice and a higher capacity to clear apoptotic cells from circulation. In contrast, depletion of macrophages with clodronate worsened fibrosis in FVB mice. The authors further revealed that mitochondria-derived DAMPs from apoptotic hepatocytes undergoing secondary necrosis drive hepatic stellate cell activation and fibrosis, highlighting the importance of the rapid removal of dying cells in the prevention of fibrosis [269]. Further, adoptive transfer of bone-marrow-derived mesenchymal (BM-MSC) cells into mice subjected to CCl_4 treatment led to the conversion of $LY6C^{HIGH}$ macrophages into pro-resolving $LY6C^{LOW}$ macrophages [270]. As expected, BM-MSC homed to the liver, however, they quickly underwent apoptosis. Efferocytosis by $LY6C^{LOW}$ macrophages induced the expression of MMP12, which counteracts fibrosis. Indeed, prevention of apoptosis via inhibition of caspases or inhibition of PS-mediated uptake prevented the anti-fibrotic effect of adoptive transfer of BM-MSC in liver fibrosis [270]. The anti-fibrotic effects of apoptosis are not restricted to the liver: Yang and colleagues report that during steady state, LPC produced by apoptotic acinar cells induces a pro-efferocytic M2a polarization in pancreatic macrophages [271]. However, during development of chronic pancreatitis, the phospholipid metabolism is altered, resulting in less production of LPC ultimately leading to decreased efferocytosis and uncleared apoptotic cells that stimulate inflammation and the development of fibrosis [271]. Another mechanism by which the clearance machinery is preventing fibrosis is through the bridging molecule MFGE8 that binds to collagen and targets it for removal by macrophages [272]. This effect is independent of apoptotic cell clearance, as MFGE8 knockout mice do not show impaired efferocytosis in the lung. Wang and colleagues report that TREM2-mediated engulfment of lipid-laden apoptotic hepatocytes occurs in the steatotic liver; however, during disease progression, TREM2 is cleaved by ADAM17, thereby impairing clearance of dying hepatocytes [273]. In consequence, secondary necrotic hepatocytes release DAMPs that exacerbate inflammation and drive progression into nonalcoholic steatohepatitis.

Linkage of Efferocytosis to Promoting Fibrosis

Contrasting the role of dead cell clearance in the pathogenesis of NASH, MERTK was shown to contribute to the disease. During steatosis, MERTK is cleaved by the protease ADAM17, although in NASH this cleavage is decreased, resulting in stronger surface expression of MERTK in liver macrophages [274]. In line with this, boosting cleavage of MERTK by ADAM17 via administration of all-trans retinoid acid ameliorated the disease. Similarly, global or myeloid-specific knockout of MERTK prevented advanced stages of NASH, whereas expression of a cleavage-resistant MERTK variant exacerbated the disease. Mechanistically, TGF-β produced via the MERTK/ERK pathway activated hepatic stellate cells that responded with increased collagen deposition [274]. This is highly interesting as cleavage of two different engulfment receptors, MERTK and TREM2, by the same protease may exacerbate or ameliorate the progression of nonalcoholic fatty liver disease, highlighting again that it is not only the uptake that matters but also which receptors are engaged [54, 273, 274]. Not only efferocytosis of hepatocytes, but also that of infiltrating neutrophils in the liver of CCL4-treated animals may contribute to fibrosis: engulfment mediated by the matricellular protein CCN1 and its receptor integrin ITGAV-CD61 on the phagocyte is able to induce a robust TGF-β response that causes transdifferentiation of hepatic stellate cells into myofibroblast-like cells [275]. Reports about the pro-fibrotic role of efferocytosis also exist for other organs than the liver: Rivati and colleagues looked at IgG4-related disease that can affect various organs [276]. It is a chronic, fibro-inflammatory condition characterized by infiltration of affected tissues by IgG4-secreting plasma cells and T lymphocytes. Interestingly, MERTK-positive, TGF-β expressing macrophages are found within lesions and they associate with apoptotic cells opsonized with the bridging molecule PROS1 and PROS1-positive B and T cells. The authors propose a model in which lymphocytes induce chronic inflammation and tissue damage, and the macrophages, which are continuously stimulated by both apoptotic cells and lymphocytes, react with the overt secretion of pro-fibrotic TGF-β [276]. This hypothesis is supported by the fact that these macrophages quickly disappear after B cell depletion with rituximab.

Efferocytosis in Fibrosis: Striking the Balance

The complexity of the question of whether efferocytosis contributes to or prevents fibrosis is especially pronounced in the lung: administration of apoptotic cells after bleomycin-induced injury prevents the development of fibrosis [277]. Mechanistically, this effect is mediated by PPARγ via STAT6, as alveolar macrophages engulfing apoptotic cells showed a sustained upregulation of this transcription factor [278, 279]. Interestingly, engulfing macrophages also secreted the pro-fibrotic molecule TGF-β; however, its concentration in the lung tissue quickly returned to baseline. Controls that did not receive apoptotic cells or where the PPARγ signaling was prevented initially showed low concentrations of TGF-β. However, a steady increase that accompanied the development of fibrosis was observed [278]. Further, combinatory treatment of mice exposed to bleomycin with apoptotic cells and the statin simvastatin prevented the development of fibrosis [280]. In these examples, efferocytosis clearly prevented the development of fibrosis, but can this finding be generalized? Kim and colleagues reported that exposing the lung to a level of apoptosis that exceeds the baseline, either via means of a diphtheria toxin-induced suicide switch in type II alveolar epithelial cells or via repeated instillation of alveolar epithelial cells, is sufficient to induce lung fibrosis [281]. Interestingly, this effect is not recapitulated by apoptotic T lymphocytes. This highlights the complexity of the role of efferocytosis in fibrosis. In the presence of an insult, such as administration of bleomycin, apoptotic stimuli from certain cells, but not others, could contribute to accelerated resolution of inflammation and thereby prevent fibrosis. In contrast, within a naïve tissue, an overt rate of apoptosis might induce a pro-fibrotic program. It should also be noted that some apoptotic cells may also release TGF-β, and the continuous presence of uncleared apoptotic cells, as seen in failure to resolve inflammation, could lead to elevated levels of TGF-β and fibrosis [282]. TGF-β, in turn, can sensitize alveolar epithelial cells to apoptosis, adding additional fuel to the fire [283]. In support of this hypothesis, Huen and colleagues have shown that deletion of TGF-β in the myeloid lineage does not prevent the development of renal fibrosis, suggesting that other sources of TGF-β drive fibrosis [284]. Therefore, the efficient removal of apoptotic cells could minimize the presence of TGF-β and disrupt the pro-fibrotic loop. Indeed, the bronchoalveolar lavage of patients with idiopathic pulmonary fibrosis shows a significantly higher burden of apoptotic cells compared to patients with pneumonitis or pneumonia, yet fewer apoptotic cells are found within alveolar macrophages [285]. Similar to what has been reported for the administration of apoptotic cells, contrasting evidence can also be found on the level of the engulfment receptors: while knockout of CD36 prevents fibrosis in mice subjected to bleomycin-induced lung injury, blockade of TIM3 increases fibrosis [286, 287]. Further upregulation of MERTK is seen in idiopathic pulmonary fibrosis and bleomycin-induced fibrosis, although potential pro-efferocytic functions might have been undermined by an increase in mitochondrial membrane potential, that is known to interfere with efferocytosis [21, 288]. Interestingly, administration of apoptotic cells decreased the expression of MERTK and ameliorated fibrosis [288]. This is in line with the work from Pastore and colleagues that suggests pro-fibrotic crosstalk of macrophages expressing MERTK and hepatic stellate cells in vitro [289]. Altogether, these findings point toward a highly context and model-dependent effect of efferocytosis in regulating fibrosis.

Concluding Thoughts

In this review, we highlight the pro-homeostatic functions of efferocytosis and consequences that may ensue when this clearance process is disturbed. We posit that efferocytosis works as a "homeostatic magnet" that pulls back any system that deviates from homeostasis. For example, efferocytosis ensures that tolerance to self is maintained. This is highlighted by the fact that autoimmune diseases are relatively rare and that it takes considerable effort to induce autoimmunity in experimental settings. Here, the role of efferocytosis is not about pulling back a system that went out of balance but rather preventing deviation from homeostasis. In situations where homeostasis is disrupted, such as sterile injury or infection, efferocytosis contributes to the resolution of inflammation and initiates return to homeostasis. However, the pro-homeostatic function of efferocytosis might also come with certain downsides, as in the case of pathogens hijacking the efferocytic machinery and tumors benefiting from the tolerogenic environment created by apoptotic cell clearance. Similarly, the role of efferocytosis is less well understood in the context of fibrosis. There is clear evidence that efferocytosis has anti-fibrotic effects, but if the efferocytic process fails to shut down inflammation completely, it may parallel chronic inflammation, with the combination driving the development of fibrosis. Further research will help delineate what tips the balance in specific contexts and which "homeostatic checkpoints" are regulated by apoptotic cell clearance.

References

1. de Cathelineau, A. M., & Henson, P. M. (2003). The final step in programmed cell death: Phagocytes carry apoptotic cells to the grave. *Essays in Biochemistry, 39*, 105–117. https://doi.org/10.1042/bse0390105
2. Kerr, J. F., Wyllie, A. H., & Currie, A. R. (1972). Apoptosis: A basic biological phenomenon with wide-ranging implications in tissue kinetics. *British Journal of Cancer, 26*, 239–257.
3. Tak, T., Tesselaar, K., Pillay, J., Borghans, J. A., & Koenderman, L. (2013). What's your age again? Determination of human neutrophil half-lives revisited. *Journal of Leukocyte Biology, 94*, 595–601. https://doi.org/10.1189/jlb.1112571
4. Vandivier, R. W., Henson, P. M., & Douglas, I. S. (2006). Burying the dead: The impact of failed apoptotic cell removal (efferocytosis) on chronic inflammatory lung disease. *Chest, 129*, 1673–1682. https://doi.org/10.1378/chest.129.6.1673
5. Medina, C. B., et al. (2020). Metabolites released from apoptotic cells act as tissue messengers. *Nature, 580*, 130–135. https://doi.org/10.1038/s41586-020-2121-3
6. Morioka, S., et al. (2018). Efferocytosis induces a novel SLC program to promote glucose uptake and lactate release. *Nature, 563*, 714–718. https://doi.org/10.1038/s41586-018-0735-5
7. Chera, S., et al. (2009). Apoptotic cells provide an unexpected source of Wnt3 signaling to drive hydra head regeneration. *Developmental Cell, 17*, 279–289. https://doi.org/10.1016/j.devcel.2009.07.014
8. Rabiller, L., et al. (2021). Driving regeneration, instead of healing, in adult mammals: The decisive role of resident macrophages through efferocytosis. *npj Regenerative Medicine, 6*, 41. https://doi.org/10.1038/s41536-021-00151-1
9. A-Gonzalez, N., et al. (2017). Phagocytosis imprints heterogeneity in tissue-resident macrophages. *The Journal of Experimental Medicine, 214*, 1281–1296. https://doi.org/10.1084/jem.20161375
10. Cummings, R. J., et al. (2016). Different tissue phagocytes sample apoptotic cells to direct distinct homeostasis programs. *Nature, 539*, 565–569. https://doi.org/10.1038/nature20138
11. Sui, B., et al. (2021). Apoptotic extracellular vesicles (ApoEVs) safeguard liver homeostasis and regeneration <em>via</em> assembling an ApoEV-golgi organelle. *bioRxiv*, 2021.2002.2024.432630. https://doi.org/10.1101/2021.02.24.432630
12. Hernández-García, D., García-Meléndrez, C., Hernández-Martínez, R., Collazo-Navarrete, O. & Covarrubias, L. (2023). Macrophages allocate before apoptosis initiation and produce reactive oxygen species during interdigital phagocytosis. *bioRxiv*, 2023.2005.2004.539494. https://doi.org/10.1101/2023.05.04.539494
13. Raymond, M. H., et al. (2022). Live cell tracking of macrophage efferocytosis during Drosophila embryo development in vivo. *Science, 375*, 1182–1187. https://doi.org/10.1126/science.abl4430
14. Hoijman, E., et al. (2021). Cooperative epithelial phagocytosis enables error correction in the early embryo. *Nature, 590*, 618–623. https://doi.org/10.1038/s41586-021-03200-3
15. Lu, L., & Osmond, D. G. (2000). Apoptosis and its modulation during B lymphopoiesis in mouse bone marrow. *Immunological Reviews, 175*, 158–174.
16. Wang, H., & Zuniga-Pflucker, J. C. (2022). Thymic microenvironment: Interactions between innate immune cells and developing thymocytes. *Frontiers in Immunology, 13*, 885280. https://doi.org/10.3389/fimmu.2022.885280
17. Surh, C. D., & Sprent, J. (1994). T-cell apoptosis detected in situ during positive and negative selection in the thymus. *Nature, 372*, 100–103. https://doi.org/10.1038/372100a0

18. Tsai, T. L., et al. (2022). Multiomics reveal the central role of pentose phosphate pathway in resident thymic macrophages to cope with efferocytosis-associated stress. *Cell Reports, 40*, 111065. https://doi.org/10.1016/j.celrep.2022.111065
19. Gerlach, B. D., et al. (2021). Efferocytosis induces macrophage proliferation to help resolve tissue injury. *Cell Metabolism, 33*, 2445–2463 e2448. https://doi.org/10.1016/j.cmet.2021.10.015
20. Wang, Y., et al. (2017). Mitochondrial fission promotes the continued clearance of apoptotic cells by macrophages. *Cell, 171*, 331–345 e322. https://doi.org/10.1016/j.cell.2017.08.041
21. Park, D., et al. (2011). Continued clearance of apoptotic cells critically depends on the phagocyte Ucp2 protein. *Nature, 477*, 220–224. https://doi.org/10.1038/nature10340
22. Grootveld, A. K., et al. (2023). Apoptotic cell fragments locally activate tingible body macrophages in the germinal center. *Cell, 186*, 1144–1161 e1118. https://doi.org/10.1016/j.cell.2023.02.004
23. van Ham, T. J., Kokel, D., & Peterson, R. T. (2012). Apoptotic cells are cleared by directional migration and elmo1- dependent macrophage engulfment. *Current Biology, 22*, 830–836. https://doi.org/10.1016/j.cub.2012.03.027
24. Sato, K., Honda, S. I., Shibuya, A., & Shibuya, K. (2018). Cutting edge: Identification of marginal reticular cells as phagocytes of apoptotic B cells in germinal centers. *Journal of Immunology, 200*, 3691–3696. https://doi.org/10.4049/jimmunol.1701293
25. Han, C. Z., et al. (2016). Macrophages redirect phagocytosis by non-professional phagocytes and influence inflammation. *Nature, 539*, 570–574. https://doi.org/10.1038/nature20141
26. Shaha, C., Tripathi, R., & Mishra, D. P. (2010). Male germ cell apoptosis: Regulation and biology. *Philosophical Transactions of the Royal Society of London. Series B, Biological Sciences, 365*, 1501–1515. https://doi.org/10.1098/rstb.2009.0124
27. Penberthy, K. K., Lysiak, J. J., & Ravichandran, K. S. (2018). Rethinking phagocytes: Clues from the retina and testes. *Trends in Cell Biology, 28*, 317–327. https://doi.org/10.1016/j.tcb.2018.01.004
28. Hai, Y., et al. (2014). The roles and regulation of Sertoli cells in fate determinations of spermatogonial stem cells and spermatogenesis. *Seminars in Cell & Developmental Biology, 29*, 66–75. https://doi.org/10.1016/j.semcdb.2014.04.007
29. Luaces, J. P., et al. (2014). Loss of Sertoli-germ cell adhesion determines the rapid germ cell elimination during the seasonal regression of the seminiferous epithelium of the large hairy armadillo Chaetophractus villosus. *Biology of Reproduction, 90*, 48. https://doi.org/10.1095/biolreprod.113.113118
30. Seco-Rovira, V., et al. (2014). The death of sertoli cells and the capacity to phagocytize elongated spermatids during testicular regression due to short photoperiod in Syrian hamster (Mesocricetus auratus). *Biology of Reproduction, 90*, 107. https://doi.org/10.1095/biolreprod.113.112649
31. Lu, Q., et al. (1999). Tyro-3 family receptors are essential regulators of mammalian spermatogenesis. *Nature, 398*, 723–728. https://doi.org/10.1038/19554
32. Xiong, W., et al. (2008). Gas6 and the Tyro 3 receptor tyrosine kinase subfamily regulate the phagocytic function of Sertoli cells. *Reproduction, 135*, 77–87. https://doi.org/10.1530/REP-07-0287
33. Chen, Y., et al. (2009). Functions of TAM RTKs in regulating spermatogenesis and male fertility in mice. *Reproduction, 138*, 655–666. https://doi.org/10.1530/REP-09-0101
34. Shi, J., et al. (2023). Absence of MerTK disrupts spermatogenesis in an age-dependent manner. *Molecular and Cellular Endocrinology, 560*, 111815. https://doi.org/10.1016/j.mce.2022.111815
35. Zhang, Y., et al. (2013). Breakdown of immune homeostasis in the testis of mice lacking Tyro3, Axl and Mer receptor tyrosine kinases. *Immunology and Cell Biology, 91*, 416–426. https://doi.org/10.1038/icb.2013.22
36. Sun, B., et al. (2010). Sertoli cell-initiated testicular innate immune response through toll-like receptor-3 activation is negatively regulated by Tyro3, Axl, and mer receptors. *Endocrinology, 151*, 2886–2897. https://doi.org/10.1210/en.2009-1498

37. Shiratsuchi, A., Kawasaki, Y., Ikemoto, M., Arai, H., & Nakanishi, Y. (1999). Role of class B scavenger receptor type I in phagocytosis of apoptotic rat spermatogenic cells by Sertoli cells. *The Journal of Biological Chemistry, 274*, 5901–5908. https://doi.org/10.1074/jbc.274.9.5901
38. Kawasaki, Y., Nakagawa, A., Nagaosa, K., Shiratsuchi, A., & Nakanishi, Y. (2002). Phosphatidylserine binding of class B scavenger receptor type I, a phagocytosis receptor of testicular sertoli cells. *The Journal of Biological Chemistry, 277*, 27559–27566. https://doi.org/10.1074/jbc.M202879200
39. Nakagawa, A., Shiratsuchi, A., Tsuda, K., & Nakanishi, Y. (2005). In vivo analysis of phagocytosis of apoptotic cells by testicular Sertoli cells. *Molecular Reproduction and Development, 71*, 166–177. https://doi.org/10.1002/mrd.20278
40. Maeda, Y., Shiratsuchi, A., Namiki, M., & Nakanishi, Y. (2002). Inhibition of sperm production in mice by annexin V microinjected into seminiferous tubules: Possible etiology of phagocytic clearance of apoptotic spermatogenic cells and male infertility. *Cell Death and Differentiation, 9*, 742–749. https://doi.org/10.1038/sj.cdd.4401046
41. Osada, Y., Sunatani, T., Kim, I. S., Nakanishi, Y., & Shiratsuchi, A. (2009). Signalling pathway involving GULP, MAPK and Rac1 for SR-BI-induced phagocytosis of apoptotic cells. *Journal of Biochemistry, 145*, 387–394. https://doi.org/10.1093/jb/mvn176
42. Elliott, M. R., et al. (2010). Unexpected requirement for ELMO1 in clearance of apoptotic germ cells in vivo. *Nature, 467*, 333–337. https://doi.org/10.1038/nature09356
43. Dong, Y. S., et al. (2016). Unexpected requirement for a binding partner of the syntaxin family in phagocytosis by murine testicular Sertoli cells. *Cell Death and Differentiation, 23*, 787–800. https://doi.org/10.1038/cdd.2015.139
44. Jin, X., et al. (2020). Testicular Lmcd1 regulates phagocytosis by Sertoli cells through modulation of NFAT1/Txlna signaling pathway. *Aging Cell, 19*, e13217. https://doi.org/10.1111/acel.13217
45. Selva, D. M., et al. (2004). The ATP-binding cassette transporter 1 mediates lipid efflux from Sertoli cells and influences male fertility. *Journal of Lipid Research, 45*, 1040–1050. https://doi.org/10.1194/jlr.M400007-JLR200
46. Wu, D., et al. (2022). Smtnl2 regulates apoptotic germ cell clearance and lactate metabolism in mouse Sertoli cells. *Molecular and Cellular Endocrinology, 551*, 111664. https://doi.org/10.1016/j.mce.2022.111664
47. Wu, D., et al. (2023). Cytoskeletal orchestration of glucose uptake in Sertoli cell to support efferocytosis of apoptotic germ cells. *Biochimica et Biophysica Acta (BBA) – Molecular Cell Research, 1870*, 119434. https://doi.org/10.1016/j.bbamcr.2023.119434
48. Xiong, H., Chen, Z., Zhao, J., Li, W., & Zhang, S. (2022). TNF-alpha/ENO1 signaling facilitates testicular phagocytosis by directly activating Elmo1 gene expression in mouse Sertoli cells. *The FEBS Journal, 289*, 2809–2827. https://doi.org/10.1111/febs.16326
49. Panneerdoss, S., et al. (2017). Cross-talk between miR-471-5p and autophagy component proteins regulates LC3-associated phagocytosis (LAP) of apoptotic germ cells. *Nature Communications, 8*, 598. https://doi.org/10.1038/s41467-017-00590-9
50. Luo, D., et al. (2020). High fat diet impairs spermatogenesis by regulating glucose and lipid metabolism in Sertoli cells. *Life Sciences, 257*, 118028. https://doi.org/10.1016/j.lfs.2020.118028
51. Xiong, W., Wang, H., Wu, H., Chen, Y., & Han, D. (2009). Apoptotic spermatogenic cells can be energy sources for Sertoli cells. *Reproduction, 137*, 469–479. https://doi.org/10.1530/REP-08-0343
52. Gong, Y., et al. (2018). Inactivation of glycogen synthase kinase-3alpha is required for mitochondria-mediated apoptotic germ cell phagocytosis in Sertoli cells. *Aging (Albany NY), 10*, 3104–3116. https://doi.org/10.18632/aging.101614
53. Ngo, J., et al. (2023). Mitochondrial morphology controls fatty acid utilization by changing CPT1 sensitivity to malonyl-CoA. *The EMBO Journal, 42*, e111901. https://doi.org/10.15252/embj.2022111901

54. Penberthy, K. K., et al. (2017). Context-dependent compensation among phosphatidylserine-recognition receptors. *Scientific Reports, 7*, 14623. https://doi.org/10.1038/s41598-017-15191-1
55. Pelletier, R. M., Yoon, S. R., Akpovi, C. D., Silvas, E., & Vitale, M. L. (2009). Defects in the regulatory clearance mechanisms favor the breakdown of self-tolerance during spontaneous autoimmune orchitis. *American Journal of Physiology. Regulatory, Integrative and Comparative Physiology, 296*, R743–R762. https://doi.org/10.1152/ajpregu.90751.2008
56. Yamashita, Y., Suzuki, C., Uchiyama, Y., & Nagata, S. (2020). Infertility caused by inefficient apoptotic germ cell clearance in Xkr8-deficient male mice. *Molecular and Cellular Biology, 40*. https://doi.org/10.1128/MCB.00402-19
57. Tay, T. W., et al. (2007). Phagocytosis plays an important role in clearing dead cells caused by mono(2-ethylhexyl) phthalate administration. *Tissue & Cell, 39*, 241–246. https://doi.org/10.1016/j.tice.2007.04.002
58. Rival, C. M., et al. (2019). Phosphatidylserine on viable sperm and phagocytic machinery in oocytes regulate mammalian fertilization. *Nature Communications, 10*, 4456. https://doi.org/10.1038/s41467-019-12406-z
59. Sandahl, M., Hunter, D. M., Strunk, K. E., Earp, H. S., & Cook, R. S. (2010). Epithelial cell-directed efferocytosis in the post-partum mammary gland is necessary for tissue homeostasis and future lactation. *BMC Developmental Biology, 10*, 122. https://doi.org/10.1186/1471-213X-10-122
60. Hanayama, R., Nagata, S., & Impaired involution of mammary glands in the absence of milk fat globule EGF factor 8. (2005). *Proceedings of the National Academy of Sciences of the United States of America, 102*, 16886–16891. https://doi.org/10.1073/pnas.0508599102
61. Rosenwald, M., et al. (2012). Serum-derived plasminogen is activated by apoptotic cells and promotes their phagocytic clearance. *Journal of Immunology, 189*, 5722–5728. https://doi.org/10.4049/jimmunol.1200922
62. Green, K. A., Nielsen, B. S., Castellino, F. J., Romer, J., & Lund, L. R. (2006). Lack of plasminogen leads to milk stasis and premature mammary gland involution during lactation. *Developmental Biology, 299*, 164–175. https://doi.org/10.1016/j.ydbio.2006.07.021
63. Stanford, J. C., et al. (2014). Efferocytosis produces a prometastatic landscape during post-partum mammary gland involution. *The Journal of Clinical Investigation, 124*, 4737–4752. https://doi.org/10.1172/JCI76375
64. Elliott, M. R., Koster, K. M., & Murphy, P. S. (2017). Efferocytosis signaling in the regulation of macrophage inflammatory responses. *Journal of Immunology, 198*, 1387–1394. https://doi.org/10.4049/jimmunol.1601520
65. Voll, R. E., et al. (1997). Immunosuppressive effects of apoptotic cells. *Nature, 390*, 350–351. https://doi.org/10.1038/37022
66. Chung, E. Y., et al. (2007). Interleukin-10 expression in macrophages during phagocytosis of apoptotic cells is mediated by homeodomain proteins Pbx1 and Prep-1. *Immunity, 27*, 952–964. https://doi.org/10.1016/j.immuni.2007.11.014
67. Savina, A., et al. (2006). NOX2 controls phagosomal pH to regulate antigen processing during crosspresentation by dendritic cells. *Cell, 126*, 205–218. https://doi.org/10.1016/j.cell.2006.05.035
68. Kumaran Satyanarayanan, S., et al. (2019). IFN-beta is a macrophage-derived effector cytokine facilitating the resolution of bacterial inflammation. *Nature Communications, 10*, 3471. https://doi.org/10.1038/s41467-019-10903-9
69. Liu, K., et al. (2002). Immune tolerance after delivery of dying cells to dendritic cells in situ. *The Journal of Experimental Medicine, 196*, 1091–1097. https://doi.org/10.1084/jem.20021215
70. Desch, A. N., et al. (2011). CD103+ pulmonary dendritic cells preferentially acquire and present apoptotic cell-associated antigen. *The Journal of Experimental Medicine, 208*, 1789–1797. https://doi.org/10.1084/jem.20110538

71. Huang, F. P., et al. (2000). A discrete subpopulation of dendritic cells transports apoptotic intestinal epithelial cells to T cell areas of mesenteric lymph nodes. *The Journal of Experimental Medicine, 191*, 435–444. https://doi.org/10.1084/jem.191.3.435
72. Steinman, R. M., Turley, S., Mellman, I., & Inaba, K. (2000). The induction of tolerance by dendritic cells that have captured apoptotic cells. *The Journal of Experimental Medicine, 191*, 411–416. https://doi.org/10.1084/jem.191.3.411
73. Bosteels, V., et al. (2023). LXR signaling controls homeostatic dendritic cell maturation. *Science Immunology, 8*, eadd3955. https://doi.org/10.1126/sciimmunol.add3955
74. Perry, J. S. A., et al. (2019). Interpreting an apoptotic corpse as anti-inflammatory involves a chloride sensing pathway. *Nature Cell Biology, 21*, 1532–1543. https://doi.org/10.1038/s41556-019-0431-1
75. Kushwah, R., Oliver, J. R., Zhang, J., Siminovitch, K. A., & Hu, J. (2009). Apoptotic dendritic cells induce tolerance in mice through suppression of dendritic cell maturation and induction of antigen-specific regulatory T cells. *Journal of Immunology, 183*, 7104–7118. https://doi.org/10.4049/jimmunol.0900824
76. Perry, J. S. A., et al. (2018). Transfer of cell-surface antigens by scavenger receptor CD36 promotes thymic regulatory T cell receptor repertoire development and allo-tolerance. *Immunity, 48*, 923–936 e924. https://doi.org/10.1016/j.immuni.2018.04.007
77. Nagata, S., Hanayama, R., & Kawane, K. (2010). Autoimmunity and the clearance of dead cells. *Cell, 140*, 619–630. https://doi.org/10.1016/j.cell.2010.02.014
78. Tian, L., et al. (2016). Enhanced efferocytosis by dendritic cells underlies memory T-cell expansion and susceptibility to autoimmune disease in CD300f-deficient mice. *Cell Death and Differentiation, 23*, 1086–1096. https://doi.org/10.1038/cdd.2015.161
79. Rovere, P., et al. (1998). Bystander apoptosis triggers dendritic cell maturation and antigen-presenting function. *Journal of Immunology, 161*, 4467–4471.
80. Rovere, P., et al. (1999). Delayed clearance of apoptotic lymphoma cells allows cross-presentation of intracellular antigens by mature dendritic cells. *Journal of Leukocyte Biology, 66*, 345–349. https://doi.org/10.1002/jlb.66.2.345
81. Birge, R. B., et al. (2016). Phosphatidylserine is a global immunosuppressive signal in efferocytosis, infectious disease, and cancer. *Cell Death and Differentiation, 23*, 962–978. https://doi.org/10.1038/cdd.2016.11
82. Ashida, H., et al. (2011). Cell death and infection: A double-edged sword for host and pathogen survival. *The Journal of Cell Biology, 195*, 931–942. https://doi.org/10.1083/jcb.201108081
83. Friedrich, A., Pechstein, J., Berens, C., & Luhrmann, A. (2017). Modulation of host cell apoptotic pathways by intracellular pathogens. *Current Opinion in Microbiology, 35*, 88–99. https://doi.org/10.1016/j.mib.2017.03.001
84. Maiti, D., Bhattacharyya, A., & Basu, J. (2001). Lipoarabinomannan from Mycobacterium tuberculosis promotes macrophage survival by phosphorylating Bad through a phosphatidylinositol 3-kinase/Akt pathway. *The Journal of Biological Chemistry, 276*, 329–333. https://doi.org/10.1074/jbc.M002650200
85. Dhiman, R., Raje, M., & Majumdar, S. (2007). Differential expression of NF-kappaB in mycobacteria infected THP-1 affects apoptosis. *Biochimica et Biophysica Acta, 1770*, 649–658. https://doi.org/10.1016/j.bbagen.2006.11.016
86. Schaible, U. E., et al. (2003). Apoptosis facilitates antigen presentation to T lymphocytes through MHC-I and CD1 in tuberculosis. *Nature Medicine, 9*, 1039–1046. https://doi.org/10.1038/nm906
87. Winau, F., et al. (2006). Apoptotic vesicles crossprime CD8 T cells and protect against tuberculosis. *Immunity, 24*, 105–117. https://doi.org/10.1016/j.immuni.2005.12.001
88. Torchinsky, M. B., Garaude, J., Martin, A. P., & Blander, J. M. (2009). Innate immune recognition of infected apoptotic cells directs T(H)17 cell differentiation. *Nature, 458*, 78–82. https://doi.org/10.1038/nature07781

89. Uderhardt, S., et al. (2012). 12/15-lipoxygenase orchestrates the clearance of apoptotic cells and maintains immunologic tolerance. *Immunity, 36*, 834–846. https://doi.org/10.1016/j.immuni.2012.03.010
90. Albert, M. L., et al. (1998). Immature dendritic cells phagocytose apoptotic cells via alphavbeta5 and CD36, and cross-present antigens to cytotoxic T lymphocytes. *The Journal of Experimental Medicine, 188*, 1359–1368. https://doi.org/10.1084/jem.188.7.1359
91. Blander, J. M., & Medzhitov, R. (2006). Toll-dependent selection of microbial antigens for presentation by dendritic cells. *Nature, 440*, 808–812. https://doi.org/10.1038/nature04596
92. Ercolini, A. M., & Miller, S. D. (2009). The role of infections in autoimmune disease. *Clinical and Experimental Immunology, 155*, 1–15. https://doi.org/10.1111/j.1365-2249.2008.03834.x
93. Noack, M., & Miossec, P. (2014). Th17 and regulatory T cell balance in autoimmune and inflammatory diseases. *Autoimmunity Reviews, 13*, 668–677. https://doi.org/10.1016/j.autrev.2013.12.004
94. Rigoni, T. S., et al. (2022). Axl receptor induces efferocytosis, dampens M1 macrophage responses and promotes heart pathology in Trypanosoma cruzi infection. *Communications Biology, 5*, 1421. https://doi.org/10.1038/s42003-022-04401-w
95. Dalboni, L. C., et al. (2021). Encephalitozoon cuniculi takes advantage of efferocytosis to evade the immune response. *PLoS One, 16*, e0247658. https://doi.org/10.1371/journal.pone.0247658
96. Salina, A. C., Souza, T. P., Serezani, C. H., & Medeiros, A. I. (2017). Efferocytosis-induced prostaglandin E2 production impairs alveolar macrophage effector functions during Streptococcus pneumoniae infection. *Innate Immunity, 23*, 219–227. https://doi.org/10.1177/1753425916684934
97. van Zandbergen, G., et al. (2004). Cutting edge: Neutrophil granulocyte serves as a vector for Leishmania entry into macrophages. *Journal of Immunology, 173*, 6521–6525. https://doi.org/10.4049/jimmunol.173.11.6521
98. Bertsias, G. K., Pamfil, C., Fanouriakis, A., & Boumpas, D. T. (2013). Diagnostic criteria for systemic lupus erythematosus: Has the time come? *Nature Reviews Rheumatology, 9*, 687–694. https://doi.org/10.1038/nrrheum.2013.103
99. Aringer, M., et al. (2019). 2019 European League Against Rheumatism/American College of Rheumatology classification criteria for systemic lupus erythematosus. *Annals of the Rheumatic Diseases, 78*, 1151–1159. https://doi.org/10.1136/annrheumdis-2018-214819
100. Munoz, L. E., Lauber, K., Schiller, M., Manfredi, A. A., & Herrmann, M. (2010). The role of defective clearance of apoptotic cells in systemic autoimmunity. *Nature Reviews Rheumatology, 6*, 280–289. https://doi.org/10.1038/nrrheum.2010.46
101. Hargraves, M. M., Richmond, H., & Morton, R. (1948). Presentation of two bone marrow elements; the tart cell and the L.E. cell. *Proceedings of the staff meetings. Mayo Clinic, 23*, 25–28.
102. Sundberg, R. D., & Lick, N. B. L. E. (1949). cells in the blood in acute disseminated lupus erythematosus. *The Journal of Investigative Dermatology, 12*, 83.
103. Hargraves, M. M. (1949). Production *in vitro* of the L.E. cell phenomenon; use of normal bone marrow elements and blood plasma from patients with acute disseminated lupus erythematosus. *Proceedings of the staff meetings. Mayo Clinic, 24*, 234–237.
104. Sabouraud, R. (1910). Sur quelques points de l'anatomie pathologique du lichen plan de Wilson. *Annales de Dermatologie et de Syphiligraphie, 5*, 491–505.
105. Crowson, A. N., & Magro, C. (2001). The cutaneous pathology of lupus erythematosus: A review. *Journal of Cutaneous Pathology, 28*, 1–23. https://doi.org/10.1034/j.1600-0560.2001.280101.x
106. Burnham, T. K., Neblett, T. R., & Fine, G. (1963). The application of the fluorescent antibody technic to the investigation of lupus erythematosus and various dermatoses. *The Journal of Investigative Dermatology, 41*, 451–456. https://doi.org/10.1038/jid.1963.140
107. Koopman, K., et al. (2017). Nuclear proteins and apoptotic bodies are found in the lupus band of patients with cutaneous lupus erythematosus. *The Journal of Investigative Dermatology, 137*, 2652–2654. https://doi.org/10.1016/j.jid.2017.08.006

108. Kuhn, A., et al. (2006). Accumulation of apoptotic cells in the epidermis of patients with cutaneous lupus erythematosus after ultraviolet irradiation. *Arthritis and Rheumatism, 54*, 939–950. https://doi.org/10.1002/art.21658
109. Agmon-Levin, N., Mosca, M., Petri, M., & Shoenfeld, Y. (2012). Systemic lupus erythematosus one disease or many? *Autoimmunity Reviews, 11*, 593–595. https://doi.org/10.1016/j.autrev.2011.10.020
110. Richard, M. L., & Gilkeson, G. (2018). Mouse models of lupus: What they tell us and what they don't. *Lupus Science & Medicine, 5*, e000199. https://doi.org/10.1136/lupus-2016-000199
111. Pau, E., Loh, C., Minty, G. E., Chang, N. H., & Wither, J. E. (2013). Identification of a lupus-susceptibility locus leading to impaired clearance of apoptotic debris on New Zealand Black chromosome 13. *Genes and Immunity, 14*, 154–161. https://doi.org/10.1038/gene.2012.64
112. Ramaswamy, M., & Siegel, R. M. (2012). Autoimmunity: Twenty years in the Fas lane. *Journal of Immunology, 189*, 5097–5100. https://doi.org/10.4049/jimmunol.1202833
113. Licht, R., Jacobs, C. W., Tax, W. J., & Berden, J. H. (2001). No constitutive defect in phagocytosis of apoptotic cells by resident peritoneal macrophages from pre-morbid lupus mice. *Lupus, 10*, 102–107. https://doi.org/10.1191/096120301672276558
114. Licht, R., Dieker, J. W., Jacobs, C. W., Tax, W. J., & Berden, J. H. (2004). Decreased phagocytosis of apoptotic cells in diseased SLE mice. *Journal of Autoimmunity, 22*, 139–145. https://doi.org/10.1016/j.jaut.2003.11.003
115. Harder, J. W., et al. (2023). Male microbiota-associated metabolite restores macrophage efferocytosis in female lupus-prone mice via activation of PPARgamma/LXR signaling pathways. *Journal of Leukocyte Biology, 113*, 41–57. https://doi.org/10.1093/jleuko/qiac002
116. Saavedra, P. H. V., et al. (2024). Broad-spectrum antibiotics disrupt homeostatic efferocytosis. *Nature Metabolism, 6*, 1682–1694. https://doi.org/10.1038/s42255-024-01107-7
117. Herrmann, M., Zoller, O. M., Hagenhofer, M., Voll, R., & Kalden, J. R. (1996). What triggers anti-dsDNA antibodies? *Molecular Biology Reports, 23*, 265–267. https://doi.org/10.1007/BF00351179
118. Perniok, A., Wedekind, F., Herrmann, M., Specker, C., & Schneider, M. (1998). High levels of circulating early apoptic peripheral blood mononuclear cells in systemic lupus erythematosus. *Lupus, 7*, 113–118. https://doi.org/10.1191/096120398678919804
119. Herrmann, M., et al. (1998). Impaired phagocytosis of apoptotic cell material by monocyte-derived macrophages from patients with systemic lupus erythematosus. *Arthritis and Rheumatism, 41*, 1241–1250. https://doi.org/10.1002/1529-0131(199807)41:7<1241::AID-ART15>3.0.CO;2-H
120. Majai, G., et al. (2014). Decreased apopto-phagocytic gene expression in the macrophages of systemic lupus erythematosus patients. *Lupus, 23*, 133–145. https://doi.org/10.1177/0961203313511557
121. Hu, C. Y., et al. (2009). Genetic polymorphism in milk fat globule-EGF factor 8 (MFG-E8) is associated with systemic lupus erythematosus in human. *Lupus, 18*, 676–681. https://doi.org/10.1177/0961203309103027
122. Lauber, K., et al. (2013). Milk fat globule-EGF factor 8 mediates the enhancement of apoptotic cell clearance by glucocorticoids. *Cell Death and Differentiation, 20*, 1230–1240. https://doi.org/10.1038/cdd.2013.82
123. Cohen, P. L., et al. (2002). Delayed apoptotic cell clearance and lupus-like autoimmunity in mice lacking the c-mer membrane tyrosine kinase. *The Journal of Experimental Medicine, 196*, 135–140. https://doi.org/10.1084/jem.20012094
124. Hanayama, R., et al. (2004). Autoimmune disease and impaired uptake of apoptotic cells in MFG-E8-deficient mice. *Science, 304*, 1147–1150. https://doi.org/10.1126/science.1094359
125. Botto, M., et al. (1998). Homozygous C1q deficiency causes glomerulonephritis associated with multiple apoptotic bodies. *Nature Genetics, 19*, 56–59. https://doi.org/10.1038/ng0598-56
126. Alperin, J. M., Ortiz-Fernandez, L., & Sawalha, A. H. (2018). Monogenic lupus: A developing paradigm of disease. *Frontiers in Immunology, 9*, 2496. https://doi.org/10.3389/fimmu.2018.02496

127. Macedo, A. C., & Isaac, L. (2016). Systemic lupus erythematosus and deficiencies of early components of the complement classical pathway. *Frontiers in Immunology, 7*, 55. https://doi.org/10.3389/fimmu.2016.00055
128. Sekine, H., et al. (2011). The benefit of targeted and selective inhibition of the alternative complement pathway for modulating autoimmunity and renal disease in MRL/lpr mice. *Arthritis and Rheumatism, 63*, 1076–1085. https://doi.org/10.1002/art.30222
129. Korb, L. C., & Ahearn, J. M. (1997). C1q binds directly and specifically to surface blebs of apoptotic human keratinocytes: Complement deficiency and systemic lupus erythematosus revisited. *Journal of Immunology, 158*, 4525–4528.
130. Taylor, P. R., et al. (2000). A hierarchical role for classical pathway complement proteins in the clearance of apoptotic cells in vivo. *The Journal of Experimental Medicine, 192*, 359–366. https://doi.org/10.1084/jem.192.3.359
131. Ramirez-Ortiz, Z. G., et al. (2013). The scavenger receptor SCARF1 mediates the clearance of apoptotic cells and prevents autoimmunity. *Nature Immunology, 14*, 917–926. https://doi.org/10.1038/ni.2670
132. Wang, F. M., et al. (2016). The dysfunctions of complement factor H in lupus nephritis. *Lupus, 25*, 1328–1340. https://doi.org/10.1177/0961203316642307
133. Martin, M., et al. (2016). Factor H uptake regulates intracellular C3 activation during apoptosis and decreases the inflammatory potential of nucleosomes. *Cell Death and Differentiation, 23*, 903–911. https://doi.org/10.1038/cdd.2015.164
134. Verbovetski, I., et al. (2002). Opsonization of apoptotic cells by autologous iC3b facilitates clearance by immature dendritic cells, down-regulates DR and CD86, and up-regulates CC chemokine receptor 7. *The Journal of Experimental Medicine, 196*, 1553–1561. https://doi.org/10.1084/jem.20020263
135. Amarilyo, G., et al. (2010). iC3b-opsonized apoptotic cells mediate a distinct anti-inflammatory response and transcriptional NF-kappaB-dependent blockade. *European Journal of Immunology, 40*, 699–709. https://doi.org/10.1002/eji.200838951
136. Parsons, H., Snyder, F., Bowen, T., Klassen, J., & Pinto, A. (1996). Immune complex disease consistent with systemic lupus erythematosus in a patient with lysinuric protein intolerance. *Journal of Inherited Metabolic Disease, 19*, 627–634. https://doi.org/10.1007/BF01799838
137. Kamoda, T., et al. (1998). Lysinuric protein intolerance and systemic lupus erythematosus. *European Journal of Pediatrics, 157*, 130–131. https://doi.org/10.1007/s004310050784
138. Tirosh, I., et al. (2019). Whole exome sequencing in childhood-onset lupus frequently detects single gene etiologies. *Pediatric Rheumatology Online Journal, 17*, 52. https://doi.org/10.1186/s12969-019-0349-y
139. Contreras, J. L., et al. (2021). Immune dysregulation mimicking systemic lupus erythematosus in a patient with lysinuric protein intolerance: Case report and review of the literature. *Frontiers in Pediatrics, 9*, 673957. https://doi.org/10.3389/fped.2021.673957
140. Demy, D. L., et al. (2020). The cationic amino acid exporter Slc7a7 is induced and vital in zebrafish tissue macrophages with sustained efferocytic activity. *Journal of Cell Science, 133*. https://doi.org/10.1242/jcs.249037
141. Geng, L., et al. (2022). Human SLE variant NCF1-R90H promotes kidney damage and murine lupus through enhanced Tfh2 responses induced by defective efferocytosis of macrophages. *Annals of the Rheumatic Diseases, 81*, 255–267. https://doi.org/10.1136/annrheumdis-2021-220793
142. Ablin, J., Verbovetski, I., Trahtemberg, U., Metzger, S., & Mevorach, D. (2005). Quinidine and procainamide inhibit murine macrophage uptake of apoptotic and necrotic cells: A novel contributing mechanism of drug-induced-lupus. *Apoptosis: An International Journal on Programmed Cell Death, 10*, 1009–1018. https://doi.org/10.1007/s10495-005-1189-4
143. Ronnblom, L., & Leonard, D. (2019). Interferon pathway in SLE: One key to unlocking the mystery of the disease. *Lupus Science & Medicine, 6*, e000270. https://doi.org/10.1136/lupus-2018-000270

144. Li, H., et al. (2015). Interferon-induced mechanosensing defects impede apoptotic cell clearance in lupus. *The Journal of Clinical Investigation, 125*, 2877–2890. https://doi.org/10.1172/JCI81059
145. Grossmayer, G. E., et al. (2017). Elevated serum lysophosphatidylcholine in patients with systemic lupus erythematosus impairs phagocytosis of necrotic cells in vitro. *Frontiers in Immunology, 8*, 1876. https://doi.org/10.3389/fimmu.2017.01876
146. Ekman, C., Jonsen, A., Sturfelt, G., Bengtsson, A. A., & Dahlback, B. (2011). Plasma concentrations of Gas6 and sAxl correlate with disease activity in systemic lupus erythematosus. *Rheumatology (Oxford), 50*, 1064–1069. https://doi.org/10.1093/rheumatology/keq459
147. Wu, J., et al. (2011). Increased plasma levels of the soluble Mer tyrosine kinase receptor in systemic lupus erythematosus relate to disease activity and nephritis. *Arthritis Research & Therapy, 13*, R62. https://doi.org/10.1186/ar3316
148. Ballantine, L., et al. (2015). Increased soluble phagocytic receptors sMer, sTyro3 and sAxl and reduced phagocytosis in juvenile-onset systemic lupus erythematosus. *Pediatric Rheumatology Online Journal, 13*, 10. https://doi.org/10.1186/s12969-015-0007-y
149. Bellan, M., et al. (2021). Increased plasma levels of Gas6 and its soluble tyrosine kinase receptors Mer and Axl are associated with immunological activity and severity of lupus nephritis. *Clinical and Experimental Rheumatology, 39*, 132–138.
150. Zhou, Z., et al. (2020). Anti-Tyro3 IgG associates with disease activity and reduces efferocytosis of macrophages in new-onset systemic lupus erythematosus. *Journal of Immunology Research, 2 180708*. https://doi.org/10.1155/2020/2180708
151. Jorge, A. M., et al. (2022). SCARF1-induced efferocytosis plays an immunomodulatory role in humans, and autoantibodies targeting SCARF1 are produced in patients with systemic lupus erythematosus. *Journal of Immunology, 208*, 955–967. https://doi.org/10.4049/jimmunol.2100532
152. Antes, U., Heinz, H. P., Hartung, K., & Loos, M. (1990). Autoantibodies against the complement component C1q in systemic lupus erythematosus. *Klinische Wochenschrift, 68*, 1066–1070. https://doi.org/10.1007/BF01649307
153. Gershov, D., Kim, S., Brot, N., & Elkon, K. B. (2000). C-Reactive protein binds to apoptotic cells, protects the cells from assembly of the terminal complement components, and sustains an antiinflammatory innate immune response: Implications for systemic autoimmunity. *The Journal of Experimental Medicine, 192*, 1353–1364. https://doi.org/10.1084/jem.192.9.1353
154. Szalai, A. J., et al. (2003). Delayed lupus onset in (NZB x NZW)F1 mice expressing a human C-reactive protein transgene. *Arthritis and Rheumatism, 48*, 1602–1611. https://doi.org/10.1002/art.11026
155. Gardai, S. J., et al. (2005). Cell-surface calreticulin initiates clearance of viable or apoptotic cells through trans-activation of LRP on the phagocyte. *Cell, 123*, 321–334. https://doi.org/10.1016/j.cell.2005.08.032
156. Scannell, M., et al. (2007). Annexin-1 and peptide derivatives are released by apoptotic cells and stimulate phagocytosis of apoptotic neutrophils by macrophages. *Journal of Immunology, 178*, 4595–4605. https://doi.org/10.4049/jimmunol.178.7.4595
157. Trendelenburg, M. (2021). Autoantibodies against complement component C1q in systemic lupus erythematosus. *Clinical & Translational Immunology, 10*, e1279. https://doi.org/10.1002/cti2.1279
158. Sjowall, C., Zickert, A., Skogh, T., Wettero, J., & Gunnarsson, I. (2009). Serum levels of autoantibodies against C-reactive protein correlate with renal disease activity and response to therapy in lupus nephritis. *Arthritis Research & Therapy, 11*, R188. https://doi.org/10.1186/ar2880
159. Bruschi, M., et al. (2014). Glomerular autoimmune multicomponents of human lupus nephritis in vivo: Alpha-enolase and annexin AI. *J Am Soc Nephrol, 25*, 2483–2498. https://doi.org/10.1681/ASN.2013090987
160. Boehm, J., Orth, T., Van Nguyen, P., & Soling, H. D. (1994). Systemic lupus erythematosus is associated with increased auto-antibody titers against calreticulin and grp94, but calreticulin

is not the Ro/SS-A antigen. *European Journal of Clinical Investigation, 24*, 248–257. https://doi.org/10.1111/j.1365-2362.1994.tb01082.x

161. Biermann, M. H. C., et al. (2018). Autoantibodies recognizing secondary NEcrotic cells promote neutrophilic phagocytosis and identify patients with systemic lupus erythematosus. *Frontiers in Immunology, 9*, 989. https://doi.org/10.3389/fimmu.2018.00989
162. Reefman, E., et al. (2007). Opsonization of late apoptotic cells by systemic lupus erythematosus autoantibodies inhibits their uptake via an Fcgamma receptor-dependent mechanism. *Arthritis and Rheumatism, 56*, 3399–3411. https://doi.org/10.1002/art.22947
163. Munoz, L. E., et al. (2009). Remnants of secondarily necrotic cells fuel inflammation in systemic lupus erythematosus. *Arthritis and Rheumatism, 60*, 1733–1742. https://doi.org/10.1002/art.24535
164. Grossmayer, G. E., et al. (2008). IgG autoantibodies bound to surfaces of necrotic cells and complement C4 comprise the phagocytosis promoting activity for necrotic cells of systemic lupus erythaematosus sera. *Annals of the Rheumatic Diseases, 67*, 1626–1632. https://doi.org/10.1136/ard.2007.081828
165. Mahajan, A., Herrmann, M., & Munoz, L. E. (2016). Clearance deficiency and cell death pathways: A model for the pathogenesis of SLE. *Frontiers in Immunology, 7*, 35. https://doi.org/10.3389/fimmu.2016.00035
166. Baumann, I., et al. (2002). Impaired uptake of apoptotic cells into tingible body macrophages in germinal centers of patients with systemic lupus erythematosus. *Arthritis and Rheumatism, 46*, 191–201. https://doi.org/10.1002/1529-0131(200201)46:1<191::AID-ART10027>3.0.CO;2-K
167. Podolska, M. J., Biermann, M. H., Maueroder, C., Hahn, J., & Herrmann, M. (2015). Inflammatory etiopathogenesis of systemic lupus erythematosus: An update. *Journal of Inflammation Research, 8*, 161–171. https://doi.org/10.2147/JIR.S70325
168. Green, D. R., Ferguson, T., Zitvogel, L., & Kroemer, G. (2009). Immunogenic and tolerogenic cell death. *Nature Reviews. Immunology, 9*, 353–363. https://doi.org/10.1038/nri2545
169. Janko, C., et al. (2011). CRP/anti-CRP antibodies assembly on the surfaces of cell remnants switches their phagocytic clearance toward inflammation. *Frontiers in Immunology, 2*, 70. https://doi.org/10.3389/fimmu.2011.00070
170. Ren, Y., et al. (2003). Increased apoptotic neutrophils and macrophages and impaired macrophage phagocytic clearance of apoptotic neutrophils in systemic lupus erythematosus. *Arthritis and Rheumatism, 48*, 2888–2897. https://doi.org/10.1002/art.11237
171. Tan, W. C., et al. (2015). Characteristics of COPD in never-smokers and ever-smokers in the general population: Results from the CanCOLD study. *Thorax, 70*, 822–829. https://doi.org/10.1136/thoraxjnl-2015-206938
172. Liu, Y., Liu, H., Li, C., Ma, C., & Ge, W. (2020). Proteome profiling of lung tissues in chronic obstructive pulmonary disease (COPD): Platelet and macrophage dysfunction contribute to the pathogenesis of COPD. *International Journal of Chronic Obstructive Pulmonary Disease, 15*, 973–980. https://doi.org/10.2147/COPD.S246845
173. Zhang, S., et al. (2015). MFG-E8, a clearance glycoprotein of apoptotic cells, as a new marker of disease severity in chronic obstructive pulmonary disease. *Brazilian Journal of Medical and Biological Research, 48*, 1032–1038. https://doi.org/10.1590/1414-431X20154730
174. Wang, Y., et al. (2017). Cigarette smoke attenuates phagocytic ability of macrophages through down-regulating Milk fat globule-EGF factor 8 (MFG-E8) expressions. *Scientific Reports, 7*, 42642. https://doi.org/10.1038/srep42642
175. Hodge, S., et al. (2007). Smoking alters alveolar macrophage recognition and phagocytic ability: Implications in chronic obstructive pulmonary disease. *American Journal of Respiratory Cell and Molecular Biology, 37*, 748–755. https://doi.org/10.1165/rcmb.2007-0025OC
176. Hodge, S., Hodge, G., Scicchitano, R., Reynolds, P. N., & Holmes, M. (2003). Alveolar macrophages from subjects with chronic obstructive pulmonary disease are deficient in their ability to phagocytose apoptotic airway epithelial cells. *Immunology and Cell Biology, 81*, 289–296. https://doi.org/10.1046/j.1440-1711.2003.t01-1-01170.x

177. Asare, P. F., et al. (2021). Inhibition of LC3-associated phagocytosis in COPD and in response to cigarette smoke. *Therapeutic Advances in Respiratory Disease, 15*, 17534666211039769. https://doi.org/10.1177/17534666211039769
178. Richens, T. R., et al. (2009). Cigarette smoke impairs clearance of apoptotic cells through oxidant-dependent activation of RhoA. *American Journal of Respiratory and Critical Care Medicine, 179*, 1011–1021. https://doi.org/10.1164/rccm.200807-1148OC
179. Hodge, S., et al. (2011). Cigarette smoke-induced changes to alveolar macrophage phenotype and function are improved by treatment with procysteine. *American Journal of Respiratory Cell and Molecular Biology, 44*, 673–681. https://doi.org/10.1165/rcmb.2009-0459OC
180. Inoue, M., Ishibashi, Y., Nogawa, H., & Yasue, T. (2012). Carbocisteine promotes phagocytosis of apoptotic cells by alveolar macrophages. *European Journal of Pharmacology, 677*, 173–179. https://doi.org/10.1016/j.ejphar.2011.12.026
181. Ito, H., et al. (2020). Cigarette smoke induces endoplasmic reticulum stress and suppresses efferocytosis through the activation of RhoA. *Scientific Reports, 10*, 12620. https://doi.org/10.1038/s41598-020-69610-x
182. Ween, M. P., et al. (2020). Effects of E-cigarette E-liquid components on bronchial epithelial cells: Demonstration of dysfunctional efferocytosis. *Respirology, 25*, 620–628. https://doi.org/10.1111/resp.13696
183. Serpa, G. L., Renton, N. D., Lee, N., Crane, M. J., & Jamieson, A. M. (2020). Electronic nicotine delivery system aerosol-induced cell death and dysfunction in macrophages and lung epithelial cells. *American Journal of Respiratory Cell and Molecular Biology, 63*, 306–316. https://doi.org/10.1165/rcmb.2019-0200OC
184. Petrache, I., et al. (2005). Ceramide upregulation causes pulmonary cell apoptosis and emphysema-like disease in mice. *Nature Medicine, 11*, 491–498. https://doi.org/10.1038/nm1238
185. Petrusca, D. N., et al. (2010). Sphingolipid-mediated inhibition of apoptotic cell clearance by alveolar macrophages. *The Journal of Biological Chemistry, 285*, 40322–40332. https://doi.org/10.1074/jbc.M110.137604
186. Barnawi, J., et al. (2015). Potential link between the Sphingosine-1-Phosphate (S1P) system and defective alveolar macrophage phagocytic function in Chronic Obstructive Pulmonary Disease (COPD). *PLoS One, 10*, e0122771. https://doi.org/10.1371/journal.pone.0122771
187. Barnawi, J., Jersmann, H., Haberberger, R., Hodge, S., & Meech, R. (2017). Reduced DNA methylation of sphingosine-1 phosphate receptor 5 in alveolar macrophages in COPD: A potential link to failed efferocytosis. *Respirology, 22*, 315–321. https://doi.org/10.1111/resp.12949
188. Barnawi, J., et al. (2016). Pro-phagocytic effects of thymoquinone on cigarette smoke-exposed macrophages occur by modulation of the sphingosine-1-phosphate signalling system. *COPD, 13*, 653–661. https://doi.org/10.3109/15412555.2016.1153614
189. Hodge, S., et al. (2008). Azithromycin improves macrophage phagocytic function and expression of mannose receptor in chronic obstructive pulmonary disease. *American Journal of Respiratory and Critical Care Medicine, 178*, 139–148. https://doi.org/10.1164/rccm.200711-1666OC
190. Hodge, S., et al. (2017). Nonantibiotic macrolides restore airway macrophage phagocytic function with potential anti-inflammatory effects in chronic lung diseases. *American Journal of Physiology. Lung Cellular and Molecular Physiology, 312*, L678–L687. https://doi.org/10.1152/ajplung.00518.2016
191. Hodge, S., & Reynolds, P. N. (2012). Low-dose azithromycin improves phagocytosis of bacteria by both alveolar and monocyte-derived macrophages in chronic obstructive pulmonary disease subjects. *Respirology, 17*, 802–807. https://doi.org/10.1111/j.1440-1843.2012.02135.x
192. Eltboli, O., et al. (2014). COPD exacerbation severity and frequency is associated with impaired macrophage efferocytosis of eosinophils. *BMC Pulmonary Medicine, 14*, 112. https://doi.org/10.1186/1471-2466-14-112
193. Huynh, M. L., et al. (2005). Defective apoptotic cell phagocytosis attenuates prostaglandin E2 and 15-hydroxyeicosatetraenoic acid in severe asthma alveolar macrophages. *American*

Journal of Respiratory and Critical Care Medicine, 172, 972–979. https://doi.org/10.1164/rccm.200501-035OC

194. Liu, Y., et al. (1999). Glucocorticoids promote nonphlogistic phagocytosis of apoptotic leukocytes. *Journal of Immunology, 162*, 3639–3646.
195. Giles, K. M., et al. (2001). Glucocorticoid augmentation of macrophage capacity for phagocytosis of apoptotic cells is associated with reduced p130Cas expression, loss of paxillin/pyk2 phosphorylation, and high levels of active Rac. *Journal of Immunology, 167*, 976–986. https://doi.org/10.4049/jimmunol.167.2.976
196. McColl, A., et al. (2009). Glucocorticoids induce protein S-dependent phagocytosis of apoptotic neutrophils by human macrophages. *Journal of Immunology, 183*, 2167–2175. https://doi.org/10.4049/jimmunol.0803503
197. Gilmour, J. S., et al. (2006). Local amplification of glucocorticoids by 11 beta-hydroxysteroid dehydrogenase type 1 promotes macrophage phagocytosis of apoptotic leukocytes. *Journal of Immunology, 176*, 7605–7611. https://doi.org/10.4049/jimmunol.176.12.7605
198. Grabiec, A. M., et al. (2017). Diminished airway macrophage expression of the Axl receptor tyrosine kinase is associated with defective efferocytosis in asthma. *The Journal of Allergy and Clinical Immunology, 140*, 1144–1146 e1144. https://doi.org/10.1016/j.jaci.2017.03.024
199. Carr, T. F., Zeki, A. A., & Kraft, M. (2018). Eosinophilic and noneosinophilic asthma. *American Journal of Respiratory and Critical Care Medicine, 197*, 22–37. https://doi.org/10.1164/rccm.201611-2232PP
200. Simpson, J. L., et al. (2013). Impaired macrophage phagocytosis in non-eosinophilic asthma. *Clinical and Experimental Allergy, 43*, 29–35. https://doi.org/10.1111/j.1365-2222.2012.04075.x
201. Gao, P., et al. (2015). Anti-inflammatory deficiencies in neutrophilic asthma: Reduced galectin-3 and IL-1RA/IL-1beta. *Respiratory Research, 16*, 5. https://doi.org/10.1186/s12931-014-0163-5
202. Sano, H., et al. (2003). Critical role of galectin-3 in phagocytosis by macrophages. *The Journal of Clinical Investigation, 112*, 389–397. https://doi.org/10.1172/JCI17592
203. Erriah, M., et al. (2019). Galectin-3 enhances monocyte-derived macrophage efferocytosis of apoptotic granulocytes in asthma. *Respiratory Research, 20*, 1. https://doi.org/10.1186/s12931-018-0967-9
204. Fernandez-Boyanapalli, R., et al. (2013). Obesity impairs apoptotic cell clearance in asthma. *The Journal of Allergy and Clinical Immunology, 131*, 1041–1047, 1047 e1041–1043. https://doi.org/10.1016/j.jaci.2012.09.028
205. Miki, H., Pei, H., Gracias, D. T., Linden, J., & Croft, M. (2021). Clearance of apoptotic cells by lung alveolar macrophages prevents development of house dust mite-induced asthmatic lung inflammation. *The Journal of Allergy and Clinical Immunology, 147*, 1087–1092 e1083. https://doi.org/10.1016/j.jaci.2020.10.005
206. Felton, J. M., et al. (2018). Mer-mediated eosinophil efferocytosis regulates resolution of allergic airway inflammation. *The Journal of Allergy and Clinical Immunology, 142*, 1884–1893 e1886. https://doi.org/10.1016/j.jaci.2018.01.029
207. Huang, Q., et al. (2023). Progranulin deficiency suppresses allergic asthma and enhances efferocytosis via PPAR-gamma/MFG-E8 regulation in macrophages. *Immunity, Inflammation and Disease, 11*, e779. https://doi.org/10.1002/iid3.779
208. Juncadella, I. J., et al. (2013). Apoptotic cell clearance by bronchial epithelial cells critically influences airway inflammation. *Nature, 493*, 547–551. https://doi.org/10.1038/nature11714
209. Morimoto, K., et al. (2006). Lovastatin enhances clearance of apoptotic cells (efferocytosis) with implications for chronic obstructive pulmonary disease. *Journal of Immunology, 176*, 7657–7665. https://doi.org/10.4049/jimmunol.176.12.7657
210. Wootton, D. G., et al. (2016). Recovery from pneumonia requires efferocytosis which is impaired in smokers and those with low body mass index and enhanced by statins. *Thorax, 71*, 1052–1054. https://doi.org/10.1136/thoraxjnl-2016-208505
211. Sekheri, M., El Kebir, D., Edner, N., & Filep, J. G. (2020). 15-Epi-LXA(4) and 17-epi-RvD1 restore TLR9-mediated impaired neutrophil phagocytosis and accelerate resolution of

lung inflammation. *Proceedings of the National Academy of Sciences of the United States of America, 117*, 7971–7980. https://doi.org/10.1073/pnas.1920193117
212. Abdulnour, R. E., et al. (2016). Aspirin-triggered resolvin D1 is produced during self-resolving gram-negative bacterial pneumonia and regulates host immune responses for the resolution of lung inflammation. *Mucosal Immunology, 9*, 1278–1287. https://doi.org/10.1038/mi.2015.129
213. Ganesan, R., et al. (2020). D-series resolvins activate phospholipase D in phagocytes during inflammation and resolution. *The FASEB Journal, 34*, 15888–15906. https://doi.org/10.1096/fj.201903025RR
214. Rymut, N., et al. (2020). Resolvin D1 promotes efferocytosis in aging by limiting senescent cell-induced MerTK cleavage. *The FASEB Journal, 34*, 597–609. https://doi.org/10.1096/fj.201902126R
215. Madenspacher, J. H., et al. (2020). Cholesterol 25-hydroxylase promotes efferocytosis and resolution of lung inflammation. *JCI Insight, 5*. https://doi.org/10.1172/jci.insight.137189
216. Hsiao, H. M., et al. (2013). A novel anti-inflammatory and pro-resolving role for resolvin D1 in acute cigarette smoke-induced lung inflammation. *PLoS One, 8*, e58258. https://doi.org/10.1371/journal.pone.0058258
217. Subramaniam, R., et al. (2016). Restoring cigarette smoke-induced impairment of efferocytosis in alveolar macrophages. *Mucosal Immunology, 9*, 873–883. https://doi.org/10.1038/mi.2015.120
218. Li, J., et al. (2022). Ulinastatin promotes macrophage efferocytosis and ameliorates lung inflammation via the ERK5/Mer signaling pathway. *FEBS Open Bio, 12*, 1498–1508. https://doi.org/10.1002/2211-5463.13461
219. Long, M. E., et al. (2017). MEK1/2 inhibition promotes macrophage reparative properties. *Journal of Immunology, 198*, 862–872. https://doi.org/10.4049/jimmunol.1601059
220. Sterner, R. C., & Sterner, R. M. (2021). CAR-T cell therapy: Current limitations and potential strategies. *Blood Cancer Journal, 11*, 69. https://doi.org/10.1038/s41408-021-00459-7
221. Klichinsky, M., et al. (2020). Human chimeric antigen receptor macrophages for cancer immunotherapy. *Nature Biotechnology, 38*, 947–953. https://doi.org/10.1038/s41587-020-0462-y
222. Morioka, S., et al. (2022). Chimeric efferocytic receptors improve apoptotic cell clearance and alleviate inflammation. *Cell, 185*, 4887–4903 e4817. https://doi.org/10.1016/j.cell.2022.11.029
223. Jenner, W., et al. (2014). Characterisation of leukocytes in a human skin blister model of acute inflammation and resolution. *PLoS One, 9*, e89375. https://doi.org/10.1371/journal.pone.0089375
224. De Maeyer, R. P. H., et al. (2020). Blocking elevated p38 MAPK restores efferocytosis and inflammatory resolution in the elderly. *Nature Immunology, 21*, 615–625. https://doi.org/10.1038/s41590-020-0646-0
225. Schloesser, D., et al. (2023). Senescent cells suppress macrophage-mediated corpse removal via upregulation of the CD47-QPCT/L axis. *The Journal of Cell Biology, 222*. https://doi.org/10.1083/jcb.202207097
226. Hu, H., et al. (2023). Defective efferocytosis by aged macrophages promotes STING signaling mediated inflammatory liver injury. *Cell Death Discovery, 9*, 236. https://doi.org/10.1038/s41420-023-01497-9
227. Khanna, S., et al. (2010). Macrophage dysfunction impairs resolution of inflammation in the wounds of diabetic mice. *PLoS One, 5*, e9539. https://doi.org/10.1371/journal.pone.0009539
228. Romana-Souza, B., Chen, L., Leonardo, T. R., Chen, Z., & DiPietro, L. A. (2021). Dermal fibroblast phagocytosis of apoptotic cells: A novel pathway for wound resolution. *The FASEB Journal, 35*, e21443. https://doi.org/10.1096/fj.202002078R
229. Maschalidi, S., et al. (2022). Targeting SLC7A11 improves efferocytosis by dendritic cells and wound healing in diabetes. *Nature, 606*, 776–784. https://doi.org/10.1038/s41586-022-04754-6

230. Ting, S. M., et al. (2020). Brain cleanup as a potential target for poststroke recovery: The role of RXR (Retinoic X receptor) in phagocytes. *Stroke, 51*, 958–966. https://doi.org/10.1161/STROKEAHA.119.027315
231. Hu, M., et al. (2021). High-salt diet downregulates TREM2 expression and blunts efferocytosis of macrophages after acute ischemic stroke. *Journal of Neuroinflammation, 18*, 90. https://doi.org/10.1186/s12974-021-02144-9
232. Nakahashi-Oda, C., et al. (2021). CD300a blockade enhances efferocytosis by infiltrating myeloid cells and ameliorates neuronal deficit after ischemic stroke. *Science Immunology, 6*, eabe7915. https://doi.org/10.1126/sciimmunol.abe7915
233. Glinton, K. E., et al. (2022). Macrophage-produced VEGFC is induced by efferocytosis to ameliorate cardiac injury and inflammation. *The Journal of Clinical Investigation, 132*. https://doi.org/10.1172/JCI140685
234. Cai, S., et al. (2023). Mitochondrial dysfunction in macrophages promotes inflammation and suppresses repair after myocardial infarction. *The Journal of Clinical Investigation, 133*. https://doi.org/10.1172/JCI159498
235. Shirakawa, K., et al. (2020). MerTK expression and ERK activation are essential for the functional maturation of osteopontin-producing reparative macrophages after myocardial infarction. *Journal of the American Heart Association, 9*, e017071. https://doi.org/10.1161/JAHA.120.017071
236. Chen, B., et al. (2019). Macrophage Smad3 protects the infarcted heart, stimulating phagocytosis and regulating inflammation. *Circulation Research, 125*, 55–70. https://doi.org/10.1161/CIRCRESAHA.119.315069
237. Zhang, S., et al. (2019). Efferocytosis fuels requirements of fatty acid oxidation and the electron transport chain to polarize macrophages for tissue repair. *Cell Metabolism, 29*, 443–456 e445. https://doi.org/10.1016/j.cmet.2018.12.004
238. Shirakawa, K., et al. (2018). IL (Interleukin)-10–STAT3–galectin-3 axis is essential for osteopontin-producing reparative macrophage polarization after myocardial infarction. *Circulation, 138*, 2021–2035. https://doi.org/10.1161/circulationaha.118.035047
239. DeBerge, M., et al. (2017). MerTK cleavage on resident cardiac macrophages compromises repair after myocardial ischemia reperfusion injury. *Circulation Research, 121*, 930–940. https://doi.org/10.1161/CIRCRESAHA.117.311327
240. Howangyin, K. Y., et al. (2016). Myeloid-epithelial-reproductive receptor tyrosine kinase and milk fat globule epidermal growth factor 8 coordinately improve remodeling after myocardial infarction via local delivery of vascular endothelial growth factor. *Circulation, 133*, 826–839. https://doi.org/10.1161/CIRCULATIONAHA.115.020857
241. Jun, J. I., Kim, K. H., & Lau, L. F. (2015). The matricellular protein CCN1 mediates neutrophil efferocytosis in cutaneous wound healing. *Nature Communications, 6*, 7386. https://doi.org/10.1038/ncomms8386
242. Del Sordo, L., et al. (2023). Impaired efferocytosis by synovial macrophages in patients with knee osteoarthritis. *Arthritis & Rhematology, 75*, 685–696. https://doi.org/10.1002/art.42412
243. Yao, Z., et al. (2023). Down-regulated GAS6 impairs synovial macrophage efferocytosis and promotes obesity-associated osteoarthritis. *eLife, 12*. https://doi.org/10.7554/eLife.83069
244. Al-Bataineh, M. M., et al. (2021). KIM-1-mediated anti-inflammatory activity is preserved by MUC1 induction in the proximal tubule during ischemia-reperfusion injury. *American Journal of Physiology. Renal Physiology, 321*, F135–F148. https://doi.org/10.1152/ajprenal.00127.2021
245. Arai, Y., Yamaoka, Y., & Morioka, S. (2021). Sweeping up dying cells during tissue injury. *Nephron, 1–4*. https://doi.org/10.1159/000517731
246. Yang, M., et al. (2023). Inhibition of retinoic acid signaling in proximal tubular epithelial cells protects against acute kidney injury. *JCI Insight*. https://doi.org/10.1172/jci.insight.173144
247. Mounier, R., et al. (2013). AMPKalpha1 regulates macrophage skewing at the time of resolution of inflammation during skeletal muscle regeneration. *Cell Metabolism, 18*, 251–264. https://doi.org/10.1016/j.cmet.2013.06.017

248. Al-Zaeed, N., Budai, Z., Szondy, Z., & Sarang, Z. (2021). TAM kinase signaling is indispensable for proper skeletal muscle regeneration in mice. *Cell Death & Disease, 12*, 611. https://doi.org/10.1038/s41419-021-03892-5
249. Liao, Z., et al. (2023). Myofiber directs macrophages IL-10-Vav1-Rac1 efferocytosis pathway in inflamed muscle following CTX myoinjury by activating the intrinsic TGF-beta signaling. *Cell Communication and Signaling: CCS, 21*, 168. https://doi.org/10.1186/s12964-023-01163-8
250. Martin-Rodriguez, O., et al. (2021). Pro-resolving factors released by macrophages after efferocytosis promote mucosal wound healing in inflammatory bowel disease. *Frontiers in Immunology, 12*, 754475. https://doi.org/10.3389/fimmu.2021.754475
251. Meriwether, D., et al. (2022). Macrophage COX2 mediates efferocytosis, resolution reprogramming, and intestinal epithelial repair. *Cellular and Molecular Gastroenterology and Hepatology, 13*, 1095–1120. https://doi.org/10.1016/j.jcmgh.2022.01.002
252. Lee, C. S., et al. (2016). boosting apoptotic cell clearance by colonic epithelial cells attenuates inflammation in vivo. *Immunity, 44*, 807–820. https://doi.org/10.1016/j.immuni.2016.02.005
253. Campana, L., et al. (2018). The STAT3-IL-10-IL-6 pathway is a novel regulator of macrophage efferocytosis and phenotypic conversion in sterile liver injury. *Journal of Immunology, 200*, 1169–1187. https://doi.org/10.4049/jimmunol.1701247
254. Zagorska, A., et al. (2020). Differential regulation of hepatic physiology and injury by the TAM receptors Axl and Mer. *Life Science Alliance, 3*. https://doi.org/10.26508/lsa.202000694
255. Ni, M., et al. (2021). T-cell immunoglobulin and mucin domain-containing protein-4 is critical for kupffer cell homeostatic function in the activation and resolution of liver ischemia reperfusion injury. *Hepatology, 74*, 2118–2132. https://doi.org/10.1002/hep.31906
256. Yurdagul, A., Jr., Doran, A. C., Cai, B., Fredman, G., & Tabas, I. A. (2017). Mechanisms and consequences of defective efferocytosis in atherosclerosis. *Frontiers in Cardiovascular Medicine, 4*, 86. https://doi.org/10.3389/fcvm.2017.00086
257. Kasikara, C., Doran, A. C., Cai, B., & Tabas, I. (2018). The role of non-resolving inflammation in atherosclerosis. *The Journal of Clinical Investigation, 128*, 2713–2723. https://doi.org/10.1172/JCI97950
258. Kiss, M. G., et al. (2023). Cell-autonomous regulation of complement C3 by factor H limits macrophage efferocytosis and exacerbates atherosclerosis. *Immunity, 56*, 1809–1824 e1810. https://doi.org/10.1016/j.immuni.2023.06.026
259. Meng, X. M., Nikolic-Paterson, D. J., & Lan, H. Y. (2016). TGF-beta: The master regulator of fibrosis. *Nature Reviews. Nephrology, 12*, 325–338. https://doi.org/10.1038/nrneph.2016.48
260. Ariel, A., & Timor, O. (2013). Hanging in the balance: Endogenous anti-inflammatory mechanisms in tissue repair and fibrosis. *The Journal of Pathology, 229*, 250–263. https://doi.org/10.1002/path.4108
261. Distler, J. H. W., et al. (2019). Shared and distinct mechanisms of fibrosis. *Nature Reviews Rheumatology, 15*, 705–730. https://doi.org/10.1038/s41584-019-0322-7
262. Wynn, T. A. (2008). Cellular and molecular mechanisms of fibrosis. *The Journal of Pathology, 214*, 199–210. https://doi.org/10.1002/path.2277
263. Lederer, D. J., & Martinez, F. J. (2018). Idiopathic pulmonary fibrosis. *The New England Journal of Medicine, 378*, 1811–1823. https://doi.org/10.1056/NEJMra1705751
264. Lech, M., & Anders, H. J. (1832). Macrophages and fibrosis: How resident and infiltrating mononuclear phagocytes orchestrate all phases of tissue injury and repair. *Biochimica et Biophysica Acta, 989–997*, 2013. https://doi.org/10.1016/j.bbadis.2012.12.001
265. Tang, P. M., Nikolic-Paterson, D. J., & Lan, H. Y. (2019). Macrophages: Versatile players in renal inflammation and fibrosis. *Nature Reviews. Nephrology, 15*, 144–158. https://doi.org/10.1038/s41581-019-0110-2
266. Wynn, T. A., & Vannella, K. M. (2016). Macrophages in tissue repair, regeneration, and fibrosis. *Immunity, 44*, 450–462. https://doi.org/10.1016/j.immuni.2016.02.015
267. Schwabe, R. F., Tabas, I., & Pajvani, U. B. (2020). Mechanisms of fibrosis development in nonalcoholic steatohepatitis. *Gastroenterology, 158*, 1913–1928. https://doi.org/10.1053/j.gastro.2019.11.311

268. Popov, Y., et al. (2010). Macrophage-mediated phagocytosis of apoptotic cholangiocytes contributes to reversal of experimental biliary fibrosis. *American Journal of Physiology. Gastrointestinal and Liver Physiology, 298*, G323–G334. https://doi.org/10.1152/ajpgi.00394.2009
269. An, P., et al. (2020). Hepatocyte mitochondria-derived danger signals directly activate hepatic stellate cells and drive progression of liver fibrosis. *Nature Communications, 11*, 2362. https://doi.org/10.1038/s41467-020-16092-0
270. Li, Y. H., et al. (2021). Mesenchymal stem cells attenuate liver fibrosis by targeting Ly6C(hi/lo) macrophages through activating the cytokine-paracrine and apoptotic pathways. *Cell Death Discovery, 7*, 239. https://doi.org/10.1038/s41420-021-00584-z
271. Yang, W. J., et al. (2022). Acinar ATP8b1/LPC pathway promotes macrophage efferocytosis and clearance of inflammation during chronic pancreatitis development. *Cell Death & Disease, 13*, 893. https://doi.org/10.1038/s41419-022-05322-6
272. Atabai, K., et al. (2009). Mfge8 diminishes the severity of tissue fibrosis in mice by binding and targeting collagen for uptake by macrophages. *The Journal of Clinical Investigation, 119*, 3713–3722. https://doi.org/10.1172/JCI40053
273. Wang, X., et al. (2023). Prolonged hypernutrition impairs TREM2-dependent efferocytosis to license chronic liver inflammation and NASH development. *Immunity, 56*, 58–77 e11. https://doi.org/10.1016/j.immuni.2022.11.013
274. Cai, B., et al. (2020). Macrophage MerTK Promotes Liver Fibrosis in Nonalcoholic Steatohepatitis. *Cell Metabolism, 31*, 406–421 e407. https://doi.org/10.1016/j.cmet.2019.11.013
275. Kim, K. H., Cheng, N., & Lau, L. F. (2022). Cellular communication network factor 1-stimulated liver macrophage efferocytosis drives hepatic stellate cell activation and liver fibrosis. *Hepatology Communications, 6*, 2798–2811. https://doi.org/10.1002/hep4.2057
276. Rovati, L., et al. (2021). Mer tyrosine kinase as a possible link between resolution of inflammation and tissue fibrosis in IgG4-related disease. *Rheumatology (Oxford), 60*, 4929–4941. https://doi.org/10.1093/rheumatology/keab096
277. Lee, Y. J., et al. (2012). Apoptotic cell instillation after bleomycin attenuates lung injury through hepatocyte growth factor induction. *The European Respiratory Journal, 40*, 424–435. https://doi.org/10.1183/09031936.00096711
278. Yoon, Y. S., et al. (2015). PPARgamma activation following apoptotic cell instillation promotes resolution of lung inflammation and fibrosis via regulation of efferocytosis and proresolving cytokines. *Mucosal Immunology, 8*, 1031–1046. https://doi.org/10.1038/mi.2014.130
279. Kim, M. J., et al. (2018). A STAT6 inhibitor AS1517499 reduces preventive effects of apoptotic cell instillation on bleomycin-induced lung fibrosis by suppressing PPARgamma. *Cellular Physiology and Biochemistry, 45*, 1863–1877. https://doi.org/10.1159/000487877
280. Lee, Y. J., et al. (2017). Simvastatin treatment boosts benefits of apoptotic cell infusion in murine lung fibrosis. *Cell Death & Disease, 8*, e2860. https://doi.org/10.1038/cddis.2017.260
281. Kim, K. K., et al. (2018). Efferocytosis of apoptotic alveolar epithelial cells is sufficient to initiate lung fibrosis. *Cell Death & Disease, 9*, 1056. https://doi.org/10.1038/s41419-018-1074-z
282. Chen, W., Frank, M. E., Jin, W., & Wahl, S. M. (2001). TGF-beta released by apoptotic T cells contributes to an immunosuppressive milieu. *Immunity, 14*, 715–725. https://doi.org/10.1016/s1074-7613(01)00147-9
283. Hagimoto, N., et al. (2002). TGF-beta 1 as an enhancer of Fas-mediated apoptosis of lung epithelial cells. *Journal of Immunology, 168*, 6470–6478. https://doi.org/10.4049/jimmunol.168.12.6470
284. Huen, S. C., Moeckel, G. W., & Cantley, L. G. (2013). Macrophage-specific deletion of transforming growth factor-beta1 does not prevent renal fibrosis after severe ischemia-reperfusion or obstructive injury. *American Journal of Physiology. Renal Physiology, 305*, F477–F484. https://doi.org/10.1152/ajprenal.00624.2012
285. Morimoto, K., Janssen, W. J., & Terada, M. (2012). Defective efferocytosis by alveolar macrophages in IPF patients. *Respiratory Medicine, 106*, 1800–1803. https://doi.org/10.1016/j.rmed.2012.08.020

286. Parks, B. W., et al. (2013). CD36, but not G2A, modulates efferocytosis, inflammation, and fibrosis following bleomycin-induced lung injury. *Journal of Lipid Research, 54*, 1114–1123. https://doi.org/10.1194/jlr.M035352
287. Isshiki, T., et al. (2017). Cutting edge: Anti-TIM-3 treatment exacerbates pulmonary inflammation and fibrosis in mice. *Journal of Immunology, 199*, 3733–3737. https://doi.org/10.4049/jimmunol.1700059
288. She, Y., et al. (2023). Elevated expression of macrophage MERTK exhibits profibrotic effects and results in defective regulation of efferocytosis function in pulmonary fibrosis. *Respiratory Research, 24*, 118. https://doi.org/10.1186/s12931-023-02424-3
289. Pastore, M., et al. (2022). Macrophage MerTK promotes profibrogenic cross-talk with hepatic stellate cells via soluble mediators. *JHEP Reports, 4*, 100444. https://doi.org/10.1016/j.jhepr.2022.100444

Messages from Beyond the Grave: How Can *Drosophila* Help Us Understand the Consequences of Apoptosis?

Alessandro Scopelliti, Luigi Zechini, and Will Wood

Abstract

Cell death is an integral part of life. Cells strategically die during the embryonic development of an organism to help sculpt developing tissues into their final form, and cells are constantly lost and replaced during homeostasis as well as through injury. In each of these scenarios, the resulting corpses must be efficiently cleared, and surviving cells must react to the loss of their dying neighbour appropriately such that the health and function of the tissue is maintained. For this to happen effectively, healthy cells need to be alerted to a death in their community rapidly and efficiently. Fortunately, cell death is not a silent process, there is no quiet drifting off into the afterlife for the departing cell but instead it announces its demise through the creation of a plethora of signals both secreted and membrane bound designed to make its neighbours aware of its imminent departure. The fruit fly *Drosophila melanogaster* through its powerful genetics and live imaging has proved an invaluable model in helping researchers understand not only the signalling that drives the process of programmed cell death but also the complex multicellular communications within a tissue that influence a cell's decision to die as well as the downstream consequences of that death. Here, we highlight how research using the humble fly can help to understand the complex multicellular coordination required to successfully deal with the loss of a cell within the complex setting of a three-dimensional living organism.

A. Scopelliti · L. Zechini · W. Wood (✉)
Centre for Inflammation Research, Institute for Regeneration and Repair, University of Edinburgh, Edinburgh, UK
e-mail: Alessandro.Scopelliti@ed.ac.uk; lzechin2@exceed.ed.ac.uk; w.wood@ed.ac.uk

C. D. Gregory et al. (eds.), *Apoptosis and Other forms of Cell Death in the Regulation of Inflammation in Health and Disease*, Advances in Experimental Medicine and Biology 1481, https://doi.org/10.1007/978-3-031-92785-0_3

Keywords

Drosophila · Apoptosis · Macrophage · Cell competition · Efferocytosis · Haemocyte

Abbreviations

AiA	Apoptosis-induced apoptosis
AiDR	Apoptosis-induced death resistance
AiP	Apoptosis-induced proliferation
Apaf1	Apoptotic protease activating factor 1
APC	Adenomatous polyposis coli
Azot	Ahuizotl
Bcl-2	B cell lymphoma 2
bZip	Basic leucine zipper
Calr	Calreticulin
CharON	Caspase and pH-activated reporter, fluorescence on
CNS	Central nervous system
Damm	Death-associated molecule related to Mch2 caspase
DARK	Death-associated APAF1-related killer
Dcp-1	Death caspase-1
Debcl	Death executioner Bcl-2
Decay	Death executioner caspase related to Apopain/Yama
DIAP1	Death-associated inhibitor of apoptosis 1
Dif	Dorsal-related immunity factor
dlg1	Discs large 1
dpp	Decapentaplegic
Dredd	Death-related ced-3/Nedd2-like caspase
DrICE	Death-related ICE-like caspase
Dronc	Death regulator Nedd2-like caspase
Duox	Dual oxidase
EGF	Epidermal growth factor
EGFR	Epidermal growth factor receptor
egr	Eiger
eIF2α	Eukaryotic translation initiation factor 2 subunit alpha
ELMO	Engulfment and cell motility
Elp3	Elongator complex protein 3
ER	Endoplasmic reticulum
ERK	Extracellular signal-regulated kinase
Fwe	Flower
Grnd	Grindelwald
hh	Hedgehog
hid	Head involution defective

IAP	Inhibitor of apoptosis proteins
IkB	Inhibitor of kappa B
Irbp18	Inverted repeat binding protein 18 kDa
JAK-STAT	Janus kinase-signal transducers and activators of transcription
JNK	c-Jun N-terminal kinase
l(2)gl	Lethal (2) giant larvae
LEC	Larval epidermal cells
M	Minute
Mahj	Mahjong
Mbc	Myoblast city
Myo1D	Myosin 1D
NADPH	Nicotinamide adenine dinucleotide phosphate hydrogen
NF-kB	Nuclear factor-kappa B
NimB4	Nimrod B4
NimC1	Nimrod C1
PERK	Pancreatic eIF-2α kinase
PI3K	Phosphatidylinositol 3-kinase
puc	Puckered
Pvf1	PDGF- and VEGF-related factor 1
Pvr	PDGF- and VEGF-receptor related
rab5	Ras-associated binding
Raf	Rapidly accelerated fibrosarcoma
Ras	Rat sarcoma virus
RHG	Reaper, hid and Grim
ROS	Reactive oxygen species
rpr	Reaper
scrib	Scribble
Shark	SH2 ankyrin repeat kinase
SIMU	Six microns under
Src42A	Src oncogene at 42A
Strica	Ser/Thr-rich caspase
Syx7	Syntaxin 7
TGF-β	Transforming growth factor-β
Tie	Tyrosine kinase with immunoglobulin and epidermal growth factor homology domains
TNF	Tumour necrosis factor
TNFR	Tumour necrosis factor receptor
TOR	Target of rapamycin
TSG101	Tumour susceptibility gene 101
wg	Wingless
Wgn	Wengen
wnt	Wingless-related integration site
Wol	Wollknaeuel
Xrp1	X-ray induced P53-dependent #1

Apoptosis Signalling in Flies

Apoptosis is the most common form of programmed cell death allowing the orderly removal of damaged or unneeded cells, a process critical for the normal development of an organism. Apoptotic death is characterised by retention of membrane integrity leading to the formation of apoptotic bodies. Dying cells produce several "find me" and "eat me" signals that direct and instruct phagocytes to clear the corpses without causing an acute inflammatory response. However, these are not the only final messages from the departing cell and the full range of secreted factors generated by apoptotic cells is only now being appreciated. These signals include metabolites, extracellular vesicles and signalling molecules that are able to influence the microenvironment such that, depending on the context, they can promote additional cell death or stimulate proliferation of neighbouring cells to compensate for the cell loss and restore tissue integrity (see also Chap. 6 of this volume). Crucially, several of these signalling pathways can also be hijacked by conditions such as tumours to promote their overgrowth and cause tissue damage.

Drosophila has proven to be a valuable model to discover and understand conserved players regulating the mechanisms of programmed cell death (Table 1). In flies, apoptosis is detected from the early stages of embryonic development where it plays an essential role in coordinating tissue growth and morphogenesis [1–3], but it is also a fundamental requirement during later stages of development operating to refine developmental fluctuations and errors [4–6].

The levels of apoptosis occurring during development vary wildly across different tissues and developmental stages from single cells to entire tissues being removed through the activation of apoptotic machinery once they have served their respective functions. For example, the entire larval salivary glands are lost and

Table 1 Core components of the apoptotic pathway in *Drosophila*

Initiator caspases	Dronc
	Dredd
	Strica
Executioner caspases	DrICE
	Dcp-1
	Decay
	Damm
Adaptor protein	DARK
IAP	DIAP1
IAP antagonists	Reaper
	Hid
	Grim
BCL-2 family members	Buffy
	Debcl
Death receptors	Grindelwald
	Wengen
Death receptor ligand	Eiger

remodelled during metamorphosis [7] while a finer selective reshaping occurs in the developing fly central nervous system, where approximately 30% of the neural population is lost during normal development. Other examples are the successive waves of apoptosis necessary for the correct patterning of the ommatidia in the developing eye, responsible for the removal of around 2000 supernumerary cells [8] or the elimination of supernumerary sensory bristles in the thorax to ensure correct spacing between sensory organs [9]. Apart from developmental activation, the apoptotic program is also activated for the selective elimination of potentially harmful cells following, for example, the accumulation of extensive DNA damage. These examples highlight the requirement of an extremely precise, context dependent, spatial and temporal coordination between autonomous and non-autonomous signals to ensure an efficient activation of the apoptotic program and a regulated response to it.

Apoptosis is triggered by the activation of a highly specialised family of cysteine proteases called Caspases. Initially discovered in the nematode *Caenorhabditis elegans* during the 1980s, Caspases have been shown to be well conserved across metazoans. Activation of Caspases is tightly controlled by a combination of transcriptional and posttranslational mechanisms to prevent undesirable activation.

Drosophila has seven Caspases that are generally divided into two classes, initiator (Dronc, Dredd, Strica) and executioner Caspases (DrICE, Dcp-1, Decay, Damm). However, only three, Dronc, DrICE and Dcp-1, seem to play a major role during apoptotic activation. Dronc is the ortholog of mammalian Caspase-2/9, while DrICE and Dcp-1 are the orthologs of Caspase-3/7, respectively. The remaining Caspases play a tissue specific or redundant function during apoptosis activation or are involved in other unrelated functions. Strica and Dcp-1, for example, are involved in mid-oogenesis cell death [10], and Dredd is involved in the activation of the transcription factor Relish, a member of the NF-kB family, during immune responses [11, 12].

Caspases are produced as catalytically inactive zymogens (pro-Caspases) that can be activated following intrinsic (e.g. oxidative stress) or extrinsic (e.g. TNF) stimuli, by association with scaffolding proteins (e.g. DARK) or cleavage by upstream Caspases. Active Caspases can act on a broad range of substrates leading to the canonical morphological hallmarks of apoptotic cell death: DNA condensation, nuclear fragmentation and membrane blebbing [13].

In flies, Caspases activation is mainly restrained by the Caspase inhibitor DIAP1, a member of the IAP family, that directly inhibits Caspases activation, while Bcl-2 family members (Buffy, Debcl) seem to play only a minor role during the apoptotic program compared to their mammalian counterparts, since the apoptosome complex can form in the absence of cytochrome c and apoptosis can still be induced in absence of Bcl-2 family members.

DIAP1 physically binds and inhibits both the initiator Caspase Dronc (Caspase-9) and the effector Caspases DrICE and Dcp-1. DIAP1 interaction with Dronc prevents the assembly of the apoptosome by preventing DARK (Apaf-1) interaction with Dronc (Fig. 1). Suppression of DIAP1 leads to Caspase activation and apoptosis in the absence of any stimulus.

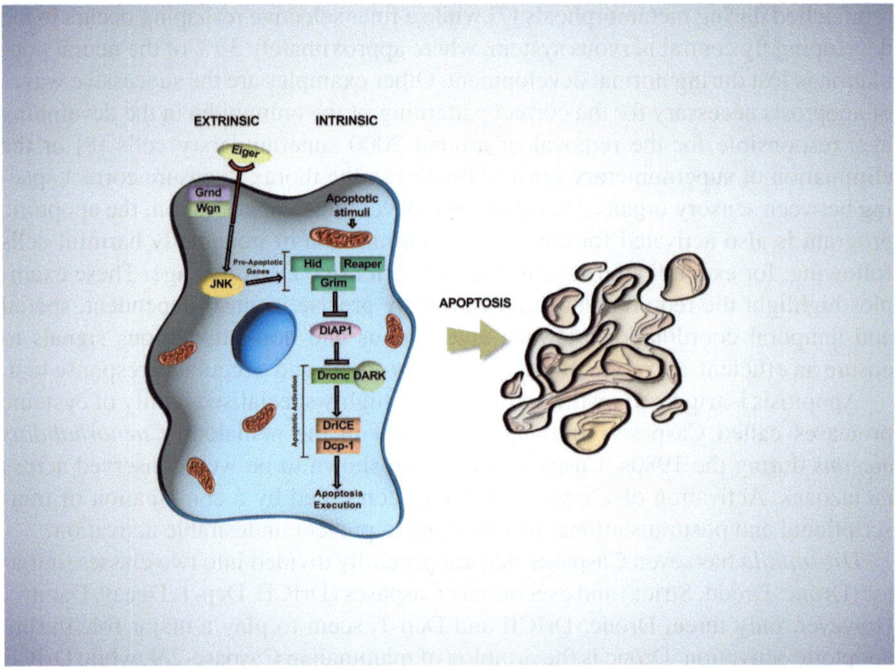

Fig. 1 Apoptotic signalling in flies. Schematic representation of the main components of the apoptotic pathway in *Drosophila*. Apoptosis is mainly controlled by the presence of the DIAP1 antagonists Reaper (Rpr), Hid and Grim (RHG genes). During intrinsic apoptotic pathway activation, mitochondria function as hubs for RHG protein activation, where they promote ubiquitination and degradation of DIAP1. DIAP1 degradation releases the initiator Caspase Dronc (Caspase-9) that associates with the scaffolding protein DARK to form an active apoptosome, leading to the activation of the executioner Caspases DrICE and Dcp-1 (Caspase 3/7). Extrinsic apoptotic activation, on the contrary, is initiated at the plasma membrane by binding of the TNF homolog Eiger to the Wengen (Wgn)/Grindelwald (Grnd) receptors, leading to JNK activation that leads to upregulation of RHG genes (Hid and Rpr) followed by Dronc activation

Upon appropriate apoptotic stimuli, the pro-apoptotic IAP antagonists Hid, Rpr (Reaper) and Grim (collectively known as RHG genes) are upregulated, localise at the mitochondria and promote proteasome degradation of DIAP1. Deletion of *hid*, *rpr* and *grim* genes completely suppresses the apoptotic machinery during embryogenesis.

Upon removal of DIAP1 inhibition, DARK can form the apoptosome complex with Dronc and induce Dronc self-cleavage and activation. Active Dronc in turn cleaves and activates DrICE and Dcp-1 effector Caspases (Fig. 1). Loss of DARK function leads to several developmental abnormalities and hyperplastic phenotypes due to supernumerary cells that fail to undergo apoptosis.

Alternatively, as is observed in mammals, apoptosis in flies can be triggered by extrinsic stimuli. *Drosophila* has a single TNF ortholog (Eiger, egr) that can engage the TNFR Grindelwald (Grnd) or Wengen (Wgn) receptors leading to the activation of the JNK pathway, triggering the upregulation of Hid and Reaper and subsequent Caspases activation (Fig. 1).

Apoptosis: A "Not-So-Silent" Way to Die

Even though apoptosis is commonly considered a silent form of cell death, the opposite appears to be true. Apoptotic cells announce their death with a "song" that has a profound ability to influence the surrounding tissue and in so doing dictate the method of their own elimination and disposal. During morphogenesis, apoptotic cells can change the mechanical properties of an epithelium driving folding, invagination and migration, and inhibition of apoptosis can lead to malformation in adult organs. Apoptotic cells can have both a local influence on the close neighbouring cells and a broader, tissue level, effect. To preserve tissue integrity, a cell committed to die can reorganise its cytoskeletal arrangement to form an intracellular actomyosin ring to induce apical constriction of the dying cell. At the same time, the apoptotic cell induces the neighbouring cells to form a supracellular acto-myosin cable around the dying cell to allow its seamless extrusion. This mechanism allows the maintenance of proper junctions during the extrusion process.

Dorsal closure is a morphogenetic event during the end of embryonic development, where the epidermal sheet converges laterally to close the space occupied by the amnioserosa, an extraembryonic tissue. Amnioserosal cells die by anoikis, a form of cell death induced by loss of adhesion. Inhibition of apoptosis in amnioserosal cells reduces the speed of dorsal closure, while induction of apoptosis accelerates it. Furthermore, it has been shown that amnioserosal apoptosis is responsible in part for the generation of the morphogenetic forces acting on the leading edge of lateral epidermal cells, required for efficient and effective closure of the dorsal epidermis [1, 14] (Fig. 2).

Another example of the long-range effect of apoptotic events comes from elegant studies of abdominal morphogenesis in the fly, an event featuring the beautifully choreographed replacement of larval epidermal cells (LECs) by different clusters of hystoblasts that proliferate, fuse and migrate to the dorsal and ventral midline. LECs are extruded from the epithelium and eliminated by apoptosis [15], mostly at the interface of LEC and hystoblasts. While hystoblast proliferation is required for LEC apoptosis, hystoblast expansion is delayed if LEC apoptosis is inhibited [16, 17]. The relationship between the dying LECs and the proliferating hystoblasts is a complex and intriguing one and is a fascinating area of ongoing research that is likely to uncover much about the dynamic balance between cell death and proliferation within a growing tissue in vivo.

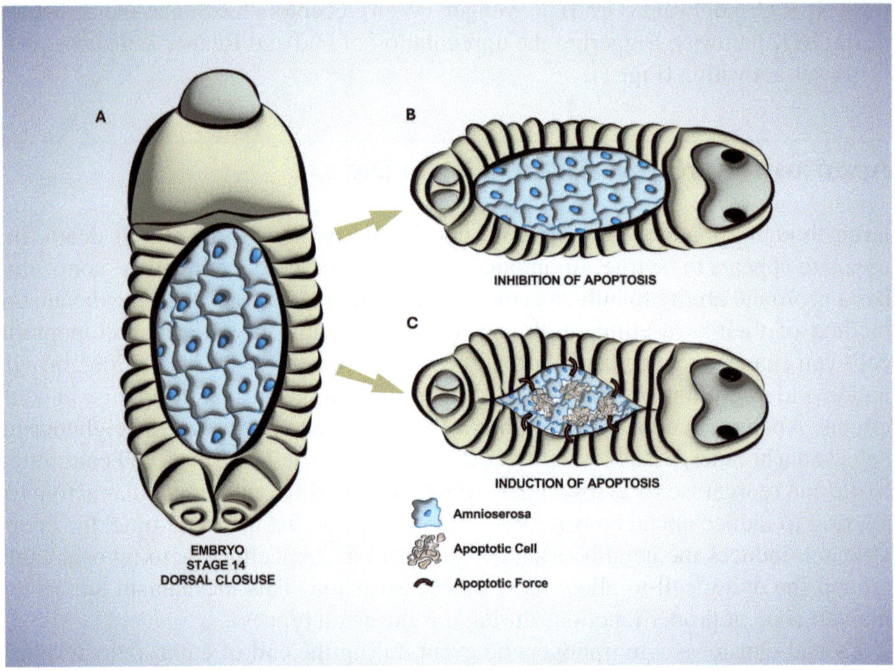

Fig. 2 Apoptosis during tissue morphogenesis. (**a**) Dorsal closure is a key morphogenetic event occurring late in embryonic development, during which the epidermal sheet converges centrally to seal the gap left by dying amnioserosa cells. (**b**) Inhibition of apoptosis in amnioserosal cells significantly slows down the dorsal closure process. (**c**) Apoptosis in these cells creates a pulling force that promotes the movement of epithelial sheets, thus speeding up the closure

Apoptosis-Induced Proliferation (AiP) and Apoptosis-Induced Apoptosis (AiA)

Despite apoptosis being widely recognised as a tumour suppressor mechanism, several clinical data associate an increase in apoptotic burden with poor prognosis [18, 19]. This paradoxical effect can be explained through the ability of dying cells to interact and affect the microenvironment via the release of mitogenic factors that can stimulate cell growth in neighbouring cells. While this communication is fundamental during organogenesis and tissue regeneration, the mechanism can be hijacked by tumour cells to grow, resist chemotherapy and relapse.

First discovered in flies, apoptosis-induced proliferation (AiP) is a compensatory mechanism of proliferation driven by apoptotic stress [20, 21] to replace injured cells by activating proliferation in the neighbour cells. X-ray ablation of a vast portion of the wing disc results in a normal adult wing due to the proliferation of the surviving cells [22]. It is now clear that apoptotic cells release mitogenic factors to promote AiP. *Drosophila* has provided many mechanistic insights into AiP with several different flavours of signal being described but all reveal a fundamental role for Caspases and most of them converge on the production of ROS and the involvement of immune cells. Due to easy genetic manipulation in flies, it is possible to uncouple the apoptotic execution from the AiP mechanisms, by combining the induction of apoptosis through Dronc activation whilst concurrently blocking the executioner Caspases through the expression of the baculovirus apoptotic inhibitor P35, thus generating an "undead" cell. As an alternative model, the apoptotic program can be activated in a temporal and spatial restricted manner in a non-vital organ (e.g. the developing eye) to observe the regenerative response. The use of the undead model has allowed researchers to understand the complex mechanisms driving AiP. In the *Drosophila* wing disc, an apoptotic stimulus triggers the activation of the unconventional Myosin Myo1D that relocalises Dronc to the plasma membrane where it activates the NADPH-oxidase Duox. Duox at the plasma membrane induces the production of extracellular ROS [23, 24]. ROS is a diffusible factor required for the recruitment of circulating macrophages (haemocytes), which once recruited, secrete Eiger that activates the JNK pathway in undead cells by engaging with the Grindelwald receptor [23, 25] leading to the secretion of mitogenic factors like Spitz (EGF), Wg (Wnt) and Dpp (TGF-β) [23, 26] (Fig. 3).

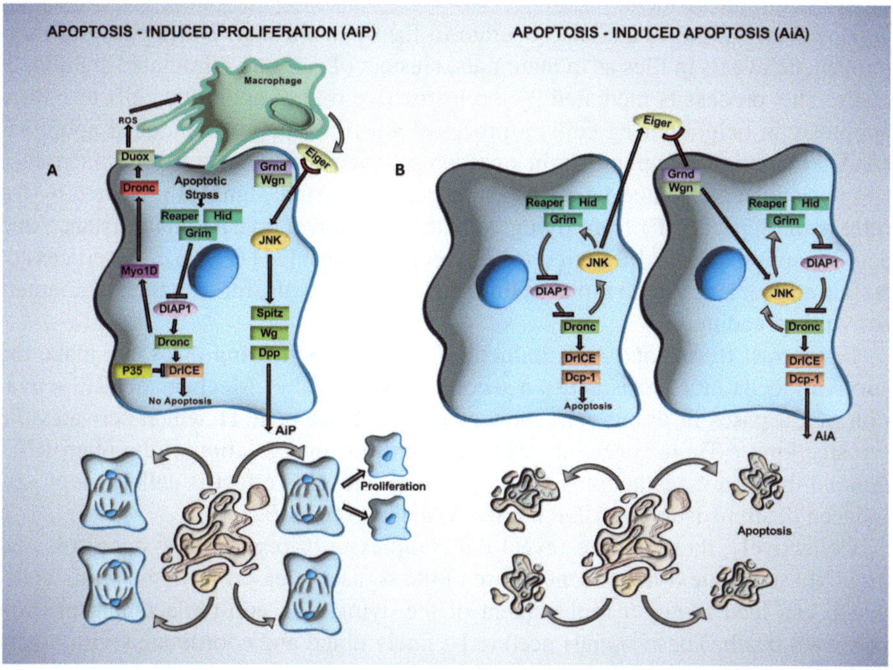

Fig. 3 Effect of apoptotic cells on their neighbours: apoptosis-induced proliferation (AiP) and apoptosis-induced apoptosis (AiA). (**a**) Apoptosis-induced proliferation (AiP): apoptotic cells can induce proliferation in neighbouring cells before they undergo cell death by releasing mitogenic signals (Wg, Dpp and Spitz), thus preserving tissue homeostasis and regeneration. During this process, dying cells recruit circulating macrophages (haemocytes) that activate the Eiger/TNF signalling axis, leading to the production and release of mitogens from dying cells. (**b**) Apoptosis-induced apoptosis (AiA): in addition to promoting cell proliferation, apoptotic cells can also emit pro-death signals, such as Eiger, that diffuse away to propagate apoptosis across a tissue

A role for effector Caspases during AiP that does not involve JNK activation has been described in a non-proliferating tissue, the *Drosophila* retina. Upon apoptosis induction in photoreceptor neurons, DrICE and Dcp-1 drive the secretion of Hedgehog (hh) [20] which drives compensatory proliferation, highlighting a context dependent mechanism of AiP.

The physiological pro-regenerative pathway of AiP may be exploited by cancer cells once they acquire one of the "hallmarks of cancer", namely the resistance to apoptosis. As observed in the undead model, the Caspase-dependent regenerative signal can drive overgrowth in transformed cells. Key to this overgrowth loop is JNK activation which plays a prominent role in several cancers [27]. JNK activation is responsible for the secretion of mitogenic factors [28, 29] and for an amplification loop being responsible for ROS production and the recruitment of macrophages, which in turn sustains JNK activity.

Like the undead model, in a *Ras*^{V12}*; scrib*$^{-/-}$-induced *Drosophila* carcinoma model, Caspase activation has been shown to be important, since both Dronc and DrICE/Dcp-1 are necessary for ROS generation and JNK activation, in an amplification loop that involves macrophage recruitment [30], highlighting the importance of AiP in tumour models.

However, apoptosis can also operate in the opposite way. During several morphogenetic events in flies as in mammals, clusters of cells are eliminated simultaneously. This process is mediated by the instructive role of apoptotic cells to induce apoptosis in neighbouring cells, a process known as apoptosis-induced apoptosis (AiA). Induction of apoptosis in one compartment of the wing disc can induce apoptosis in cells located far away. This death instructive signal is mediated by the release of Eiger (TNF) from dying cells that can travel considerable distance and activate apoptosis in a different wing disc compartment [31] (Fig. 3). The process of AiA is an efficient way to rapidly eliminate groups of cells, for example, to counteract viral spreading.

In contrast to AiP and AiA, induction of apoptosis in wing discs can make the surviving cells more resistant to a second death stimulus. Mechanistically, activation of Caspases in dying cells induces the secretion of Pvf1 which activates the tyrosine kinase Tie in surviving cells leading to the upregulation of the microRNA *bantam* that represses apoptosis by repressing hid. This effect is called apoptosis-induced death resistance (AiDR) or the Mahakali effect [32].

Collectively, these studies reveal the complexity of responses to apoptosis and highlight the context-dependent nature of the signals released from the dying cells, which can lead to either replacement of the dying cell, non-replacement or even increased death. These signals need to be finely tuned and coordinated with others in the microenvironment to ensure cell number within a tissue is maintained during homeostasis and to prevent excessive cell loss or gain which can lead to pathology including neoplastic growth in tumours. What is clear is that these findings in the fly challenge the simplistic view of apoptosis as a tumour-suppressive or tumour-preventive mechanism.

Apoptosis in Cell Competition and Cancer

Cell competition is a homeostatic mechanism of active elimination, via extrusion and/or cell death of viable but unfit (or loser) cells within a tissue by direct interaction with fitter (or winner) neighbouring cells. This mechanism affects clone dynamics within a tissue and is crucial to maintain tissue homeostasis by removing damaged or unfit cells to improve tissue functionality (Fig. 4).

The first description of cell competition came from clonal analysis of *Minute* (M) mutants in flies [33, 34], a group of ribosome subunit mutants affecting growth rate. When the whole imaginal disc carries the mutation, the flies develop normally, but when the tissue is made up from a mosaic of mutant clones interspersed with wild-type tissue, the *M/+* clones are eliminated from the tissue by a JNK-dependent apoptotic program. Typically, the elimination of loser cells happens by activation of the apoptotic pathway at the boundary of the clones, implying paracrine interaction between winner, wild type cells and loser, *M/+* clones. Over the years, similar competitive interactions leading to apoptotic elimination have been described for mutants in polarity genes (*dlg*, *scrib*) [35, 36], ER-stress (*wol*, *Elp3*, *Calr*) [37, 38] endocytosis (Syx7, TSG101, *rab5*) [39–43], E3 ubiquitin ligase (*Mahj*) [38], to mention a few. Contrary to *Minute* clones that are characterised by a reduced growth ability, some mutations that confer a loser status generate over-proliferating tissues but are still eliminated if in contact with normal cells. These mutants show that

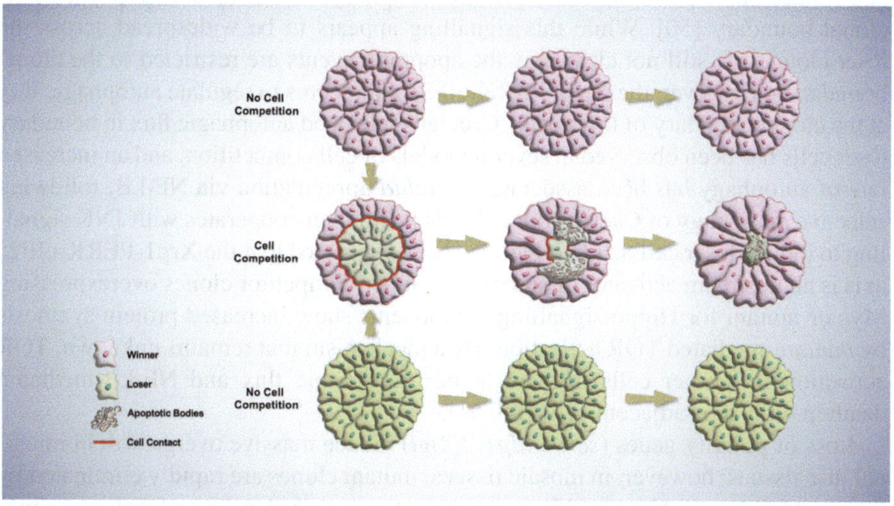

Fig. 4 Apoptosis in cell competition. Competitive interaction between heterologous cell populations may result in the selective elimination of relatively less fit populations (losers) by fitter (winner) ones. When a tissue is composed of a homogeneous population composed entirely of either loser or winner cells (top and bottom panels), no competition occurs. However, when the two populations come into contact, the less fit population is outcompeted and specifically eliminated at the population boundary by the fitter one through apoptosis and/or extrusion. The process enables winner cells to proliferate at the expenses of the loser cells and take over the tissue (middle panel)

winner status is not necessarily dictated by an increased proliferative ability and support the notion that cell competition can act as a tumour suppressive event that eliminates pre-cancerous, unfit cells from normal tissues. However, the winner or loser status of a cell is more dynamic than initially appreciated. Winner or loser status can be gained or lost upon changes in the fitness advantage with respect to the neighbouring population. Wild-type cells can lose their winner status over mutant clones if confronted with clones bearing oncogenic mutations (Myc [5, 44], Wnt [45], JAK-STAT [46], Hippo [47]), an event called super-competition.

Despite different genetic backgrounds giving rise to unique types of cell competition, some common underlying mechanisms have been identified and studies in flies have provided several clues as to how a cell acquires its winner/loser status and the mechanisms by which the winner cell induces apoptosis in loser clones.

Although our understanding of the process of cell competition is still fragmentary, a distinctive trait that distinguishes loser over winner cells is the reduced rate of protein synthesis [37] with an enhanced protein synthesis rate associated with the super-competitor status [48].

Several stressors converge on the upregulation of Xrp1/Irbp18 and the phosphorylation and inhibition of the eIF2α translation initiation factor. ER stress induces PERK-dependent phosphorylation of eIF2α, inhibiting protein synthesis. Xrp1/Irbp18 is a complex of two bZip transcription factors involved in the upregulation of genes associated with the loser status (*flower*, *cactus*, *Dif*, *puc*) and driving apoptosis (*hid*, *rpr*) in loser cells [49]; this complex is upregulated upon ER stress and is necessary and sufficient for eIF2α phosphorylation [37] and apoptosis at the clonal boundary [50]. While this signalling appears to be widespread across the loser clone, it is still not clear how the apoptotic events are restricted to the clonal boundaries. However, the Xrp1-PERK-eIF2α axis seems to regulate autophagic flux at the clonal boundary of loser cells. Crucially, elevated autophagic flux in boundary loser cells has been observed in several models of cell competition, and an increased rate of autophagy has been associated with *hid* upregulation via NF-kB, following selective autophagy of Cactus (IkB) [48, 51]. Hid then cooperates with JNK signalling to induce loser cell death. It has also been suggested that the Xrp1-PERK-eIF2a axis is an upstream activator of autophagy. Super-competitor clones overexpressing Myc or mutant for Hippo signalling components show increased protein synthesis by *bantam*-mediated TOR activation. By a mechanism that remains unknown, TOR activation in winner cells triggers higher autophagic flux and NF-kB-mediated death in wild-type adjacent cells [48, 51].

Loss of polarity genes (*scrib*, *dlg1*, *l(2)gl*) induce massive overgrowth in imaginal disc tissues; however, in mosaic tissues, mutant clones are rapidly eliminated by their wild-type neighbours [35], supporting the notion that loss of polarity genes induces a loser condition. The elimination of *scrib*$^{-/-}$ clones occurs by a combination of JNK-induced Caspase-dependent apoptosis in loser cells and Caspase-independent elimination. Despite the source of egr being context dependent in different loss of polarity models, it appears that recruited macrophages are the main source of egr responsible for JNK activation in loser clones [52].

Polarity deficient clones are eliminated by a JNK-dependent mechanism [35, 53], and inhibition of the receptor Grnd or the ligand Eiger results in tumour overgrowth of the clones [36, 54]. Mechanistically, loss of polarity mislocalises to the basolateral side of the wing disk epithelium where it is activated by the circulating Eiger to induce cell death and extrusion. Interestingly, activation of Eiger/JNK in winning cells was also shown to induce engulfment of loser scrib$^{-/-}$ cells by neighbouring wild-type cells via a Pvr/ELMO/Mbc signalling axis [55] in one of very few studies to address the fate of the dying cell in this system.

Another factor crucial to eliminate loser clones is the Flower (Fwe)/Azot signalling axis, discovered in supercompetitor Myc overexpressing clones. Flower is a plasma membrane Ca^{2+} channel; particular splice isoforms Fwe$^{Lose\text{-}A/\text{-}B}$ are selectively overexpressed in loser cells while the Fwe^{Ubi} isoform is associated with winner clones. Fwe^{Lose} upregulation is responsible for the upregulation of Azot, an EF-hand protein that mediates the upregulation of *hid* and therefore the elimination of loser clones [56–58]. It has been proposed that the interaction of Flower isoforms at the clone boundary can contribute to the elimination of loser cells at this specific location. The ortholog of Fwe^{Ubi} (hFWE-Win) is upregulated in a subset of malignant cancers suggesting that a cell competition mechanism similar to that identified in the fly may be involved in clonal selection during cancer development [59]. Cell competition is considered a tumour suppressive mechanism; however, it is now clear that this mechanism can be hijacked by cancer cells to promote their own proliferation and clonal selection at the expense of surrounding normal cells [60]. As mentioned before, Myc overexpression promotes the elimination of wild-type cells not just in wing discs but also contributes to the establishment of APC mutant clones in adult fly models of adenoma [61].

Recent evidence suggests that cell competition and its resulting death can be triggered in the absence of mutations. When two cell populations with different growth rate, either due to differences in proliferation or cell growth, come into contact, the two populations are subjected to mechanical deformation, stretching and compression. In some instances, this cell deformation can lead to loser cell elimination by apoptosis, a phenomenon called “mechanical competition” [62]. In the *Drosophila* notum, Myc overexpressing clones eliminate wild-type cells as a result of a PI3K-dependent accumulation of F-Actin in winner clones, and the resulting differences in tension between clones result in loser elimination [62]. Similarly, tissue overcrowding induces cell delamination and death during normal development of the notum [63]. Mechanical compression or stretching in the pupal notum can respectively suppress or activate EGFR/Ras/Raf/ERK signalling to regulate death of the notum cells by directly modulating Hid activity [64] and a recent elegant study utilised live imaging to show that the pattern of cell death in this developing tissue can be predicted by the geometry of individual cells with a small apical area being predictive of an apoptotic commitment [65]. Such studies provide insight into our understanding of the spatial regulation of apoptosis within a tissue but what role (if any) the dying cell plays in helping to coordinate this tissue wide coordination of apoptosis once it has committed to die remains to be seen.

As highlighted above, research in the fly has shed light on the signalling and mechanisms that initiate apoptosis in the loser cells during cell competition. However, we know surprisingly little about what happens after death in this context. Based on our knowledge of apoptosis in other settings it seems unlikely that the dying cell goes quietly. What additional factors are produced by dying cells and how they communicate with surviving neighbours (both winners and future losers) are still poorly understood. Furthermore, our understanding of how corpses are cleared and how tissue integrity is maintained during this process is very limited. Future work is needed to address these and related questions, but it can be assumed that the dying cell may well be playing a more active role in this fascinating process than is currently appreciated.

Clearance of Apoptosis

The swift clearance of apoptotic corpses is crucial to avert tissue damage caused by an excessive inflammatory response. One area where researchers have made remarkable advances in understanding the downstream consequences of cell death in *Drosophila* concerns the mechanisms by which apoptotic corpses are disposed of during the fly's development. The recognition, uptake and digestion of apoptotic corpses, a series of events collectively termed "efferocytosis" is a swift and highly efficient process primarily executed by specialist phagocytic cells, haemocytes and glial cells, and in some instances by non-professional phagocytes, which typically do not take part in apoptotic cell clearance, but can occasionally develop this capability. Haemocytes are the fly equivalent of mammalian macrophages and from the embryonic stage onward, they patrol and rapidly detect and engulf apoptotic corpses wherever they come across them [66]. Most of the embryonic apoptosis occurs in the developing central nervous system (CNS) with up to 30% of neurons dying through apoptosis. These corpses are cleared both by circulating macrophages that lie in close proximity to the CNS and specialist phagocytic glial cells that reside within the developing nerve cord. These phagocytic glia share functional and molecular characteristics to vertebrate microglia and their phagocytic ability is comparable to the circulating macrophages. Unlike macrophages, however, they are immotile and remain resident in the CNS throughout development.

Efficient removal of apoptotic corpses requires a coordinated set of events that involve the recruitment of phagocytes by secretion of "find me" signals from the apoptotic bodies. In flies, only a limited number of "find me" signals have been identified. Evidence shows that H_2O_2 released from damaged tissue is crucial for the recruitment of haemocytes to the site of injury by activating a Src42A/Draper/Shark signalling pathway [67, 68]. Once recruited, the phagocyte needs to engage with specific "eat me" molecules (e.g. phosphatidylserine) exposed by the apoptotic corpses. This recognition is mediated by a specific set of phagocytic receptors and secreted bridging molecules. This interaction leads to rearrangements of the phagocyte cytoskeletal components and the formation of the phagosome which undergoes a process of maturation and degradation within the phagocyte.

Not only does the fly employ different types of phagocytes to clear apoptosis in different anatomical locations but individual phagocytes can adapt their engulfment strategies depending on the specific environment in which they find the corpse. Macrophages can engulf apoptotic corpses through two distinct processes termed "lamellipodial phagocytosis" and "filopodial phagocytosis." These two mechanisms require a different set of actin regulatory proteins and remarkably macrophages will rapidly switch between the two modalities when they find themselves spatially constrained [69]. Phagocytic glia appear to exclusively rely on "filopodial phagocytosis" due to their permanent spatial constraint being confined within the ventral nerve cord. This strategy also allows these cells to scan a wide area without the need for migration. Despite these morphological differences and the distinct phagocytic strategy, both phagocytic cell types employ the same suite of receptors to efficiently perform efferocytosis. Both macrophages and phagocytic glia express the Nimrod-family receptors SIMU and Draper, which are utilised by both phagocytes during the clearance of developmental apoptosis. Both receptors have been reported to bind phosphatidylserine which is exposed on the outer leaflet of apoptotic corpse membranes and acts as a robust "eat me" signal. Although Draper seems to be not essential for apoptotic corpse engulfment, it rather appears important for subsequent corpse digestion and processing [70, 71]. Another Nimrod family member NimB4 has been recently shown to play a role in the clearance of apoptotic corpses in embryos and larvae. NimB4 acts as a secreted bridging molecule that binds apoptotic corpses and enhances efferocytosis. Furthermore, the same study revealed a role for two further members of the same gene family, NimC1 and Eater, in corpse uptake highlighting the importance of this intriguing family of receptors in controlling *Drosophila* developmental efferocytosis [72].

The extraordinary speed of efferocytosis has historically presented a challenge for researchers aiming to study this process, since capturing in vivo the moments of uptake are almost impossible. However, the use of *Drosophila* with its relative ease of live imaging has opened new avenues to study this process in real-time within the complex setting of a three-dimensional living animal. A recent study generated transgenic flies carrying a novel efferocytic fluorescent biosensor called CharON (Caspase and pH activated reporter, fluorescence on) [73]. This advancement enabled the visualisation of the entire efferocytic process, from activation of apoptosis in the dying cell, through its detection, engulfment and digestion by macrophages live within a living embryo. This study was able to, for the first time, quantify the dynamics of in vivo efferocytosis. It demonstrated that macrophages prioritise uptake of corpses over their digestion and revealed a striking heterogeneity in the number of apoptotic corpses internalised per macrophage (apoptotic burden) across the phagocyte population. Engulfment of an apoptotic corpse has a profound effect on the macrophage, transforming them from cells that cannot adequately detect damage signals from wounds to fully primed macrophages able to initiate a robust inflammatory response to sites of tissue damage [74]. Interestingly, ingestion of just one apoptotic corpse is sufficient to trigger this "priming" response but the amount a phagocyte eats has fascinating consequences for the cell. Whilst macrophages containing a high number of cell corpses (high burden) are able to detect and migrate to a wound as effectively as their low burden counterparts, they are unable to make the switch to engulf necrotic debris once at the injury site [73]. Such experiments highlight how the downstream consequences of a dying cell can be far-reaching and accomplished indirectly through affecting the behaviour of surrounding cells—in this case, the phagocyte that engulfed it.

Conclusions and Future Directions

Drosophila has played an essential role in advancing our knowledge of the fundamental components of cell death pathways and their intricate biochemical interactions. Despite the progress made over recent years, however, cell death remains a research field with several challenges and key questions that remain unanswered.

The complete understanding of how different cell death and cell survival signals converge in different pathophysiological contexts within a complex multicellular organism is still lacking. For instance, the influence of apoptotic cells on the development and function of immune cells or the effect of metabolic pathways on apoptotic signals or the nature and effects, both local and systemic, of the molecules released by apoptotic bodies are all aspects ripe for further research (Fig. 5). A

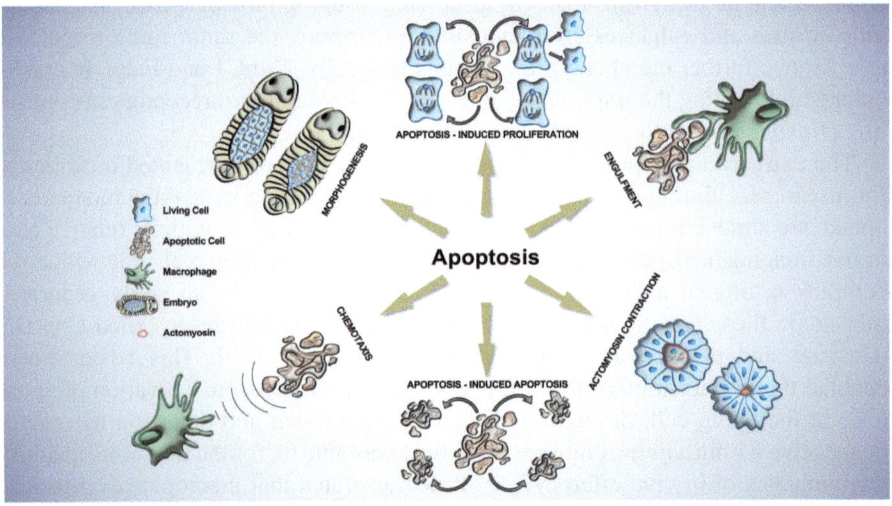

Fig. 5 Apoptosis: a "not-so-silent" way to die. Apoptotic cells play an active and dynamic role in shaping the behaviours of neighbouring cells through a variety of mechanisms. These dying cells influence processes such as engulfment, proliferation, chemotaxis and the modulation of mechanical forces. Importantly, these interactions are often bidirectional, with apoptotic cells both affecting and being affected by the surrounding cellular environment. This crosstalk can lead to the non-autonomous induction of apoptosis within a broader cell community, extending the influence of cell death beyond the individual cell. The active communication between apoptotic and living cells is critical for maintaining tissue homeostasis and is integral to physiological processes such as tissue remodelling, regeneration and morphogenesis. Through these interactions, apoptotic cells help coordinate complex cellular responses that ensure proper development and tissue integrity

deeper understanding of these processes is essential to advance not only our knowledge of basic biology but to also develop new therapeutic strategies for pathologies where apoptosis plays a critical role.

Due to the wide and constantly increasing availability of sophisticated genetic tools together with an ease of experimental manipulation that includes the potential for live imaging, *Drosophila* is in a prime position to overcome the complexities inherent in studying apoptosis and make a significant contribution to tackle these outstanding challenges.

Acknowledgements The research underlying this work is funded by a Wellcome Trust Invesigator Award to WW (22460/Z/21/Z) and an MRC Programme Grant to WW (MR/W019264/1). We would like to thank C. La Floresta (carlottalf@gmail.com) for graphical assistance.

Competing Interests The authors declare no competing interests.

References

1. Toyama, Y., Peralta, X. G., Wells, A. R., Kiehart, D. P., & Edwards, G. S. (2008). Apoptotic force and tissue dynamics during Drosophila embryogenesis. *Science, 321*, 1683–1686.
2. Monier, B., Gettings, M., Gay, G., Mangeat, T., Schott, S., Guarner, A., & Suzanne, M. (2015). Apico-basal forces exerted by apoptotic cells drive epithelium folding. *Nature, 518*, 245–248.
3. Matamoro-Vidal, A., Cumming, T., Davidovic, A., Levillayer, F., & Levayer, R. (2024). Patterned apoptosis has an instructive role for local growth and tissue shape regulation in a fast-growing epithelium. *Current Biology, 34*, 376–388 e377.
4. Merino, M. M., Seum, C., Dubois, M., & Gonzalez-Gaitan, M. (2022). A role for flower and cell death in controlling morphogen gradient scaling. *Nature Cell Biology, 24*, 424–433.
5. de la Cova, C., Abril, M., Bellosta, P., Gallant, P., & Johnston, L. A. (2004). Drosophila myc regulates organ size by inducing cell competition. *Cell, 117*, 107–116.
6. Baker, N. E. (2020). Emerging mechanisms of cell competition. *Nature Reviews. Genetics, 21*, 683–697.
7. Lee, C. Y., & Baehrecke, E. H. (2001). Steroid regulation of autophagic programmed cell death during development. *Development, 128*, 1443–1455.
8. Cagan, R. L., & Ready, D. F. (1989). The emergence of order in the Drosophila pupal retina. *Developmental Biology, 136*, 346–362.
9. Koto, A., Kuranaga, E., & Miura, M. (2011). Apoptosis ensures spacing pattern formation of Drosophila sensory organs. *Current Biology, 21*, 278–287.
10. Baum, J. S., Arama, E., Steller, H., & McCall, K. (2007). The Drosophila caspases Strica and Dronc function redundantly in programmed cell death during oogenesis. *Cell Death and Differentiation, 14*, 1508–1517.
11. Stoven, S., Ando, I., Kadalayil, L., Engstrom, Y., & Hultmark, D. (2000). Activation of the Drosophila NF-kappaB factor relish by rapid endoproteolytic cleavage. *EMBO Reports, 1*, 347–352.
12. Leulier, F., Rodriguez, A., Khush, R. S., Abrams, J. M., & Lemaitre, B. (2000). The Drosophila caspase Dredd is required to resist gram-negative bacterial infection. *EMBO Reports, 1*, 353–358.
13. Crawford, E. D., & Wells, J. A. (2011). Caspase substrates and cellular remodeling. *Annual Review of Biochemistry, 80*, 1055–1087.
14. Gorfinkiel, N., Blanchard, G. B., Adams, R. J., & Martinez Arias, A. (2009). Mechanical control of global cell behaviour during dorsal closure in Drosophila. *Development, 136*, 1889–1898.

15. Bischoff, M. (2012). Lamellipodia-based migrations of larval epithelial cells are required for normal closure of the adult epidermis of Drosophila. *Developmental Biology, 363*, 179–190.
16. Nakajima, Y., Kuranaga, E., Sugimura, K., Miyawaki, A., & Miura, M. (2011). Nonautonomous apoptosis is triggered by local cell cycle progression during epithelial replacement in Drosophila. *Molecular and Cellular Biology, 31*, 2499–2512.
17. Ninov, N., Chiarelli, D. A., & Martin-Blanco, E. (2007). Extrinsic and intrinsic mechanisms directing epithelial cell sheet replacement during Drosophila metamorphosis. *Development, 134*, 367–379.
18. de Jong, J. S., van Diest, P. J., & Baak, J. P. (2000). Number of apoptotic cells as a prognostic marker in invasive breast cancer. *British Journal of Cancer, 82*, 368–373.
19. Gregory, C. D., & Pound, J. D. (2010). Microenvironmental influences of apoptosis in vivo and in vitro. *Apoptosis, 15*, 1029–1049.
20. Fan, Y., & Bergmann, A. (2008). Apoptosis-induced compensatory proliferation. The cell is dead. Long live the cell! *Trends in Cell Biology, 18*, 467–473.
21. Mollereau, B., Perez-Garijo, A., Bergmann, A., Miura, M., Gerlitz, O., Ryoo, H. D., Steller, H., & Morata, G. (2013). Compensatory proliferation and apoptosis-induced proliferation: A need for clarification. *Cell Death and Differentiation, 20*, 181.
22. Haynie, J. L., & Bryant, P. J. (1977). The effects of X-rays on the proliferation dynamics of cells in the imaginal wing disc of Drosophila melanogaster. *Wilehm Roux Arch Developmental Biology, 183*, 85–100.
23. Fogarty, C. E., Diwanji, N., Lindblad, J. L., Tare, M., Amcheslavsky, A., Makhijani, K., Bruckner, K., Fan, Y., & Bergmann, A. (2016). Extracellular reactive oxygen species drive apoptosis-induced proliferation via drosophila macrophages. *Current Biology, 26*, 575–584.
24. Amcheslavsky, A., Wang, S., Fogarty, C. E., Lindblad, J. L., Fan, Y., & Bergmann, A. (2018). Plasma membrane localization of apoptotic caspases for non-apoptotic functions. *Developmental Cell, 45*, 450–464 e453.
25. Diwanji, N., & Bergmann, A. (2017). The beneficial role of extracellular reactive oxygen species in apoptosis-induced compensatory proliferation. *Fly (Austin), 11*, 46–52.
26. Shlevkov, E., & Morata, G. (2012). A dp53/JNK-dependant feedback amplification loop is essential for the apoptotic response to stress in Drosophila. *Cell Death and Differentiation, 19*, 451–460.
27. Bubici, C., & Papa, S. (2014). JNK signalling in cancer: In need of new, smarter therapeutic targets. *British Journal of Pharmacology, 171*, 24–37.
28. Fan, Y., Wang, S., Hernandez, J., Yenigun, V. B., Hertlein, G., Fogarty, C. E., Lindblad, J. L., & Bergmann, A. (2014). Genetic models of apoptosis-induced proliferation decipher activation of JNK and identify a requirement of EGFR signaling for tissue regenerative responses in Drosophila. *PLoS Genetics, 10*, e1004131.
29. Ryoo, H. D., Gorenc, T., & Steller, H. (2004). Apoptotic cells can induce compensatory cell proliferation through the JNK and the Wingless signaling pathways. *Developmental Cell, 7*, 491–501.
30. Perez, E., Lindblad, J. L., & Bergmann, A. (2017). Tumor-promoting function of apoptotic caspases by an amplification loop involving ROS, macrophages and JNK in Drosophila. *eLife, 6*, e26747.
31. Perez-Garijo, A., Fuchs, Y., & Steller, H. (2013). Apoptotic cells can induce non-autonomous apoptosis through the TNF pathway. *eLife, 2*, e01004.
32. Bilak, A., Uyetake, L., & Su, T. T. (2014). Dying cells protect survivors from radiation-induced cell death in Drosophila. *PLoS Genetics, 10*, e1004220.
33. Morata, G., & Ripoll, P. (1975). Minutes: Mutants of drosophila autonomously affecting cell division rate. *Developmental Biology, 42*, 211–221.
34. Katsukawa, M., Ohsawa, S., Zhang, L., Yan, Y., & Igaki, T. (2018). Serpin facilitates tumor-suppressive cell competition by blocking toll-mediated Yki activation in Drosophila. *Current Biology, 28*, 1756–1767 e1756.
35. Brumby, A. M., & Richardson, H. E. (2003). Scribble mutants cooperate with oncogenic Ras or Notch to cause neoplastic overgrowth in Drosophila. *The EMBO Journal, 22*, 5769–5779.

36. Igaki, T., Pastor-Pareja, J. C., Aonuma, H., Miura, M., & Xu, T. (2009). Intrinsic tumor suppression and epithelial maintenance by endocytic activation of Eiger/TNF signaling in Drosophila. *Developmental Cell, 16*, 458–465.
37. Ochi, N., Nakamura, M., Nagata, R., Wakasa, N., Nakano, R., & Igaki, T. (2021). Cell competition is driven by Xrp1-mediated phosphorylation of eukaryotic initiation factor 2alpha. *PLoS Genetics, 17*, e1009958.
38. Tamori, Y., Bialucha, C. U., Tian, A. G., Kajita, M., Huang, Y. C., Norman, M., Harrison, N., Poulton, J., Ivanovitch, K., Disch, L., et al. (2010). Involvement of Lgl and Mahjong/VprBP in cell competition. *PLoS Biology, 8*, e1000422.
39. Moberg, K. H., Schelble, S., Burdick, S. K., & Hariharan, I. K. (2005). Mutations in erupted, the Drosophila ortholog of mammalian tumor susceptibility gene 101, elicit non-cell-autonomous overgrowth. *Developmental Cell, 9*, 699–710.
40. Thompson, B. J., Mathieu, J., Sung, H. H., Loeser, E., Rorth, P., & Cohen, S. M. (2005). Tumor suppressor properties of the ESCRT-II complex component Vps25 in Drosophila. *Developmental Cell, 9*, 711–720.
41. Vaccari, T., & Bilder, D. (2005). The Drosophila tumor suppressor vps25 prevents nonautonomous overproliferation by regulating notch trafficking. *Developmental Cell, 9*, 687–698.
42. Menut, L., Vaccari, T., Dionne, H., Hill, J., Wu, G., & Bilder, D. (2007). A mosaic genetic screen for Drosophila neoplastic tumor suppressor genes based on defective pupation. *Genetics, 177*, 1667–1677.
43. Ballesteros-Arias, L., Saavedra, V., & Morata, G. (2014). Cell competition may function either as tumour-suppressing or as tumour-stimulating factor in Drosophila. *Oncogene, 33*, 4377–4384.
44. Moreno, E., & Basler, K. (2004). dMyc transforms cells into super-competitors. *Cell, 117*, 117–129.
45. Vincent, J. P., Kolahgar, G., Gagliardi, M., & Piddini, E. (2011). Steep differences in wingless signaling trigger Myc-independent competitive cell interactions. *Developmental Cell, 21*, 366–374.
46. Rodrigues, A. B., Zoranovic, T., Ayala-Camargo, A., Grewal, S., Reyes-Robles, T., Krasny, M., Wu, D. C., Johnston, L. A., & Bach, E. A. (2012). Activated STAT regulates growth and induces competitive interactions independently of Myc, Yorkie, Wingless and ribosome biogenesis. *Development, 139*, 4051–4061.
47. Abrams, J. M. (2002). Competition and compensation: Coupled to death in development and cancer. *Cell, 110*, 403–406.
48. Nagata, R., Akai, N., Kondo, S., Saito, K., Ohsawa, S., & Igaki, T. (2022). Yorkie drives supercompetition by non-autonomous induction of autophagy via bantam microRNA in Drosophila. *Current Biology, 32*, 1064–1076 e1064.
49. Baillon, L., Germani, F., Rockel, C., Hilchenbach, J., & Basler, K. (2018). Xrp1 is a transcription factor required for cell competition-driven elimination of loser cells. *Scientific Reports, 8*, 17712.
50. Lee, C. H., Kiparaki, M., Blanco, J., Folgado, V., Ji, Z., Kumar, A., Rimesso, G., & Baker, N. E. (2018). A regulatory response to ribosomal protein mutations controls translation, growth, and cell competition. *Developmental Cell, 46*, 807.
51. Nagata, R., Nakamura, M., Sanaki, Y., & Igaki, T. (2019). Cell competition is driven by autophagy. *Developmental Cell, 51*, 99–112 e114.
52. Cordero, J. B., Macagno, J. P., Stefanatos, R. K., Strathdee, K. E., Cagan, R. L., & Vidal, M. (2010). Oncogenic Ras diverts a host TNF tumor suppressor activity into tumor promoter. *Developmental Cell, 18*, 999–1011.
53. Igaki, T., Pagliarini, R. A., & Xu, T. (2006). Loss of cell polarity drives tumor growth and invasion through JNK activation in Drosophila. *Current Biology, 16*, 1139–1146.
54. Andersen, D. S., Colombani, J., Palmerini, V., Chakrabandhu, K., Boone, E., Rothlisberger, M., Toggweiler, J., Basler, K., Mapelli, M., Hueber, A. O., et al. (2015). The Drosophila TNF receptor Grindelwald couples loss of cell polarity and neoplastic growth. *Nature, 522*, 482–486.

55. Ohsawa, S., Sugimura, K., Takino, K., Xu, T., Miyawaki, A., & Igaki, T. (2011). Elimination of oncogenic neighbors by JNK-mediated engulfment in drosophila. *Developmental Cell, 20*, 315–328.
56. Merino, M. M., Rhiner, C., Lopez-Gay, J. M., Buechel, D., Hauert, B., & Moreno, E. (2015). Elimination of unfit cells maintains tissue health and prolongs lifespan. *Cell, 160*, 461–476.
57. Rhiner, C., Lopez-Gay, J. M., Soldini, D., Casas-Tinto, S., Martin, F. A., Lombardia, L., & Moreno, E. (2010). Flower forms an extracellular code that reveals the fitness of a cell to its neighbors in Drosophila. *Developmental Cell, 18*, 985–998.
58. Yao, C. K., Lin, Y. Q., Ly, C. V., Ohyama, T., Haueter, C. M., Moiseenkova-Bell, V. Y., Wensel, T. G., & Bellen, H. J. (2009). A synaptic vesicle-associated Ca2+ channel promotes endocytosis and couples exocytosis to endocytosis. *Cell, 138*, 947–960.
59. Madan, E., Pelham, C. J., Nagane, M., Parker, T. M., Canas-Marques, R., Fazio, K., Shaik, K., Yuan, Y., Henriques, V., Galzerano, A., et al. (2019). Flower isoforms promote competitive growth in cancer. *Nature, 572*, 260–264.
60. Moreno, E. (2008). Is cell competition relevant to cancer? *Nature Reviews. Cancer, 8*, 141–147.
61. Suijkerbuijk, S. J., Kolahgar, G., Kucinski, I., & Piddini, E. (2016). Cell competition drives the growth of intestinal adenomas in Drosophila. *Current Biology, 26*, 428–438.
62. Levayer, R., Hauert, B., & Moreno, E. (2015). Cell mixing induced by myc is required for competitive tissue invasion and destruction. *Nature, 524*, 476–480.
63. Marinari, E., Mehonic, A., Curran, S., Gale, J., Duke, T., & Baum, B. (2012). Live-cell delamination counterbalances epithelial growth to limit tissue overcrowding. *Nature, 484*, 542–545.
64. Moreno, E., Valon, L., Levillayer, F., & Levayer, R. (2019). Competition for space induces cell elimination through compaction-driven ERK downregulation. *Current Biology, 29*, 23–34 e28.
65. Cachoux, V. M. L., Balakireva, M., Gracia, M., Bosveld, F., Lopez-Gay, J. M., Maugarny, A., Gaugue, I., di Pietro, F., Rigaud, S. U., Noiret, L., et al. (2023). Epithelial apoptotic pattern emerges from global and local regulation by cell apical area. *Current Biology, 33*, 4807–4826 e4806.
66. Weavers, H., & Wood, W. (2016). Creating a buzz about macrophages: The fly as an in vivo model for studying immune cell behavior. *Developmental Cell, 38*, 129–132.
67. Moreira, S., Stramer, B., Evans, I., Wood, W., & Martin, P. (2010). Prioritization of competing damage and developmental signals by migrating macrophages in the Drosophila embryo. *Current Biology, 20*, 464–470.
68. Evans, I. R., Rodrigues, F. S., Armitage, E. L., & Wood, W. (2015). Draper/CED-1 mediates an ancient damage response to control inflammatory blood cell migration in vivo. *Current Biology, 25*, 1606–1612.
69. Davidson, A. J., & Wood, W. (2020). Macrophages use distinct actin regulators to switch engulfment strategies and ensure phagocytic plasticity in vivo. *Cell Reports, 31*, 107692.
70. Kurant, E., Axelrod, S., Leaman, D., & Gaul, U. (2008). Six-microns-under acts upstream of Draper in the glial phagocytosis of apoptotic neurons. *Cell, 133*, 498–509.
71. Shklyar, B., Sellman, Y., Shklover, J., Mishnaevski, K., Levy-Adam, F., & Kurant, E. (2014). Developmental regulation of glial cell phagocytic function during Drosophila embryogenesis. *Developmental Biology, 393*, 255–269.
72. Petrignani, B., Rommelaere, S., Hakim-Mishnaevski, K., Masson, F., Ramond, E., Hilu-Dadia, R., Poidevin, M., Kondo, S., Kurant, E., & Lemaitre, B. (2021). A secreted factor NimrodB4 promotes the elimination of apoptotic corpses by phagocytes in Drosophila. *EMBO Reports, 22*, e52262.
73. Raymond, M. H., Davidson, A. J., Shen, Y., Tudor, D. R., Lucas, C. D., Morioka, S., Perry, J. S. A., Krapivkina, J., Perrais, D., Schumacher, L. J., et al. (2022). Live cell tracking of macrophage efferocytosis during Drosophila embryo development in vivo. *Science, 375*, 1182–1187.
74. Weavers, H., Evans, I. R., Martin, P., & Wood, W. (2016). Corpse engulfment generates a molecular memory that primes the macrophage inflammatory response. *Cell, 165*, 1658–1671.

Metabolic Implications of Efferocytosis During Inflammation

Dmitry Namgaladze, Bernhard Brüne, and Andreas Weigert

Abstract

Every day billions of cells are removed in the human body by apoptosis. These apoptotic cells need to be rapidly and safely removed in order to avoid inflammation. This is ensured by a process called efferocytosis, where professional phagocytes such as macrophages take up apoptotic cells. Efferocytosis not only ensures the safe removal of cellular debris but also endows phagocytes with the capacity to actively contribute to the resolution of inflammation by altering their phenotype. Therefore, actively promoting efferocytosis may contribute to the resolution of auto-inflammatory disease conditions. While the initial steps of the recognition of dying cells by macrophages are well investigated, it is less clear what happens to the ingested material inside the phagocyte and how it influences their pro-resolving properties. In this chapter, we therefore briefly discuss the impact of efferocytosis during inflammation. We focus on the metabolic challenge macrophages face when removing multiple dying cells, e.g., during inflam-

D. Namgaladze
Goethe University Frankfurt, Faculty of Medicine, Institute of Biochemistry I, Frankfurt, Germany
e-mail: namgaladze@biochem.uni-frankfurt.de

B. Brüne · A. Weigert (✉)
Goethe University Frankfurt, Faculty of Medicine, Institute of Biochemistry I, Frankfurt, Germany

Frankfurt Cancer Institute, Goethe University Frankfurt, Frankfurt, Germany

German Cancer Consortium (DKTK), Partner Site Frankfurt, Frankfurt, Germany

Mannheim Institute for Innate Immunoscience, Department for Immunity of Inflammation, Medical Faculty Mannheim of Heidelberg University, Mannheim, Germany
e-mail: b.bruene@biochem.uni-frankfurt.de; weigert@biochem.uni-frankfurt.de

C. D. Gregory et al. (eds.), *Apoptosis and Other forms of Cell Death in the Regulation of Inflammation in Health and Disease*, Advances in Experimental Medicine and Biology 1481, https://doi.org/10.1007/978-3-031-92785-0_4

mation, and how the knowledge resulting from understanding the molecular programs that promote corpse digestion might be used in the future to combat inflammatory diseases in humans.

Keywords

Inflammation · Macrophages · Efferocytosis · Metabolism

Abbreviations

Adora2a	Adenosine receptor A_{2A}
ALOX5	5-Lipoxygenase
ALOX15	Arachidonate 15-lipoxygenase
AMPK	AMP-activated protein kinase
ApoE	Apolipoprotein E
Arg	Arginase
BAI	Brain angiogenesis inhibitor
BMDM	Bone-marrow-derived macrophages
CCL	C-C motif chemokine ligand
COPD	Chronic obstructive pulmonary disease COX, cyclooxygenase
CX_3CL1	Chemokine (C-X3-C motif) ligand 1/fractalkine
DAMPs	Danger-associated molecular patterns
Dbl	Diffuse B cell lymphoma
Drp1	Dynamin-related protein 1
ELMO1	Engulfment and cell motility protein 1
EMAPII	Endothelial monocyte-activating polypeptide-II
FAO	Fatty acid oxidation
GAS	Growth arrest-specific
Gls1	Glutaminase
GLUT	Glucose transporter
GPR	G-protein coupled receptor
GSH	Glutathione
GVHD	Graft-versus-host disease
ICAM	Intercellular adhesion molecule
IL	Interleukin
IFN	Interferon
IRF	Interferon regulatory factor
LAP	LC3-associated phagocytosis
LC3	Microtubule-associated proteins 1A/1B light chain 3B
LPC	Lysophosphatidylcholine
LPS	Lipopolysaccharide
LXR	Liver X receptor
MHC	Major histocompatibility complex
mTORC2	Mammalian target of rapamycin complex 2

NF-κB	Nuclear factor kappa-light-chain-enhancer of activated B cells
OXPHOS	Oxidative phosphorylation
PAF	Platelet-activating factor
PBX1	Pre-B cell leukemia transcription factor-1
PG	Prostaglandin
PFK1	Phosphofructokinase-1
PFKFB2	Phosphofructokinase-2
PPAR	Peroxisome proliferator and regulator of transcription
PPP	Pentose phosphate pathway
PROS1	Protein S
PROTAC	Proteolysis targeting chimera
PS	Phosphatidylserine
PTX	Pentraxin
Rac1	Ras-related C3 botulinum toxin substrate 1
RNS	Reactive nitrogen species
ROS	Reactive oxygen species
S1P	Sphingosine-1-phosphate
SAM	S-adenosylmethionine
SIRP	Signal-regulatory protein
SIRT1	Sirtuin 1
SLC16A1	Monocarboxylate transporter 1
SLE	Systemic lupus erythematosus
SPMs	Specialized pro-resolving lipid mediators
SREBP2	Sterol regulatory element-binding protein 2
STAT	Signal transducer and activator of transcription
TAM	TYRO3, AXL, and MER receptor tyrosine kinases
TAMs	Tumor-associated macrophages
TGF	Transforming growth factor
TIM	T cell/transmembrane, immunoglobulin, and mucin
TNF	Tumor necrosis factor
TXNIP	Thioredoxin-interacting protein
UCP2	Mitochondrial uncoupling protein 2
VEGF	Vascular endothelial growth factor
Xkr	Xk-related protein

Introduction: The Consequences of Inadequately Orchestrated Inflammation

Acute inflammation is the body's reaction to various noxious agents. Contrary to intuition, inflammation, when it is properly executed, is not a disease itself, but rather the physiological program of our body to deal with threats to homeostasis. The salutary intention of inflammation implies that its harmful components that we have all experienced, and which, among others, serve to eliminate pathogens, must be terminated. Inflammation is therefore not only initiated by a multitude of intricately encoded mechanisms, but likely its resolution programs are equally well

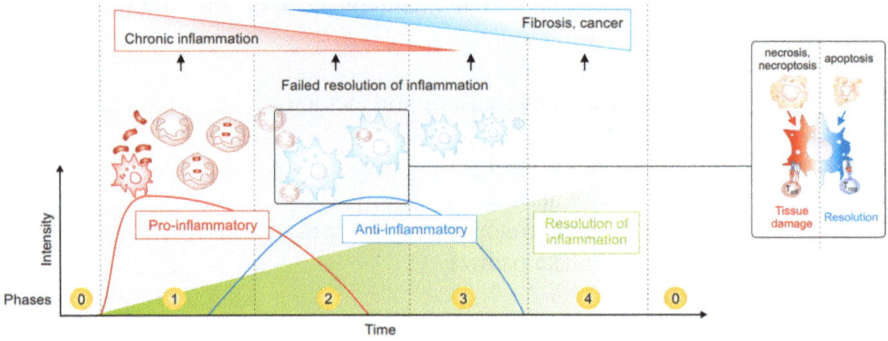

Fig. 1 Efferocytosis as a key decision point for the resolution of inflammation. Inflammatory reactions can be separated into discrete, but overlapping phases. Resolution of inflammation starts during the pro-inflammatory phase (phase 1) with clearance of the inflammatory stimulus, followed by removal of the initial inflammatory effectors and triggering of a counter-balancing anti-inflammatory response (phase 2). Anti-inflammatory effectors are subsequently removed (phase 3) and tissue repair pathways are terminated (phase 4) to return to homeostasis (phase 0). Depending on the phase, failure of resolution can lead to the pathological states indicated. The clearance of dying cells is a key checkpoint to decide if inflammation progresses or is resolved. Hereby, interaction of phagocytes with apoptotic cells induces resolution programs, where interaction with cells undergoing spontaneous or programmed necrosis, e.g., due to failed apoptotic cell clearance leading to secondary necrosis, perpetuates inflammation

coordinated. However, we still lack basic understanding of mechanisms of the resolution of inflammation. This is partly due to the fact that, for a long time, resolution of inflammation was interpreted as a merely passive phenomenon including the exhaustion or demise of leukocytes and the degradation or dilution of inflammatory mediators. The hypothesis that resolution of inflammation could be an active process is therefore still relatively new [1–4]. We even lack the understanding at which phase of the inflammatory reaction the resolution of inflammation is triggered. For the sake of simplicity, acute inflammatory reactions can be separated into different phases (Fig. 1) [5, 6].

Starting from a phase 0, in which the tissue/organ is in homeostasis, disruption of homeostasis by a noxious agent that is sensed by local immune effectors initiates a pro-inflammatory reaction (phase 1) with the goal to clear the noxious agent. For instance, in acute bacterial infections, this occurs through the recruitment of granulocytes to phagocytose and kill pathogens. Without removal of the pathogen, inflammation cannot resolve, illustrating that mechanisms of the initiation of inflammation overlap with those of its resolution [7]. In phase 2 of an inflammatory reaction, primary, pro-inflammatory immune effectors must leave the inflamed area. This occurs, for example, through cell death of activated granulocytes and their removal by phagocytes such as macrophages via efferocytosis. This process does not only remove inflammatory effectors but also induces anti-inflammatory activation of macrophages and therefore an active inhibition of further inflammatory reactions. Recruited anti-inflammatory immune effectors must also subsequently leave the tissue in phase 3. One example is the emigration of macrophages into the lymphatic system. In phase 4, cellular signaling processes, such as proliferation or matrix formation, are terminated in the parenchyma and stroma of the tissue in order to return to homeostasis (phase 0). If only one of these phases does not proceed adequately, tissue homeostasis may be permanently disturbed. If the source of inflammation and/or primary immune effectors are not eliminated in phases 1 and 2, excessive inflammation with organ failure may occur when strong immune-activating noxa are involved, while chronic smoldering inflammation manifests itself in the case of relatively weak immune activators. Persistence of anti-inflammatory signals in phase 3 or cellular processes in phase 4 may cause fibrosis and/or hyperproliferation, which may contribute, e.g., to carcinogenesis [5, 8]. Indeed, a dysfunctional resolution program has been associated with carcinogenesis [9] and the activation of resolution programs by specialized pro-resolving lipid mediators (SPMs) during surgical tumor removal reduces post-surgical metastasis [10].

As indicated above, macrophages take center stage during the proper execution of an inflammatory reaction. Macrophages are ubiquitous tissue-resident immune cells, whose primary task is maintaining tissue integrity and function. They command a remarkable sensory repertoire to actively respond to disturbances in tissue homeostasis by initiating inflammation [11, 12]. However, they are also involved in repairing the damage caused by the inflammatory reaction [13–16]. This is reflected in their remarkable trophic repertoire which ranges from shaping the extracellular matrix, phagocytosing and recycling debris, to tissue-specific functions such as modulating tissue innervation or promoting conductance in the heart [12, 17–19]. The ability of macrophages to properly respond to the changing temporo-spatial demands during an acute inflammatory reaction suggests a high functional plasticity. This property is due to a broad range of germ-line encoded receptors expressed on macrophages, whose interplay activates transcriptional enhancers or repressors to generate a large number of activation states [20–23]. Discrete activation (also called polarization) states of macrophages were postulated during inflammation. They are reminiscent of the different phases of inflammation discussed above [20, 24–26]. The most common activation states are summarized in the M1/M2 paradigm. Macrophages acquire an M1 or classically activated phenotype when being stimulated with interferon-γ (IFN-γ) and lipopolysaccharide (LPS). This is thought to mimic the situation of acute infection or more generally, type 1 inflammation, corresponding to phase 1 of the inflammatory cascade. M1 macrophages accordingly display an anti-microbial, pro-inflammatory profile, and are characterized by activation of transcription factors nuclear factor kappa-light-chain-enhancer of activated B cells (NF-κB), signal transducer and activator of transcription 1 (STAT1), and interferon regulatory factor 5 (IRF 5). M1 macrophages produce pro-inflammatory mediators such as the cytokines tumor necrosis factor (TNF)-α, interleukin (IL)-1β, IL-6, and IL-12, as well as reactive oxygen and nitrogen species (ROS/RNS) [27, 28]. Continuous activation of such inflammatory macrophages may support auto-inflammation [29]. A polarized M2 or alternative activated phenotype is produced by stimulation with the TH2 cytokines IL-4 or IL-13, which mimics conditions of type 2 inflammation, but also corresponds to the anti-inflammatory phase (phase 2) of the inflammatory cascade. Such M2 macrophages are characterized by the activation of the transcription factors STAT6 and IRF4 and production of chemokines such as CCL17 and CCL18, phagocytic receptors such as CD206, arginases (Arg1/2) to limit NO production, and the capacity to modulate the extracellular matrix to aid in tissue repair [20, 23, 28, 30, 31]. However, prolonged activation of such cells may result in fibrosis [32]. As the M1/M2 nomenclature is derived from experiments subjecting macrophages to the isolated stimuli indicated above in vitro, it has natural limitations. Macrophages in tissues during inflammation will probably never face M1 or M2 stimuli in isolation [33], and the transcriptional programs induced by IFN-γ or IL-4 in macrophages are transient rather than stable [33–35]. This allows macrophages to adapt to different functional needs in local niches and allows macrophages to switch their phenotype, e.g., to contribute to the resolution of inflammation. Pro-resolving macrophage phenotypes are, among others, triggered by efferocytosis [36, 37]. These macrophages are enhanced in their capacity for debris removal, indicating a feed-forward response to interaction of macrophages with dying cells, and tissue repair [36, 37]. Interestingly, IL-4 is required [38] and synergizes with efferocytosis [39] to establish the formation of pro-resolving macrophage phenotypes. This indicates, on the one hand, the aforementioned plasticity of macrophages and, on the other hand, suggests a fluid transition from early to later phases in the inflammatory cascade, dependent on the presence of dying cells. At the molecular level, pro-resolving macrophages induced by efferocytosis therefore share features with alternatively activated M2 macrophages and lose M1 characteristics, even though they retain certain M1 features including anti-microbial effector functions. For instance, they do not readily produce inflammatory cytokines due to suppression of NF-κB activation [40, 41], at least partly due to the action of the transcription factor peroxisome proliferator and regulator of transcription (PPAR)-γ [42–45], and produce pro-resolving cytokines such as transforming growth factor (TGF)-β [46, 47]. They upregulate arginases and have a limited capacity for ROS/RNS formation [48–52]. However, they are capable of producing prostaglandins via expression of cyclooxygenase (COX)-2 [53, 54] and produce lipocalin-2, which can scavenge iron to prevent microbial expansion [55, 56]. Macrophages taking up dying cells not only scavenge but also actively redistribute iron as well as cholesterol recovered from dying cells to, e.g., aid in tissue repair. Therefore, the clearance and recycling of cellular debris is a crucial function of macrophages during inflammation [57, 58]. Consequently, defects in efferocytosis are associated with improper execution of an inflammatory response in chronic inflammatory and auto-immune settings, but also affect an adequate response to infection and cancer [59, 60]. Therefore, understanding the mechanisms that shape specific macrophage polarization and the associated functional responses during inflammation per se, and the response to dying cells in particular, is important. Hereby, efferocytosis, as compared to stimulation with cytokines, is associated with a specific challenge, which is the handling of the apoptotic cell constituents by the phagocytes. In the following paragraphs, we therefore outline how dying cells affect the macrophage phenotype during inflammation, with a focus on macrophage metabolism during efferocytosis.

Multiple Mechanisms During Efferocytosis Avoid or Dampen Inflammation

To maintain tissue homeostasis, damaged or aged cells need to be removed, which usually occurs by apoptosis. This happens frequently in a number of tissues with a high cellular turnover such as barrier tissues and the hematopoietic system, with estimates reaching up to 1% of the cells that make up our body being removed by efferocytosis every day [61]. A striking example is neutrophils, which have a lifespan of approximately 1 day and physiologically die by apoptosis. Between 10^{10} and 10^{11} neutrophils alone need to be removed daily by efferocytosis [62]. Apoptosis is a strictly regulated ATP-dependent process that ensures that cells are dismantled broken up into discrete packages without losing cell membrane integrity to avoid the leakage of intracellular danger-associated molecular patterns (DAMPs) into the extracellular space [63]. This prevents inflammatory reactions in response to such DAMPs, indicating that apoptotic cells need to be removed by efferocytosis swiftly before secondary necrosis occurs once the remaining ATP pool is depleted. Moreover, apoptotic cells provide signals to induce tolerance and to prevent a systemic auto-immune response, besides being just passively removed to avoid inflammation [60, 64, 65]. Efferocytosis is a highly coordinated and mechanistically redundant process that includes phagocyte recruitment, apoptotic cell recognition, uptake, and digestion. We summarize these steps and their relevance for avoiding or dampening inflammation below (Fig. 2). For details, we refer to recent excellent reviews [59, 61, 65–67].

Apoptotic cells can be taken up by a variety of cells including nonprofessional phagocytes such as epithelial cells [68, 69]. However, the capacity of these cells to take up and digest apoptotic debris is limited. Therefore, professional phagocytes such as macrophages [70] need to be recruited toward the site of cell death. To ensure professional phagocyte recruitment, apoptotic cells release a number of molecules into the extracellular space, which have been called "find-me signals." These include the nucleotides ATP and UTP, lipids such as sphingosine-1-phosphate (S1P) and lysophosphatidylcholine (LPC), and proteins such as the chemokines fractalkine (CX_3CL1) and C-C motif chemokine ligand (CCL)12, the ribosomal protein S19, and endothelial monocyte-activating polypeptide-II (EMAPII), a fragment of human tyrosyl tRNA synthetase [65, 71–74]. Find-me signals serve not only to recruit phagocytes but also to initiate first steps to avoid inflammation and even trigger anti-inflammatory programs in macrophages. For instance, there are indications that nucleotides released from apoptotic cells can be broken down into immune-suppressive molecules such as adenosine, which signal through adenosine receptors such as Adora2a on macrophages [75, 76]. Accordingly, genetic interference with Adora2a impaired the resolution of inflammation in murine zymosan-induced peritonitis [75]. S1P signaling through S1P receptors not only shifts the cytokine balance from pro- to anti-inflammatory, modulates T cell activation by macrophages and dendritic cells, and endows macrophages with tissue repair capacity [54, 76–79], but also primes macrophages for further uptake of cell debris [80]. Selective metabolite cocktails released from dying cells, rather than individual metabolites, were shown

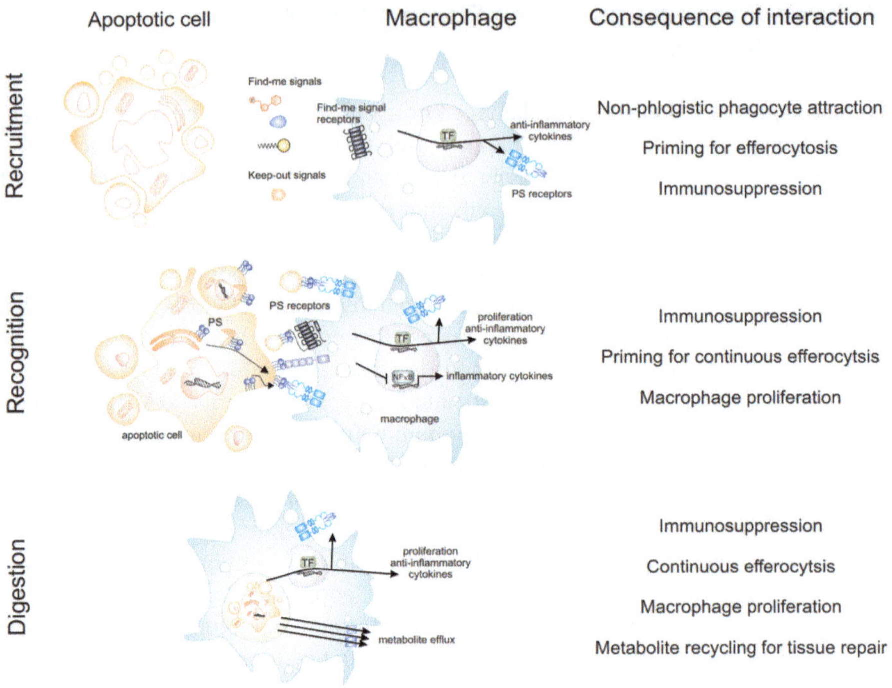

Fig. 2 Pro-resolving impact of the different phases of efferocytosis. Efferocytosis is initiated by the release of find-me signals by apoptotic cells. These recruit phagocytes but also prime them for efferocytosis by inducing efferocytic receptors and initiate a first wave of anti-inflammatory mediator production. Upon recognition of eat-me signals on apoptotic cells by efferocytic receptors (phosphatidylserine (PS) and its receptors shown as the prototypic example), signaling pathways are activated that induce a transcriptional response to sustain immunosuppression, enable continuous uptake of cell debris, and induce macrophage proliferation. These responses are sustained by metabolites generated during apoptotic cell digestion, while additionally, recycled metabolites are provided to bystander cells to enable tissue repair

to dampen inflammation in arthritis models and avoid organ rejection during lung transplantation in mice, suggesting tolerance induction [81]. Beyond local immune modulation, CX_3CL1 has the capacity to attract blood monocytes to sites of tissue damage to increase the phagocytic capacity by enhancing numbers of phagocytes, again avoiding secondary necrosis [82]. Finally, lactoferrin is released from dying cells to limit the recruitment of granulocytes toward dying cells [83]. This "keep-out signal" therefore avoids granulocyte-dependent amplification of inflammation and contributes to non-inflammatory removal of dying cells. Besides "find-me" and "keep-out signals," apoptotic cells release molecules that signal the occurrence of tissue damage with the goal to avoid the build-up of detrimental inflammation. One of these signals is the IL-1 family cytokine IL-38. Many cytokines of this family are produced in their active form in the context of inflammation and cell death [84, 85]. Hereby, as opposed to most IL-1 family cytokines, IL-38 is not an alarmin that is released when cells loose plasma membrane integrity but is produced by apoptotic cells [86]. Upon release from dying cells, IL-38 particularly limits IL-17 production by T cells directly and indirectly, which is a common trigger of auto-inflammation [86, 87]. Thus, apoptotic cells serve as an early signal to limit the extent of auto-inflammatory reactions and to promote the resolution of inflammation.

Once phagocytes have been recruited toward the apoptotic cells, specific recognition of dying self is initiated by the exposure of "eat-me signals" on the outer leaflet of the plasma membrane of the apoptotic cells, which are recognized by a broad array of receptors on phagocytes. This occurs either directly by ligand-receptor interaction, or through specific bridging molecules, which add another layer of adjustability to the system [67]. Through these multiple interactions, a platform known as the phagocytic synapse is formed. Profound changes occur at the plasma membrane of apoptotic cells in order to discriminate living from apoptotic cells, including alterations in lipid, sugar, and protein composition [71]. Proteins such as calreticulin, annexin I, intercellular adhesion molecule (ICAM)3, and the long pentraxin PTX3 are exposed on dying cells and are recognized by phagocytes either directly or indirectly [71]. The most significant changes occur at the level of lipids. Oxidized lipid moieties are recognized by CD36 on phagocytes [88]. However, the most prominent signal that serves to distinguish living from dying cells is phosphatidylserine (PS). PS is usually found at the intracellular leaflet of the plasma membrane, but is shuttled to the extracellular leaflet during apoptosis [89]. This requires caspase-dependent inactivation of P4 ATPase family flippases and activation of XK-related protein family scramblases, specifically Xkr8 [90]. PS is also exposed on cells succumbing to other forms of cell death, including necrotic or necroptotic cells [91], but specifically in the context of apoptosis, recognition of PS by phagocytic receptors counteracts inflammatory signaling and activates tolerogenic pathways [91, 92]. First experiments investigating the immunomodulatory role of PS on apoptotic cells found that PS recognition by macrophages induced the production of the anti-inflammatory cytokine TGF-β, and lipids such as platelet-activating factor (PAF) and prostaglandin (PG)E_2, which in turn reduced the production of pro-inflammatory cytokines including IL-1β, IL-8, and TNF-α [47]. PS receptors are numerous, highlighting its importance for apoptotic cell removal. PS receptors thus act redundantly, but context-specific PS receptor interactions have also been described that allow tailored reactions to dying cells [65]. Prominent PS receptors are brain angiogenesis inhibitor 1 (BAI1), stabilin-1/2, αVβ3–5 integrins, CD300 family proteins, T cell/transmembrane, immunoglobulin, and mucin (TIM) family proteins, and TYRO3, AXL, and MERTK (TAM) receptor tyrosine kinases. Among these, BAI1, stabilin-1/2, and the TIM family proteins directly recognize PS, while TAM and integrin receptors require bridging molecules. TAM receptors recognize PS via the Gla-domain-containing proteins growth arrest-specific 6 (GAS6) and protein S (PROS1), and integrin receptors recognize MFGE8 that is bound to PS [65, 92]. Besides recognizing PS for removal of dying cells, most of these PS receptors have been implicated in regulating inflammation in efferocytosing macrophages. However, nonredundant signals emanating from individual receptors have been reported. For instance, for TAM receptors, different modes of interaction with GAS6 and PROS1 have been observed that are likely the basis for the activation of different signaling pathways [93, 94]. While all three TAM receptors promote efferocytosis, TYRO3-deficient mice show neurological disorders, mice lacking AXL exhibit vascular complications, while MERTK-deficient mice display an autoimmune phenotype. In macrophages, MERTK signaling plays a role in both inflammatory disease and cancer, inhibiting the former and promoting the latter, while being generally associated with resolution of inflammation and tissue repair. In contrast, AXL signaling on macrophages may have diametrically opposed functional outcomes compared to MERTK [95, 96]. Besides directly affecting inflammatory signaling pathways, recognition by MERTK also promoted the proliferation of macrophages in cancer and inflammation models [39, 97, 98], probably serving as another means to increase the phagocytic capacity. Among TIM family receptors, mainly TIM-3 and TIM-4 are expressed on macrophages. PS recognition by TIM-4 expressed by liver macrophages prevented NLRP3 inflammasome activation to attenuate nonalcoholic fatty liver disease [99]. TIM-3 has been connected to a range of anti-inflammatory actions in transplant tolerance, autoimmunity, viral infection, and cancer, but its comparatively broad expression among immune cells and its ligand promiscuity make it difficult to attribute specific functions to PS recognition by macrophages [100]. Among stabilins, stabilin-2 is mainly expressed in macrophages and participates in TGF-β production to regulate immune responses [101]. Thus, specific recognition of apoptotic cells triggers another round of signaling pathways that help to avoid inflammation. Following the recognition of dying cells, macrophages engulf their bait and are now faced with a new challenge. How to deal with and distribute the molecular load they just received?

Metabolic Challenges During Efferocytosis

Phagocytes that have taken up dying cells face a significant metabolic burden, as they need to cope with the amount of molecules from the cells they devoured. This is particularly challenging for professional phagocytes such as macrophages that can take up multiple cells simultaneously or sequentially and need to digest a diverse range of biomolecules and redistribute them or their breakdown products to neighboring cells [102]. There are limitations to the phagocytic capacity of macrophages. Moreover, ingested metabolites during efferocytosis profoundly affect the inflammatory profile of phagocytes. In the following paragraphs, we first describe the impact of efferocytosis on metabolic pathways in macrophages and vice versa (Fig. 3) and then illustrate how the impact of efferocytosis on macrophage metabolism might be utilized to avoid dysfunctional inflammation.

A Requirement for Mitochondrial Metabolism for Efferocytosis

Mitochondrial activity is essential for a multitude of macrophage functions, not only for ATP supply but also by providing various mitochondrial metabolites to support epigenetic regulation of gene expression or biosynthesis of inflammatory

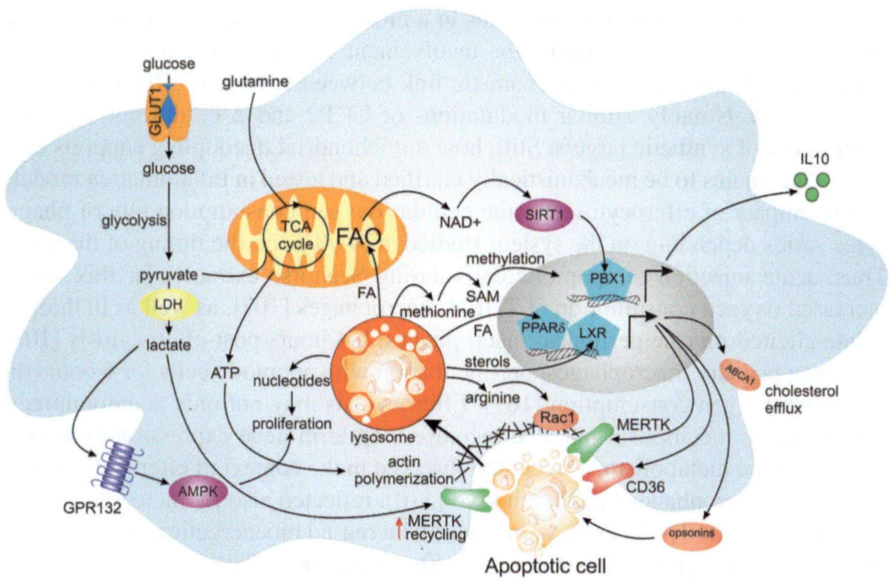

Fig. 3 Connections between metabolism and efferocytosis. Efferocytosis of apoptotic cells releases metabolites, which regulate various aspects of the efferocytic process and the macrophage pro-resolving phenotype. Lysosomal degradation of apoptotic cells generates amino acids, such as arginine and methionine. Subsequent metabolism of those promotes epigenetic remodeling via methylation and supports phagocytic cup formation through Rac1 activity. Nucleotides from digested apoptotic cells support macrophage proliferation, while fatty acids and sterols activate PPARδ and LXR-dependent gene expression of phagocytic receptors and opsonins. Fatty acid oxidative metabolism provides NAD^+ supporting IL-10 expression via the SIRT1/PBX axis. Mitochondrial oxidative metabolism also provides ATP necessary for actin polymerization during efferocytosis. Glucose catabolism generates lactate, which can support efferocytosis via calcium-dependent MERTK recycling, but also signals extracellularly via GPR132, supporting efferocytic macrophage proliferation in an AMPK-dependent manner

mediators [103]. Oxidative metabolism increases during IL-4-induced polarization of macrophages [104], and it was suggested that it is necessary to support anti-inflammatory effects of IL-4, promoting pro-resolving and tissue reparative macrophage phenotypes.

First investigations on the impact of mitochondrial metabolism on efferocytosis date back to 2011 [105]. Using a Chinese hamster ovary fibroblast cell line LR73 [106] engulfing apoptotic murine thymocytes as an efferocytosis model, Park et al. observed an increased mitochondrial membrane potential ($\Delta\Psi_m$) within 1 h of dying cell engulfment. At the same time, efferocytosis increased protein expression of the mitochondrial uncoupling protein 2 (UCP2), functionally counteracting mitochondrial hyperpolarization. Silencing UCP2 decreased the efferocytic capacity, while UCP2 overexpression increased it. This effect was mimicked by mitochondrial uncouplers and respiratory chain complex inhibitors (at concentrations not affecting cellular ATP levels). Deficiency of UCP2 in murine bone-marrow-derived macrophages (BMDM) also attenuated efferocytosis, and UCP2 knockout mice showed impaired clearance of dead thymocytes in a model of dexamethasone-induced thymic involution. While discarding the involvement of fatty acid oxidation or mitochondrial ROS, a definitive mechanistic link between $\Delta\Psi_m$ and efferocytosis was not provided. Notably, similar modulations of UCP2 and $\Delta\Psi_m$ did not affect the engulfment of synthetic targets. Still, how mitochondrial uncoupling supports efferocytosis remains to be mechanistically clarified and tested in inflammation models.

The impact of efferocytosis on the cellular oxygen consumption rate of phagocytes varies depending on the system studied, as well as on the timing of the assay. Thus, acute injection of apoptotic cells during Seahorse extracellular flux assays increased oxygen consumption of THP-1 macrophages [107], as well as in thioglycolate-elicited murine peritoneal macrophages 2–4 hours post-efferocytosis [108]. In human primary macrophages, pre-incubation with apoptotic cells for 6 hours did not affect oxygen consumption [109]. Efferocytosis may not only acutely upregulate oxidative metabolism but also increase long-term gene expression associated with oxidative metabolism, as recently observed in the context of efferocytic tumor-associated macrophages (TAMs) in vivo [110], reflected also by increased respiration of efferocytic TAMs ex vivo. In general, increased bioenergetic requirements of efferocytic macrophages to support engulfment and degradation of apoptotic cells appear to be well-balanced, since macrophage ATP levels do not drop upon efferocytosis [111, 112]. Interestingly, in peritoneal macrophages undergoing efferocytosis of apoptotic thymocytes, AMP-activated protein kinase (AMPK), a sensor of reduced cellular energy charge, was reported to be activated [113]. Inhibiting AMPK reduced efferocytosis in vitro and in vivo. It was suggested that the "find-me signal" LPC [114] causes reduction of $\Delta\Psi_m$ and mitochondrial ATP generation in macrophages, thus inducing canonical AMP-dependent AMPK activation. However, later work suggested protein kinase A-dependent mechanism of AMPK activation in phagocytes, independent of energy loss [111] (see below).

Efferocytosis of multiple targets or continual efferocytosis apparently has divergent metabolic requirements as compared to uptake of a single apoptotic body [112]. Wang et al. observed that BMDM taking up apoptotic Jurkat cells showed altered mitochondrial morphology, with increased fragmentation (or fission) of mitochondrial networks, accompanied by increased protein levels of a pro-fission protein dynamin-related protein 1 (Drp1). Drp1 deficiency attenuated efferocytosis, especially at high apoptotic cell burden, preventing a second round of apoptotic cell uptake. Mechanistically, fission reduced calcium sequestration by mitochondria, provoking increased cytosolic Ca^{2+} levels. This supported phagosome sealing and NADPH oxidase activity, promoting microtubule-associated proteins 1A/1B light chain 3B (LC3)-assisted phagolysosomal degradation of engulfed material [112]. Increased calcium also supported endocytic membrane flow to the plasma membrane, which is necessary for high capacity efferocytosis. In this case, bioenergetic function of mitochondria was secondary to their calcium-buffering properties.

Several studies addressed specific fuel requirements of efferocytic macrophages. Glutamine metabolism is considered particularly relevant for interleukin-4-stimulated macrophages [115], which also exhibit increased efferocytosis [39]. The role of glutamine metabolism in the context of efferocytosis was addressed by Merlin et al. [116]. Studying glutaminase (Gls1)-deficient macrophages, the authors showed that while the absence of Gls1 did not compromise macrophage viability, it impaired efferocytosis by BMDM or peritoneal macrophages in vitro and by Kupffer cells and spleen red pulp macrophages in vivo. Gls1 deficiency also impaired efferocytosis in atherosclerotic plaques in the apolipoprotein E (ApoE)-knockout hyperlipidemic mouse model. Gls1-deficient macrophages showed reduced glutamine incorporation into the tricarboxylic acid cycle and decreased oxygen consumption, mitochondrial respiratory capacity, and impaired ATP production. Inhibiting complexes III and II of the respiratory chain attenuated efferocytosis in this system (in contrast to [105]). At the same time, modulation of ROS did not affect efferocytosis. Interestingly, glutamate dehydrogenase-deficient macrophages showed the opposite phenotype, i.e., increased respiration, and mitochondrial ATP production as well as elevated efferocytosis. At the same time, respiration and efferocytosis were attenuated by inhibiting or silencing transaminases (Got1 and Got2), suggesting the activity of the malate/aspartate shuttle to be critical for bioenergetic support of efferocytosis. In addition, Gls1 deficiency caused a drop of the NADPH/$NADP^+$ ratio, restoration of which also recovered the efferocytic capacity. Mitochondrial ATP appeared to support actin polymerization necessary for phagocytic cup formation, and preventing mitochondrial ATP release by blocking the activity of the adenine nucleotide translocase impaired efferocytosis.

Macrophages can also use fatty acid oxidation (FAO) to fuel oxidative phosphorylation (OXPHOS), although FAO per se does not appear to determine macrophage phenotype changes in response to various stimuli [117]. FAO was also investigated in the context of efferocytosis. In a murine model of myocardial infarction, Zhang et al. observed that efferocytic macrophages exhibited increased OXPHOS [108]. Similarly, efferocytic peritoneal macrophages increased oxygen consumption, which was sensitive to FAO inhibition. FAO was necessary to produce the anti-inflammatory cytokine interleukin-10 in efferocytic macrophages. Compromising the mitochondrial respiratory chain through deletion of complex III subunit Rieske iron-sulfur protein did not affect apoptotic cell uptake but abolished IL-10 production by efferocytic macrophages. Although Rieske protein deficiency did not affect cellular ATP, it decreased the NAD^+/NADH ratio as well as total NAD, and supplementation of the NAD precursor NMN restored IL-10 secretion. Reduced NAD^+ was linked to an attenuated activity of the NAD^+-dependent protein deacetylase sirtuin 1 (SIRT1), which in turn was necessary to support IL-10 transcription via transcription factor pre-B cell leukemia transcription factor-1 (PBX1).

In summary, these data indicate a requirement of functional mitochondria for initial as well as continuous efferocytosis and link specific mitochondrial metabolic pathways, particularly fatty acid metabolism, to the anti-inflammatory effects of efferocytosis. Interestingly, ATP levels appear to be well balanced during efferocytosis even though the dying cells provide a rich source of fuel for ATP synthesis. This balance appears to be critical given that mitochondrial uncoupling is observed and required for efferocytosis.

Multiple Roles of Glucose Metabolism During Efferocytosis

Glucose is a major fuel to support innate immune responses [118]. Glucose catabolism is unique compared to other fuels in that it provides ATP both through cytosolic steps (often called anaerobic glycolysis) and via OXPHOS. Glucose is taken up into the cells using a family of facilitative glucose transporters (GLUT). In immune cells, GLUT1 (SLC2A1) is the major glucose transporter. Regarding the role of glucose transport in efferocytosis, Morioka et al. [119] observed that efferocytosis caused up-regulation of GLUT1 mRNA in professional and nonprofessional (LR73) phagocytes. Efferocytosis increased glucose uptake and the extracellular acidification rate (a marker for glycolysis) already within 1 h. This correlated with increased GLUT1 surface expression, which was linked to GLUT1 phosphorylation by serum and glucocorticoid-regulated kinase 1. Interfering with GLUT1 function reduced efferocytosis in vitro and in vivo, accompanied by decreased actin polymerization in phagocytes with compromised GLUT1 function. Interestingly, increasing glucose oxidation by inhibiting negative regulators of the pyruvate dehydrogenase complex pyruvate dehydrogenase kinase 1 and 4 also attenuated efferocytosis, suggesting opposing roles of glycolysis and OXPHOS in the LR73 model. In addition to the importance of glucose uptake, lactate release through monocarboxylate transporter 1 (SLC16A1) was also necessary to support efferocytosis, and interfering with its function reduced apoptotic cell uptake. Lactate released by efferocytic macrophages supported an anti-inflammatory gene expression program with increased expression of IL-10, TGF-β, or vascular endothelial growth factor (VEGF) elicited in macrophages by supernatants of efferocytic cells in a SLC16A1-dependent fashion.

The involvement of glycolysis in continuous efferocytosis was investigated later in detail [120]. Here, efferocytosis induced a transient (1–6 h) increase in glycolysis in BMDMs or human macrophages associated with increased GLUT1 surface expression (but no changes in GLUT1 mRNA). Efferocytosis induced Akt-dependent phosphorylation and degradation of a promoter of GLUT1 endocytosis, thioredoxin-interacting protein (TXNIP). In addition, efferocytosis elevated glycolysis by increasing activity of the rate-limiting glycolytic enzyme phosphofructokinase-1 (PFK1). This, in turn, was due to activation of phosphofructokinase-2 (PFKFB2), an enzyme providing PFK1 allosteric activator fructose-2,6-bisphosphate. PFKFB2 was phosphorylated at the activating Ser485 site in an AKT-dependent manner. Thus, AKT activation provided two hits to activate glycolysis in efferocytic macrophages: via TXNIP and PFKB2. Investigating functional consequences of increased glycolysis, as reported by Schilperoort et al., showed that inhibiting glycolysis impaired continuous efferocytosis. Specifically, lactate produced by glycolytic efferocytes supported continuous efferocytosis by promoting recycling of the efferocytic receptors MERTK and LRP1 in an intracellular calcium-dependent manner. This was dependent on intracellular lactate, since suppressing lactate release by silencing SLC16A1 further increased continuous efferocytosis. Intracellular lactate may also promote expression of pro-resolution genes via histone lactylation [121], although specific roles of histone lactylation in the context of efferocytosis were not studied so far. Intracellular lactate also supports histone acetylation during IL-4-induced macrophage polarization [122]. In addition to intracellular effects, extracellularly released lactate may affect macrophages in an autocrine manner through stimulation of the lactate-sensing G protein-coupled receptors GPR81 or GPR132 [123]. It was suggested that lactate-dependent activation of GPR132 stimulates proliferation of efferocytic macrophages [111]. This occurs through stabilization of Myc via a mechanism involving activation of Gs-coupled GPR132. Elevation of cyclic AMP and activation of protein kinase A downstream of GPR132 activated AMPK, which in turn stimulated protein deacetylase SIRT1. SIRT1 deacetylated Myc, attenuating its proteasomal degradation.

Recent studies addressed metabolic alterations of efferocytic cells in vivo or ex vivo. In addition to increased OXPHOS, efferocytic TAMs also displayed increased glycolysis [110], suggestive of a generally elevated energetic state. Whether this extends to other (patho)physiological states of efferocytes in vivo remains to be investigated. Tissue-resident macrophages adapt to specific microenvironments of their respective locations by adjusting metabolism [124]. This includes adaptation to chronic hypoxia, which is often present in tissues exhibiting high cell turnover, such as the bone marrow. Wang et al. [125] explored how macrophage metabolism changes upon exposure to chronic hypoxia, and how this impacts efferocytosis. Macrophages adapting to chronic (7 d) hypoxia exhibit enhanced efferocytosis. Adaptation to chronic hypoxia also elicited reductions of OXPHOS and lactate production but elevated glucose flow into the pentose phosphate pathway (PPP). This supported NADPH production, which is necessary for maintaining low levels of oxidative stress and preventing excessive lysosomal acidification. Targeting PPP by interfering with the activity of glucose 6-phosphate dehydrogenase (G6PD) reduced efferocytosis.

Erythrocytes are the most abundant cells in the body, and the proper clearance of aged erythrocytes by reticuloendothelial macrophages in the spleen and liver (erythrophagocytosis) is critical to prevent iron toxicity. Studying the metabolome of macrophages following erythrophagocytosis, Catala et al. noticed not only changes common with efferocytosis but also a switch in the usage of glucose, with a reduced flux through lower glycolysis with a concomitantly increased PPP [126], probably to counterbalance increased oxidative stress and NADPH demand after phagocytosis of iron-loaded erythrocytes. In contrast to these observations, peritoneal macrophages or BMDMs engulfing apoptotic thymocytes or Jurkat cells decreased the expression of PPP-associated mRNAs as well as reduced levels of PPP intermediates, NADPH, and glutathione (GSH) 24 h post-engulfment [127]. A pharmacological activator of G6PD reduced efferocytosis and increased levels of pro-inflammatory cytokines while reducing IL-10 and TGF-β expression, suggesting a negative influence of the PPP on the pro-resolving phenotype of efferocytic macrophages.

The role of glucose metabolism during efferocytosis, thus, appears to be context-dependent. It may vary dependent on the nature of the dying cell that is engulfed and on the overall tissue context. These features suggest a certain degree of specificity to modulate efferocytosis in different disease contexts that may be targetable in the future.

Handling of Peptides, Amino Acids, and Nucleotides During Efferocytosis

Macrophage amino acid metabolism has diverse roles in macrophage physiology. In particular, arginine metabolism and polyamine production are important during anti-inflammatory interleukin-4-induced polarization of macrophages [128]. Yurdagul et al. investigated the impact of arginine metabolism on efferocytic macrophages [129]. Efferocytic macrophages specifically accumulate cargo-derived arginine and its metabolite ornithine. In IL-4-stimulated macrophages, or in pro-resolving macrophages isolated from the peritoneum of mice subjected to zymosan-induced peritonitis, deficiency of the arginine-to-ornithine converting enzyme arginase attenuated continuous efferocytosis. Arginase loss also impaired efferocytosis in vivo in the model of glucocorticoid-induced thymic involution. In the context of atherosclerosis, the absence of arginase reduced efferocytosis in regressing plaques, which increased plaque necrosis and raised the levels of the inflammatory cytokines IL-1β and TNF-α in regressing plaques. Mechanistically, ornithine conversion to putrescine by ornithine decarboxylase was required to support continuous efferocytosis. Putrescine increased mRNA expression of the Ras-related C3 botulinum toxin substrate 1 (Rac1) guanosine triphosphate exchange factor diffuse B cell lymphoma (Dbl), which promoted Rac1 activity necessary for actin remodeling to support the formation of phagocytic cups. This occurred due to Dbl mRNA stabilization by putrescine, which promoted Dbl mRNA association with the mRNA-stabilizing protein HuR. In human macrophages lacking arginase, apoptotic cells provided ornithine, and silencing ornithine decarboxylase reduced continuous efferocytosis in this model as well. Polyamines in efferocytic cells may not only be derived from arginine metabolism but also directly imported in an endocytic process demanding Rac1 activation in phagocytes [130]. Interestingly, in this model using LPS-primed murine peritoneal macrophages engulfing apoptotic Jurkat cells, ornithine to putrescine conversion was substantially decreased in phagocytes, suggesting model-dependent differences in polyamine metabolism may be due to divergent arginine metabolism upon LPS vs. IL-4-priming [130]. Methionine is an amino acid with known roles in epigenetic regulation of gene expression through its role as a substrate for synthesis of S-adenosylmethionine (SAM), a methyl group donor supporting histone and DNA methylation. Efferocytosis causes an increase in cellular methionine levels [129, 130], and blocking methionine conversion into SAM by inhibiting methionine adenosyltransferase 2A prevented induction of pro-resolution genes such as cyclooxygenase 2 (PTGS2) and TGF-β in the phagocytes [131]. This response was also dependent on the activity of DNA methyltransferase 3A, which methylated a negative regulator of the extracellular signal-regulated kinase (ERK) pathway that controls PTGS2 expression, namely dual specificity phosphatase 4. Thus, overload with amino acids resulting from efferocytosis appears to serve as a signal to phagocytosing macrophages to further increase their phagocytic capacity, but may also deliver signals to actively pursue resolution of inflammation.

Besides amino acids, macrophages engulfing dying cells need to adequately deal with small peptides upon digestion of apoptotic cell-derived proteins. This is of particular importance since macrophages are antigen-presenting cells that may present peptides derived from their prey to adaptive immunity. In order to prevent inflammation in the context of efferocytosis, the presentation of potential self-antigens needs to be avoided [132, 133]. While there are similarities between the formation of phagosomes upon pathogen uptake by macrophages with the formation of an "efferosome" containing the apoptotic cell, certain differences exist [132, 133]. Altered trafficking patterns of efferosome-derived vesicles dependent on the small GTPase Rab17, for instance, prevent loading of potential auto-antigens on major histocompatibility complex (MHC)-II molecules [134]. Moreover, there is evidence for a faster turnover of efferosomes compared to phagosomes, which may avoid efficient antigen processing and presentation [135]. Finally, efferocytosis can employ parts of the machinery of canonical autophagy in a process called LC3-associated phagocytosis (LAP) [136]. LAP enables rapid processing of the ingested material derived from apoptotic cells, which limits antigen presentation and inflammation [136, 137]. Accordingly, depletion of a central component of LAP, the Beclin1-interacting protein Rubicon, aggravated disease in a murine model of the auto-immune disease systemic lupus erythematosus (SLE) [138]. A requirement for efficient removal of dying cells is supported by findings in mice and humans suggesting that altered expression of efferocytic receptors is associated with auto-immune conditions including SLE, rheumatoid arthritis, and multiple sclerosis [139–142].

Intracellular nucleotides derived from engulfed cells may also participate in shaping the phenotype of the phagocytes. Efferocytic macrophages show increased proliferation [143], which is partly dependent on the digestion of engulfed DNA by lysosomal DNase2a. However, its role was not only a provision of nucleotides to supply proliferation. Rather, DNase2a-liberated oligonucleotides activated DNA-dependent protein kinase (DNA-PK). DNA-PK, in turn, activated the mammalian target of rapamycin complex 2 (mTORC2)-Akt signaling pathway necessary for proliferation, through increasing mRNA expression of Myc. This mechanism synergized with Myc protein stabilization promoted by efferocyte-derived lactate as described above [111]. Again, intracellular metabolites derived from dying cells appear to signal to macrophages to increase their capacity for the uptake of dying material. While amino acid-derived polyamines reach this goal by a qualitative improvement of cell uptake, supporting continuous efferocytosis, nucleotides may increase the phagocytic capacity quantitatively by supporting macrophage proliferation. These mechanisms are required for efficient efferocytosis in tissues, which ultimately avoids presentation of self-antigens to adaptive immunity.

Lipids in Establishing the Phenotype of Efferocytic Macrophages

Macrophage engulfment of apoptotic cells causes substantial increases in the cellular lipid content. Macrophages respond to these increases by activating transcriptional programs of lipid handling, orchestrated in the first line by transcription factors such as peroxisome proliferator-activated receptors (PPARs) as well as liver X receptors (LXRs) [144]. The former respond to intracellular fatty acids and their oxidized derivatives, whereas the latter respond to oxysterols or sterol intermediates of the cholesterol biosynthetic pathway. Activation of PPARδ was observed in efferocytic BMDMs [145], and PPARδ deficiency suppressed the expression of various efferocytic receptors or opsonins (MerTK, Gas6, Mfge8, C1q). PPARδ-deficient macrophages also failed to suppress inflammatory responses upon apoptotic cell engulfment. Efferocytic macrophages also activate PPARγ [52]. Treating human macrophages during differentiation with a PPARγ antagonist reduced expression of proteins with pro-efferocytic functions (CD36, transglutaminase 2, pentraxin 3 or TAM family member AXL) and attenuated efferocytosis [146]. PPARγ-deficient peritoneal macrophages exhibited defects in efferocytosis, associated with decreased expression of multiple efferocytosis effector molecules (CD36, MERTK, AXL, C1Q, TGM2) [147]. PPARγ deficiency also led to impaired efferocytosis in vivo, which may have contributed to development of autoinflammatory glomerulonephritis in these mice. In addition to defects of efferocytosis, PPARγ-deficient efferocytic macrophages also showed impaired suppression of inflammatory cytokine expression in response to LPS, as well as reduced expression of anti-inflammatory cytokines TGF-β and IL-10. Thus, both PPARγ and PPARδ may orchestrate overlapping gene expression programs in efferocytic macrophages. Relative contribution of PPARγ and PPARδ may be dependent on the cellular system studied. Thus, in human primary macrophages engulfing apoptotic cells, PPARδ activation was predominant, apparently in response to lysosomal degradation of apoptotic cell-derived neutral lipids and phospholipids [109].

Lysosomal degradation of engulfed apoptotic cell membranes generates not only fatty acids but also cholesterol. Cholesterol is sensed by LXRs as well as by the endoplasmic reticulum-located transcription factor sterol regulatory element-binding protein 2 (SREBP2). Whereas LXRs are activated upon efferocytosis, SREBP2 activity is suppressed [107, 109, 144, 148, 149]. In human efferocytic macrophages, these changes are associated with accumulation of several sterols, such as lathosterol, desmosterol, lanosterol, and dihydrolanosterol, likely originating from engulfed cells [149]. On the other hand, using THP-1 macrophages as phagocytes, Viaud et al. suggested oxysterols derived from lysosomal hydrolysis and oxidation of engulfed cholesteryl esters to activate LXR and suppress SREBP2 in the phagocytes [107]. A canonical response to LXR activation is up-regulation of cholesterol efflux via transporters ABCA1 and ABCG1, and efferocytic macrophages show increased LXR-dependent cholesterol efflux [150]. LXR-deficient macrophages showed impaired efferocytosis [144], which was attributed to decreased expression of MERTK. MERTK up-regulation was also observed in human efferocytic macrophages [149]. LXR-deficient mice also displayed increased levels of apoptotic cells in spleen or thymus, and impaired efferocytosis in a model of dexamethasone-induced thymic involution. Aged LXR-deficient mice developed lupus-like autoimmune disease. LXR-deficient efferocytic macrophages also displayed attenuated induction of TGF-β and IL-10 as well as compromised suppression of the inflammatory response to LPS. In human macrophages, LXR may synergize with another nuclear receptor, retinoic acid receptor, to induce the expression of transglutaminase 2, which supported efferocytosis [151]. Oxysterols derived from engulfed apoptotic cells were also suggested to promote mitochondrial metabolism by elevating mitochondrial calcium via mitochondrial-ER contacts [107]. In IL-4-polarized human macrophages, efferocytosis potentiated expression of arachidonate 15-lipoxygenase (ALOX15) and IL-1 receptor antagonist in an LXR-dependent fashion, while not affecting the expression of other canonical IL-4 target genes [149]. Increased ALOX15 contributed to enhanced generation of oxygenated derivatives of arachidonic and docosahexaenoic acids in IL-4-polarized efferocytic macrophages [149]. These substances are thought to be important for the generation of SPMs that cause inflammation resolution [152]. SPM synthesis requires sequential processing of polyunsaturated fatty acids by ALOX15 and 5-lipoxygenase (ALOX5). Interestingly, efferocytosis was reported to promote translocation of ALOX5 from nucleus to cytoplasm in a MERTK-dependent fashion [153], which was proposed to promote SPM generation. Prolonged incubation with apoptotic tumor cells may, however, downregulate ALOX5 expression, which may be relevant for the immunosuppressive phenotype of tumor-associated macrophages [154]. It should be noted that the physiological significance of SPMs has recently come into question due to potential methodological problems with SPM detection [155], necessitating further investigation toward the roles of oxygenated fatty acid derivatives in the context of efferocytosis. In contrast to observations in the murine system, in human macrophages an attenuated inflammatory response to LPS was apparently independent of LXR/PPARδ agonism [109]. Thus, lipids taken up from apoptotic cells and their metabolic products appear to critically shape macrophage functions during inflammation. They boost the capacity of macrophages for efferocytosis by inducing uptake receptors and promote anti-inflammatory and pro-resolving programs. Whether they support continuous efferocytosis remains to be determined in the future.

Conclusions and Outlook: Capitalizing on Metabolism of Dying Cells During Inflammation

During the last years, a considerable amount of knowledge regarding changes of metabolism in efferocytosing macrophages has been gained. This includes several comprehensive metabolome studies as well as detailed mechanistic work uncovering pathways linking metabolic alterations to the phenotype of the phagocytes. At the same time, a vast variety of different cellular models of efferocytosis raises the question whether the findings are of general relevance. It is apparent that the nature of engulfed material (e.g., red blood cell versus tumor cells) may greatly influence the metabolic outcome of dying cell engulfment. In addition, the time course of efferocytosis may have different metabolic consequences. A critical aspect is the use of the efferocytic cell model. Most studies have been conducted in murine macrophages (either BMDM or peritoneal), while limited information exists on metabolic consequences of efferocytosis in human macrophages. While there are a variety of models for investigating efferocytosis in vivo in the murine system and in less complex animals, we have very limited abilities to investigate efferocytosis in the human system in vivo. Investigating this phenomenon would need solid biomarkers of efferocytosis and efferocytosis-dependent metabolism in macrophages of human tissue sections. These may include some of the metabolites discussed above. Rapidly evolving tools such as spatial metabolomics might therefore be instrumental to investigate the role of efferocytosis in human inflammatory diseases.

Even if biomarkers of appropriate versus failed metabolism of efferocytosed materials in tissue macrophages have been defined, the question remains why we should care. The argument has been made that we have observed definite evidence for defective efferocytosis only in a limited set of human diseases or in patient subgroups such as in SLE, chronic obstructive pulmonary disease (COPD), asthma, and atherosclerosis, but not in a large number of auto-inflammatory conditions per se [156–159]. Instead, inappropriate handling of the metabolic load derived from dying cell uptake, e.g., through deficiency in metabolite digestion of apoptotic cells, may be of essence to allow clearance of harmful necrotic material as observed in a *Drosophila* model [102]. As discussed above, a number of metabolic changes in macrophages upon efferocytosis are linked to continuous dying cell uptake, either qualitatively by enabling continuous efferocytosis or quantitatively by increasing the number of macrophages, which in turn may deliver anti-inflammatory and pro-resolving signals. Thus, improving the digestion of apoptotic corpses may be useful to restore homeostasis in auto-inflammatory diseases.

There are already approaches to test the power of apoptotic cell clearance concerning dampening inflammation in early preclinical and clinical development. The infusion of donor apoptotic cells as prophylaxis for acute graft-versus-host disease (GVHD) after HLA-matched myeloablative allogeneic hematopoietic stem cell transplantation was tested in an early clinical trial, found to be safe, and showed signs of tolerogenic activity [160]. The same principle was used to demonstrate a dampening impact of apoptotic cell injection on inflammation in models of sepsis and arthritis [161, 162]. Interestingly, part of the efficacy of therapeutic approaches using mesenchymal stem cells may be attributable to these cells undergoing apoptosis [163]. Consequently, apoptotic cell injection is currently tested in phase 1 and phase 2 clinical trials in patients with sepsis, osteoarthritis, and psoriatic arthritis (e.g., NCT04612413, NCT06522035, NCT06233474). Instead of using the dying cell per se, approaches to mimic apoptotic cells have been explored. For instance, PS-containing liposomes have been shown to promote the resolution of inflammation by acting on local macrophages in inflammation models to restrict inflammatory leukocyte recruitment [164]. Another strategy that has been tested in a number of pre-clinical settings including models of arthritis and colitis is the injection of supernatants of macrophages that have previously taken up apoptotic cells. This approach capitalizes on anti-inflammatory mediators such as TGF-β produced by the macrophages, as well as, potentially, on the recycled and secreted metabolites derived from apoptotic cell digestion [37].

Compared to these strategies that rely on the anti-inflammatory and tolerogenic effects of apoptotic cells acting on macrophages, it is poorly understood whether boosting the apoptotic cell uptake and recycling capacity in macrophages would be beneficial under conditions of distorted inflammation. Improving the uptake of dying cells was recently approached by various strategies. First, besides "eat me" signals, cells can expose signals to prevent their uptake, so-called don't eat me signals. One of these is CD47, which is recognized by its ligand signal-regulatory protein (SIRP)-α on macrophages to limit cell uptake [165]. Blocking CD47 with an antibody augmented phagocytosis, attenuated disease development in a murine atherosclerosis model [166], and reduced vascular inflammation in a cohort of cancer patients [167]. Second, the principle of chimeric efferocytosis receptors has been established. Expression of fusion proteins consisting of the intracellular domain of engulfment and cell motility protein 1 (ELMO1), a protein involved in actin skeleton re-arrangement during efferocytosis, and the extracellular domain of the PS receptors BAI1 or TIM-4 increased efferocytosis [168]. This was connected to reduced inflammation in multiple inflammatory disease models, including colitis. Next, macrophages taking up dying cells release extracellular vesicles, which, when acting on macrophages in a feed-forward manner, increased efferocytosis and decreased plaque necrosis in atherosclerosis models [169]. Finally, extracting efferocytosis signals from apoptotic neutrophil membranes and loading them onto a short fiber biomimetic matrix generated biomimetic short fibers that activated continuous efferocytosis to aid in muscle repair [170]. Besides these strategies, recognition of PS via MERTK promoted the proliferation of macrophages in cancer and inflammation models [39, 97, 98, 171], which could be utilized to increase the phagocytic capacity. Similarly, lactate was shown to promote proliferation of efferocytic macrophages [111]. Increased numbers of macrophages may, however, come with an increased fibrosis risk [172].

While these approaches aid in improving efferocytosis efficacy, they still do not address the need for proper corpse digestion. Several strategies could be of relevance. Instead of increasing macrophage numbers, improving apoptotic corpse turnover qualitatively may be envisioned. For instance, LAP, which can speed-up apoptotic cell digestion, may be activatable by proteolysis targeting chimera (PROTAC)-like approaches [173] to connect efferocytosis receptors to the autophagy machinery. Another approach to augment efferocytosis via LAP might be boosting mitochondrial fission [112]. Promoting mitochondrial fission via introducing Drp1 in *Drosophila* was feasible and improved healthy aging in these animals [174]. The means to induce mitochondrial fission specifically in phagocytes would need to be established. Apoptotic cell-derived arginine and ornithine conversion to putrescine promoted continuous efferocytosis via activation of Rac1 and altered actin dynamics [129]. However, a potential connection of putrescine to activation of LAP remains to be determined. To improve turnover of dying cells, macrophages need to effectively export metabolites and increasing the efflux of such metabolites may also aid in corpse digestion. The efflux of cholesterol derived from dying cells depends on the activation of LXR and/or PPARγ to upregulate cholesterol efflux via ABCA1 [107, 171]. Thus, LXR or PPARγ agonists might be useful to improve apoptotic cell turnover [65]. Further studies are required to identify the flux of cellular metabolites within and out of macrophages during efferocytosis to potentially identify new targets for improving the digestion of apoptotic cells. These strategies, as outlined above, would likely not only improve the removal of debris in inflamed tissues, but also endow macrophages with other pro-resolving functions such as the production of anti-inflammatory cytokines. Such an approach would be of major interest, since triggering the resolution of inflammation may hold advantages toward counteracting auto-inflammation over current anti-inflammatory strategies, since the latter affect the magnitude rather than the duration of inflammation [8].

Acknowledgments We apologize to researchers whose important primary contributions to the matter were cited indirectly by referring to current reviews.

Author Contributions DN, BB, and AW wrote the manuscript. All authors contributed to the article and approved the submitted version.

Conflict of Interest The authors declare that the research was conducted in the absence of any commercial or financial relationships that could be construed as a potential conflict of interest.

Funding The authors are supported by Deutsche Forschungsgemeinschaft (SFB 1039 A05, B04, B06; GRK 2336, TP1, TP6), the LOEWE Center Frankfurt Cancer Institute (FCI), and the EnABLE Cluster funded by the Hessen State Ministry for Higher Education, Research and the Arts.

References

1. Henson, P. M. (1991). Resolution of inflammation. *A perspective. Chest, 99*(3 Suppl), 2S–6S.
2. Perretti, M. (1997). Endogenous mediators that inhibit the leukocyte-endothelium interaction. *Trends in Pharmacological Sciences, 18*(11), 418–425.
3. Lawrence, T., Willoughby, D. A., & Gilroy, D. W. (2002). Anti-inflammatory lipid mediators and insights into the resolution of inflammation. *Nature Reviews. Immunology, 2*(10), 787–795.
4. Levy, B. D., et al. (2001). Lipid mediator class switching during acute inflammation: Signals in resolution. *Nature Immunology, 2*(7), 612–619.
5. Buckley, C. D., et al. (2013). The resolution of inflammation. *Nature Reviews. Immunology, 13*(1), 59–66.
6. Nathan, C., & Ding, A. (2010). Nonresolving inflammation. *Cell, 140*(6), 871–882.
7. Sugimoto, M. A., et al. (2016). Resolution of inflammation: What controls its onset? *Frontiers in Immunology, 7*, 160.
8. Fullerton, J. N., & Gilroy, D. W. (2016). Resolution of inflammation: A new therapeutic frontier. *Nature Reviews. Drug Discovery, 15*(8), 551–567.
9. Fishbein, A., et al. (2020). Carcinogenesis: Failure of resolution of inflammation? *Pharmacology & Therapeutics*, 107670.
10. Panigrahy, D., et al. (2019). Preoperative stimulation of resolution and inflammation blockade eradicates micrometastases. *The Journal of Clinical Investigation, 129*(7), 2964–2979.
11. Lavin, Y., & Merad, M. (2013). Macrophages: Gatekeepers of tissue integrity. *Cancer Immunology Research, 1*(4), 201–209.
12. Wynn, T. A., Chawla, A., & Pollard, J. W. (2013). Macrophage biology in development, homeostasis and disease. *Nature, 496*(7446), 445–455.
13. Corliss, B. A., et al. (2016). Macrophages: An inflammatory link between angiogenesis and Lymphangiogenesis. *Microcirculation, 23*(2), 95–121.

14. Minutti, C. M., et al. (2019). A macrophage-Pericyte Axis directs tissue restoration via Amphiregulin-induced transforming growth factor Beta activation. *Immunity, 50*(3), 645–654 e6.
15. Hu, M. S., et al. (2017). Delivery of monocyte lineage cells in a biomimetic scaffold enhances tissue repair. JCI. *Insight, 2*(19).
16. Minutti, C. M., et al. (2017). Tissue-specific contribution of macrophages to wound healing. *Seminars in Cell & Developmental Biology, 61*, 3–11.
17. Hulsmans, M., et al. (2017). Macrophages facilitate electrical conduction in the heart. *Cell, 169*(3), 510–522 e20.
18. Wolf, Y., et al. (2017). Brown-adipose-tissue macrophages control tissue innervation and homeostatic energy expenditure. *Nature Immunology, 18*(6), 665–674.
19. Pollard, J. W. (2009). Trophic macrophages in development and disease. *Nature Reviews. Immunology, 9*(4), 259–270.
20. Murray, P. J., et al. (2014). Macrophage activation and polarization: Nomenclature and experimental guidelines. *Immunity, 41*(1), 14–20.
21. Murray, P. J., & Wynn, T. A. (2011). Protective and pathogenic functions of macrophage subsets. *Nature Reviews. Immunology, 11*(11), 723–737.
22. Stefater, J. A., 3rd, et al. (2011). Metchnikoff's policemen: Macrophages in development, homeostasis and regeneration. *Trends in Molecular Medicine, 17*(12), 743–752.
23. Brune, B., et al. (2013). Redox control of inflammation in macrophages. *Antioxidants & Redox Signaling, 19*(6), 595–637.
24. Mosser, D. M., & Edwards, J. P. (2008). Exploring the full spectrum of macrophage activation. *Nature Reviews. Immunology, 8*(12), 958–969.
25. Nahrendorf, M., & Swirski, F. K. (2016). Abandoning M1/M2 for a network model of macrophage function. *Circulation Research, 119*(3), 414–417.
26. Mantovani, A., et al. (2004). The chemokine system in diverse forms of macrophage activation and polarization. *Trends in Immunology, 25*(12), 677–686.
27. Krausgruber, T., et al. (2011). IRF5 promotes inflammatory macrophage polarization and TH1-TH17 responses. *Nature Immunology, 12*(3), 231–238.
28. Hu, S., et al. (2015). Posttranslational modification of Sirt6 activity by peroxynitrite. *Free Radical Biology & Medicine, 79*, 176–185.
29. Yang, S., et al. (2023). Regulatory factor X1 induces macrophage M1 polarization by promoting DNA demethylation in autoimmune inflammation. JCI. *Insight, 8*(20).
30. Lawrence, T., & Natoli, G. (2011). Transcriptional regulation of macrophage polarization: Enabling diversity with identity. *Nature Reviews. Immunology, 11*(11), 750–761.
31. Martinez, F. O., et al. (2013). Genetic programs expressed in resting and IL-4 alternatively activated mouse and human macrophages: Similarities and differences. *Blood, 121*(9), e57–e69.
32. Long, H., et al. (2023). Macrophages and fibrosis: How resident and infiltrating mononuclear phagocytes account for organ injury, regeneration or atrophy. *Frontiers in Immunology, 14*, 1194988.
33. Piccolo, V., et al. (2017). Opposing macrophage polarization programs show extensive epigenomic and transcriptional cross-talk. *Nature Immunology, 18*(5), 530–540.
34. Cassetta, L., Cassol, E., & Poli, G. (2011). Macrophage polarization in health and disease. *ScientificWorldJournal, 11*, 2391–2402.
35. Stout, R. D., et al. (2005). Macrophages sequentially change their functional phenotype in response to changes in microenvironmental influences. *Journal of Immunology, 175*(1), 342–349.
36. Rodriguez-Morales, P., & Franklin, R. A. (2023). Macrophage phenotypes and functions: Resolving inflammation and restoring homeostasis. *Trends in Immunology, 44*(12), 986–998.
37. Saas, P., et al. (2022). Resolution therapy: Harnessing efferocytic macrophages to trigger the resolution of inflammation. *Frontiers in Immunology, 13*, 1021413.

38. Miyake, K., et al. (2024). Single-cell transcriptomics identifies the differentiation trajectory from inflammatory monocytes to pro-resolving macrophages in a mouse skin allergy model. *Nature Communications, 15*(1), 1666.
39. Bosurgi, L., et al. (2017). Macrophage function in tissue repair and remodeling requires IL-4 or IL-13 with apoptotic cells. *Science, 356*(6342), 1072–1076.
40. Cvetanovic, M., & Ucker, D. S. (2004). Innate immune discrimination of apoptotic cells: Repression of proinflammatory macrophage transcription is coupled directly to specific recognition. *Journal of Immunology, 172*(2), 880–889.
41. Bonizzi, G., & Karin, M. (2004). The two NF-kappaB activation pathways and their role in innate and adaptive immunity. *Trends in Immunology, 25*(6), 280–288.
42. Jennewein, C., et al. (2008). Sumoylation of peroxisome proliferator-activated receptor gamma by apoptotic cells prevents lipopolysaccharide-induced NCoR removal from kappaB binding sites mediating transrepression of proinflammatory cytokines. *Journal of Immunology, 181*(8), 5646–5652.
43. Ramos, G. C., et al. (2007). Apoptotic mimicry: Phosphatidylserine liposomes reduce inflammation through activation of peroxisome proliferator-activated receptors (PPARs) in vivo. *British Journal of Pharmacology, 151*(6), 844–850.
44. von Knethen, A., et al. (2013). 5-lipoxygenase contributes to PPARgamma activation in macrophages in response to apoptotic cells. *Cellular Signalling, 25*(12), 2762–2768.
45. Sen, P., et al. (2006). Apoptotic cells induce Mer tyrosine kinase-dependent blockade of NF-{kappa}B activation in dendritic cells. *BloodJournal.*
46. Stanford, J. C., et al. (2014). Efferocytosis produces a prometastatic landscape during postpartum mammary gland involution. *The Journal of Clinical Investigation, 124*(11), 4737–4752.
47. Fadok, V. A., et al. (1998). Macrophages that have ingested apoptotic cells in vitro inhibit proinflammatory cytokine production through autocrine/paracrine mechanisms involving TGF-beta, PGE2, and PAF. *The Journal of Clinical Investigation, 101*(4), 890–898.
48. Weigert, A., & Brune, B. (2008). Nitric oxide, apoptosis and macrophage polarization during tumor progression. *Nitric Oxide, 19*(2), 95–102.
49. Freire-de-Lima, C. G., et al. (2006). Apoptotic cells, through transforming growth factor-beta, coordinately induce anti-inflammatory and suppress pro-inflammatory eicosanoid and NO synthesis in murine macrophages. *The Journal of Biological Chemistry, 281*(50), 38376–38384.
50. Johann, A. M., et al. (2007). Apoptotic cells induce arginase II in macrophages, thereby attenuating NO production. *The FASEB Journal, 21*(11), 2704–2712.
51. Serinkan, B. F., et al. (2005). Apoptotic cells quench reactive oxygen and nitrogen species and modulate TNF-alpha/TGF-beta1 balance in activated macrophages: Involvement of phosphatidylserine-dependent and -independent pathways. *Cell Death and Differentiation, 12*(8), 1141–1144.
52. Johann, A. M., et al. (2006). Recognition of apoptotic cells by macrophages activates the peroxisome proliferator-activated receptor-gamma and attenuates the oxidative burst. *Cell Death and Differentiation, 13*(9), 1533–1540.
53. Freire-de-Lima, C. G., et al. (2000). Uptake of apoptotic cells drives the growth of a pathogenic trypanosome in macrophages. *Nature, 403*(6766), 199–203.
54. Brecht, K., et al. (2011). Macrophages programmed by apoptotic cells promote angiogenesis via prostaglandin E2. *The FASEB Journal, 25*(7), 2408–2417.
55. Jung, M., et al. (2017). Iron handling in tumor-associated macrophages-is there a new role for Lipocalin-2? *Frontiers in Immunology, 8*, 1171.
56. Jung, M., et al. (2016). Lipocalin 2 from macrophages stimulated by tumor cell-derived sphingosine 1-phosphate promotes lymphangiogenesis and tumor metastasis. *Science Signaling, 9*(434), ra64.
57. Kourtzelis, I., et al. (2019). DEL-1 promotes macrophage efferocytosis and clearance of inflammation. *Nature Immunology, 20*(1), 40–49.
58. Alivernini, S., et al. (2020). Distinct synovial tissue macrophage subsets regulate inflammation and remission in rheumatoid arthritis. *Nature Medicine, 26*(8), 1295–1306.

59. Doran, A. C., Yurdagul, A., Jr., & Tabas, I. (2020). Efferocytosis in health and disease. *Nature Reviews. Immunology, 20*(4), 254–267.
60. Weigert, A., et al. (2016). Killing is not enough: How apoptosis hijacks tumor-associated macrophages to promote cancer progression. *Advances in Experimental Medicine and Biology, 930*, 205–239.
61. Moon, B., et al. (2023). After cell death: The molecular machinery of efferocytosis. *Experimental & Molecular Medicine, 55*(8), 1644–1651.
62. Tak, T., et al. (2013). What's your age again? Determination of human neutrophil half-lives revisited. *Journal of Leukocyte Biology, 94*(4), 595–601.
63. Murao, A., et al. (2021). Release mechanisms of major DAMPs. *Apoptosis, 26*(3–4), 152–162.
64. Willems, J. J., Arnold, B. P., & Gregory, C. D. (2014). Sinister self-sacrifice: The contribution of apoptosis to malignancy. *Frontiers in Immunology, 5*, 299.
65. Mehrotra, P., & Ravichandran, K. S. (2022). Drugging the efferocytosis process: Concepts and opportunities. *Nature Reviews. Drug Discovery, 21*(8), 601–620.
66. Elliott, M. R., & Ravichandran, K. S. (2016). The dynamics of apoptotic cell clearance. *Developmental Cell, 38*(2), 147–160.
67. Boada-Romero, E., et al. (2020). The clearance of dead cells by efferocytosis. *Nature Reviews. Molecular Cell Biology, 21*(7), 398–414.
68. Elliott, M. R., et al. (2010). Unexpected requirement for ELMO1 in clearance of apoptotic germ cells in vivo. *Nature, 467*(7313), 333–337.
69. Juncadella, I. J., et al. (2013). Apoptotic cell clearance by bronchial epithelial cells critically influences airway inflammation. *Nature, 493*(7433), 547–551.
70. Thorp, E., Subramanian, M., & Tabas, I. (2011). The role of macrophages and dendritic cells in the clearance of apoptotic cells in advanced atherosclerosis. *European Journal of Immunology, 41*(9), 2515–2518.
71. Gregory, C. D., & Pound, J. D. (2010). Microenvironmental influences of apoptosis in vivo and in vitro. *Apoptosis, 15*(9), 1029–1049.
72. Gregory, C. D., & Pound, J. D. (2011). Cell death in the neighbourhood: Direct microenvironmental effects of apoptosis in normal and neoplastic tissues. *The Journal of Pathology, 223*(2), 177–194.
73. Medina, C. B., & Ravichandran, K. S. (2016). Do not let death do us part: 'find-me' signals in communication between dying cells and the phagocytes. *Cell Death and Differentiation, 23*(6), 979–989.
74. Poon, I. K. H., & Ravichandran, K. S. (2024). Targeting Efferocytosis in Inflammaging. *Annual Review of Pharmacology and Toxicology, 64*, 339–357.
75. Yamaguchi, H., et al. (2014). Immunosuppression via adenosine receptor activation by adenosine monophosphate released from apoptotic cells. *eLife, 3*, e02172.
76. Weis, N., et al. (2009). Heme oxygenase-1 contributes to an alternative macrophage activation profile induced by apoptotic cell supernatants. *Molecular Biology of the Cell, 20*(5), 1280–1288.
77. Weigert, A., et al. (2007). Tumor cell apoptosis polarizes macrophages role of sphingosine-1-phosphate. *Molecular Biology of the Cell, 18*(10), 3810–3819.
78. Sekar, D., et al. (2012). Apoptotic tumor cells induce IL-27 release from human DCs to activate Treg cells that express CD69 and attenuate cytotoxicity. *European Journal of Immunology, 42*(6), 1585–1598.
79. Dillmann, C., et al. (2016). S1PR4 signaling attenuates ILT 7 internalization to limit IFN-alpha production by human Plasmacytoid dendritic cells. *Journal of Immunology, 196*(4), 1579–1590.
80. Luo, B., et al. (2016). Erythropoeitin signaling in macrophages promotes dying cell clearance and immune tolerance. *Immunity, 44*(2), 287–302.
81. Medina, C. B., et al. (2020). Metabolites released from apoptotic cells act as tissue messengers. *Nature, 580*(7801), 130–135.
82. Truman, L. A., et al. (2008). CX3CL1/fractalkine is released from apoptotic lymphocytes to stimulate macrophage chemotaxis. *Blood, 112*(13), 5026–5036.

83. Bournazou, I., et al. (2009). Apoptotic human cells inhibit migration of granulocytes via release of lactoferrin. *The Journal of Clinical Investigation, 119*(1), 20–32.
84. Han, Y., et al. (2020). IL-36 family cytokines in protective versus destructive inflammation. *Cellular Signalling, 75*, 109773.
85. Mora, J., & Weigert, A. (2016). IL-1 family cytokines in cancer immunity – a matter of life and death. *Biological Chemistry, 397*(11), 1125–1134.
86. Mora, J., et al. (2016). Interleukin-38 is released from apoptotic cells to limit inflammatory macrophage responses. *Journal of Molecular Cell Biology, 8*(5), 426–438.
87. Han, Y., et al. (2019). IL-38 ameliorates skin inflammation and limits IL-17 production from gammadelta T cells. *Cell Reports, 27*(3), 835–846 e5.
88. Greenberg, M. E., et al. (2006). Oxidized phosphatidylserine-CD36 interactions play an essential role in macrophage-dependent phagocytosis of apoptotic cells. *The Journal of Experimental Medicine, 203*(12), 2613–2625.
89. Kur, I. M., & Weigert, A. (2024). Phosphatidylserine externalization as immune checkpoint in cancer. *Pflügers Archiv*.
90. Nagata, S., Sakuragi, T., & Segawa, K. (2020). Flippase and scramblase for phosphatidylserine exposure. *Current Opinion in Immunology, 62*, 31–38.
91. Shlomovitz, I., Speir, M., & Gerlic, M. (2019). Flipping the dogma – phosphatidylserine in non-apoptotic cell death. *Cell Communication and Signaling: CCS, 17*(1), 139.
92. Birge, R. B., et al. (2016). Phosphatidylserine is a global immunosuppressive signal in efferocytosis, infectious disease, and cancer. *Cell Death and Differentiation, 23*(6), 962–978.
93. Tsou, W. I., et al. (2014). Receptor tyrosine kinases, TYRO3, AXL, and MER, demonstrate distinct patterns and complex regulation of ligand-induced activation. *The Journal of Biological Chemistry, 289*(37), 25750–25763.
94. Lew, E. D., et al. (2014). Differential TAM receptor-ligand-phospholipid interactions delimit differential TAM bioactivities. *eLife, 3*.
95. Zagorska, A., et al. (2020). Differential regulation of hepatic physiology and injury by the TAM receptors Axl and Mer. *Life Science Alliance, 3*(8).
96. DeBerge, M., et al. (2021). Macrophage AXL receptor tyrosine kinase inflames the heart after reperfused myocardial infarction. *The Journal of Clinical Investigation, 131*(6).
97. Sekar, D., et al. (2022). Phosphatidylserine synthase PTDSS1 shapes the tumor Lipidome to maintain tumor-promoting inflammation. *Cancer Research, 82*(8), 1617–1632.
98. Knuth, A. K., et al. (2021). Apoptotic cells induce proliferation of peritoneal macrophages. *International Journal of Molecular Sciences, 22*(5).
99. Liu, W., et al. (2019). Tim-4 inhibits NLRP3 Inflammasome via the LKB1/AMPKalpha pathway in macrophages. *Journal of Immunology, 203*(4), 990–1000.
100. Wolf, Y., Anderson, A. C., & Kuchroo, V. K. (2020). TIM3 comes of age as an inhibitory receptor. *Nature Reviews. Immunology, 20*(3), 173–185.
101. Park, S. Y., & Kim, I. S. (2019). Stabilin receptors: Role as phosphatidylserine receptors. *Biomolecules, 9*(8).
102. Raymond, M. H., et al. (2022). Live cell tracking of macrophage efferocytosis during drosophila embryo development in vivo. *Science, 375*(6585), 1182–1187.
103. Marques, E., Kramer, R., & Ryan, D. G. (2024). Multifaceted mitochondria in innate immunity. *NPJ Metab Health Dis, 2*(1), 6.
104. Mills, E. L., & O'Neill, L. A. (2016). Reprogramming mitochondrial metabolism in macrophages as an anti-inflammatory signal. *European Journal of Immunology, 46*(1), 13–21.
105. Park, D., et al. (2011). Continued clearance of apoptotic cells critically depends on the phagocyte Ucp2 protein. *Nature, 477*(7363), 220–224.
106. Pollard, J. W., & Stanners, C. P. (1979). Characterization of cell lines showing growth control isolated from both the wild type and a leucyl-tRNA synthetase mutant of Chinese hamster ovary cells. *Journal of Cellular Physiology, 98*(3), 571–585.
107. Viaud, M., et al. (2018). Lysosomal cholesterol hydrolysis couples Efferocytosis to anti-inflammatory oxysterol production. *Circulation Research, 122*(10), 1369–1384.

108. Zhang, S., et al. (2019). Efferocytosis fuels requirements of fatty acid oxidation and the electron transport chain to polarize macrophages for tissue repair. *Cell Metabolism, 29*(2), 443–456 e5.
109. Mota, A. C., et al. (2021). Lysosome-dependent LXR and PPARdelta activation upon Efferocytosis in human macrophages. *Frontiers in Immunology, 12*, 637778.
110. Gonzalez, M. A., et al. (2023). Phagocytosis increases an oxidative metabolic and immune suppressive signature in tumor macrophages. *The Journal of Experimental Medicine, 220*(6).
111. Ngai, D., Schilperoort, M., & Tabas, I. (2023). Efferocytosis-induced lactate enables the proliferation of pro-resolving macrophages to mediate tissue repair. *Nature Metabolism, 5*(12), 2206–2219.
112. Wang, Y., et al. (2017). Mitochondrial fission promotes the continued clearance of apoptotic cells by macrophages. *Cell, 171*(2), 331–345 e22.
113. Jiang, S., et al. (2013). Mitochondria and AMP-activated protein kinase-dependent mechanism of efferocytosis. *The Journal of Biological Chemistry, 288*(36), 26013–26026.
114. Lauber, K., et al. (2003). Apoptotic cells induce migration of phagocytes via caspase-3-mediated release of a lipid attraction signal. *Cell, 113*(6), 717–730.
115. Jha, A. K., et al. (2015). Network integration of parallel metabolic and transcriptional data reveals metabolic modules that regulate macrophage polarization. *Immunity, 42*(3), 419–430.
116. Merlin, J., et al. (2021). Non-canonical glutamine transamination sustains efferocytosis by coupling redox buffering to oxidative phosphorylation. *Nature Metabolism, 3*(10), 1313–1326.
117. Namgaladze, D., & Brune, B. (2016). Macrophage fatty acid oxidation and its roles in macrophage polarization and fatty acid-induced inflammation. *Biochimica et Biophysica Acta, 1861*(11), 1796–1807.
118. Namgaladze, D., & Brune, B. (2023). Rapid glycolytic activation accompanying innate immune responses: Mechanisms and function. *Frontiers in Immunology, 14*, 1180488.
119. Morioka, S., et al. (2018). Efferocytosis induces a novel SLC program to promote glucose uptake and lactate release. *Nature, 563*(7733), 714–718.
120. Schilperoort, M., et al. (2023). PFKFB2-mediated glycolysis promotes lactate-driven continual efferocytosis by macrophages. *Nature Metabolism, 5*(3), 431–444.
121. Zhang, D., et al. (2019). Metabolic regulation of gene expression by histone lactylation. *Nature, 574*(7779), 575–580.
122. Noe, J. T., et al. (2021). Lactate supports a metabolic-epigenetic link in macrophage polarization. *Science Advances, 7*(46), eabi8602.
123. Manoharan, I., et al. (2021). Lactate-dependent regulation of immune responses by dendritic cells and macrophages. *Frontiers in Immunology, 12*, 691134.
124. Wculek, S. K., et al. (2024). Metabolic homeostasis of tissue macrophages across the lifespan. *Trends in Endocrinology and Metabolism.*
125. Wang, Y. T., et al. (2023). Metabolic adaptation supports enhanced macrophage efferocytosis in limited-oxygen environments. *Cell Metabolism, 35*(2), 316–331 e6.
126. Catala, A., et al. (2020). Metabolic reprogramming of mouse bone marrow derived macrophages following Erythrophagocytosis. *Frontiers in Physiology, 11*, 396.
127. He, D., et al. (2021). Pentose phosphate pathway regulates Tolerogenic apoptotic cell clearance and immune tolerance. *Frontiers in Immunology, 12*, 797091.
128. Puleston, D. J., et al. (2019). Polyamines and eIF5A Hypusination modulate mitochondrial respiration and macrophage activation. *Cell Metabolism, 30*(2), 352–363 e8.
129. Yurdagul, A., Jr., et al. (2020). Macrophage metabolism of apoptotic cell-derived arginine promotes continual Efferocytosis and resolution of injury. *Cell Metabolism, 31*(3), 518–533 e10.
130. McCubbrey, A. L., et al. (2022). Polyamine import and accumulation causes immunomodulation in macrophages engulfing apoptotic cells. *Cell Reports, 38*(2), 110222.
131. Ampomah, P. B., et al. (2022). Macrophages use apoptotic cell-derived methionine and DNMT3A during efferocytosis to promote tissue resolution. *Nature Metabolism, 4*(4), 444–457.

132. Yin, C., & Heit, B. (2021). Cellular responses to the Efferocytosis of apoptotic cells. *Frontiers in Immunology, 12*, 631714.
133. Kinchen, J. M., & Ravichandran, K. S. (2008). Phagosome maturation: Going through the acid test. *Nature Reviews. Molecular Cell Biology, 9*(10), 781–795.
134. Yin, C., et al. (2016). Rab17 mediates differential antigen sorting following efferocytosis and phagocytosis. *Cell Death & Disease, 7*(12), e2529.
135. Erwig, L. P., et al. (2006). Differential regulation of phagosome maturation in macrophages and dendritic cells mediated by rho GTPases and ezrin-radixin-moesin (ERM) proteins. *Proceedings of the National Academy of Sciences of the United States of America, 103*(34), 12825–12830.
136. Asare, P. F., et al. (2020). LC3-associated phagocytosis (LAP): A potentially influential mediator of Efferocytosis-related tumor progression and aggressiveness. *Frontiers in Oncology, 10*, 1298.
137. Wan, J., et al. (2020). LC3-associated phagocytosis protects against inflammation and liver fibrosis via immunoreceptor inhibitory signaling. *Science Translational Medicine, 12*(539).
138. Wong, S. W., Sil, P., & Martinez, J. (2018). Rubicon: LC3-associated phagocytosis and beyond. *The FEBS Journal, 285*(8), 1379–1388.
139. Kimani, S. G., et al. (2014). Contribution of defective PS recognition and Efferocytosis to chronic inflammation and autoimmunity. *Frontiers in Immunology, 5*, 566.
140. Zizzo, G., et al. (2013). Circulating levels of soluble MER in lupus reflect M2c activation of monocytes/macrophages, autoantibody specificities and disease activity. *Arthritis Research & Therapy, 15*(6), R212.
141. Waterborg, C. E. J., et al. (2018). Protective role of the MER tyrosine kinase via Efferocytosis in rheumatoid arthritis models. *Frontiers in Immunology, 9*, 742.
142. Binder, M. D., et al. (2016). Common and low frequency variants in MERTK are independently associated with multiple sclerosis susceptibility with discordant association dependent upon HLA-DRB1*15:01 status. *PLoS Genetics, 12*(3), e1005853.
143. Gerlach, B. D., et al. (2021). Efferocytosis induces macrophage proliferation to help resolve tissue injury. *Cell Metabolism, 33*(12), 2445–2463 e8.
144. Noelia, A. G., et al. (2009). Apoptotic cells promote their own clearance and immune tolerance through activation of the nuclear receptor LXR. *Immunity, 31*(2), 245–258.
145. Mukundan, L., et al. (2009). PPAR-delta senses and orchestrates clearance of apoptotic cells to promote tolerance. *Nature Medicine, 15*(11), 1266–1272.
146. Majai, G., et al. (2007). PPARgamma-dependent regulation of human macrophages in phagocytosis of apoptotic cells. *European Journal of Immunology, 37*(5), 1343–1354.
147. Roszer, T., et al. (2011). Autoimmune kidney disease and impaired engulfment of apoptotic cells in mice with macrophage peroxisome proliferator-activated receptor gamma or retinoid X receptor alpha deficiency. *Journal of Immunology, 186*(1), 621–631.
148. Gharib, S. A., et al. (2019). Transcriptional and functional diversity of human macrophage repolarization. *The Journal of Allergy and Clinical Immunology, 143*(4), 1536–1548.
149. Snodgrass, R. G., et al. (2021). Efferocytosis potentiates the expression of arachidonate 15-lipoxygenase (ALOX15) in alternatively activated human macrophages through LXR activation. *Cell Death and Differentiation, 28*(4), 1301–1316.
150. Kiss, R. S., et al. (2006). Apoptotic cells induce a phosphatidylserine-dependent homeostatic response from phagocytes. *Current Biology, 16*(22), 2252–2258.
151. Rebe, C., et al. (2009). Induction of transglutaminase 2 by a liver X receptor/retinoic acid receptor alpha pathway increases the clearance of apoptotic cells by human macrophages. *Circulation Research, 105*(4), 393–401.
152. Schilperoort, M., et al. (2023). The role of efferocytosis-fueled macrophage metabolism in the resolution of inflammation. *Immunological Reviews, 319*(1), 65–80.
153. Cai, B., et al. (2016). MerTK cleavage limits proresolving mediator biosynthesis and exacerbates tissue inflammation. *Proceedings of the National Academy of Sciences of the United States of America, 113*(23), 6526–6531.

154. Ringleb, J., et al. (2018). Apoptotic cancer cells suppress 5-lipoxygenase in tumor-associated macrophages. *Journal of Immunology, 200*(2), 857–868.
155. O'Donnell, V. B., et al. (2023). Failure to apply standard limit-of-detection or limit-of-quantitation criteria to specialized pro-resolving mediator analysis incorrectly characterizes their presence in biological samples. *Nature Communications, 14*(1), 7172.
156. Asare, P. F., et al. (2021). Inhibition of LC3-associated phagocytosis in COPD and in response to cigarette smoke. *Therapeutic Advances in Respiratory Disease, 15*, 17534666211039769.
157. Martinez, J., & Cook, D. N. (2021). What's the deal with efferocytosis and asthma? *Trends in Immunology, 42*(10), 904–919.
158. Back, M., et al. (2019). Inflammation and its resolution in atherosclerosis: Mediators and therapeutic opportunities. *Nature Reviews Cardiology, 16*(7), 389–406.
159. Trzeciak, A., Wang, Y. T., & Perry, J. S. A. (2021). First we eat, then we do everything else: The dynamic metabolic regulation of efferocytosis. *Cell Metabolism, 33*(11), 2126–2141.
160. Mevorach, D., et al. (2014). Single infusion of donor mononuclear early apoptotic cells as prophylaxis for graft-versus-host disease in myeloablative HLA-matched allogeneic bone marrow transplantation: A phase I/IIa clinical trial. *Biology of Blood and Marrow Transplantation, 20*(1), 58–65.
161. Karbian, N., et al. (2020). Apoptotic cell therapy for cytokine storm associated with acute severe sepsis. *Cell Death & Disease, 11*(7), 535.
162. Bonnefoy, F., et al. (2016). Apoptotic cell infusion treats ongoing collagen-induced arthritis, even in the presence of methotrexate, and is synergic with anti-TNF therapy. *Arthritis Research & Therapy, 18*(1), 184.
163. Pang, S. H. M., et al. (2021). Mesenchymal stromal cell apoptosis is required for their therapeutic function. *Nature Communications, 12*(1), 6495.
164. Sordi, R., et al. (2020). The apoptosis clearance signal phosphatidylserine inhibits leukocyte migration and promotes inflammation resolution in vivo. *European Journal of Pharmacology*, 173095.
165. Oronsky, B., et al. (2020). Just eat it: A review of CD47 and SIRP-alpha antagonism. *Seminars in Oncology, 47*(2–3), 117–124.
166. Kojima, Y., et al. (2016). CD47-blocking antibodies restore phagocytosis and prevent atherosclerosis. *Nature, 536*(7614), 86–90.
167. Jarr, K. U., et al. (2021). Effect of CD47 blockade on vascular inflammation. *The New England Journal of Medicine, 384*(4), 382–383.
168. Morioka, S., et al. (2022). Chimeric efferocytic receptors improve apoptotic cell clearance and alleviate inflammation. *Cell, 185*(26), 4887–4903 e17.
169. Bhattacharya, P., et al. (2023). Efferocytes release extracellular vesicles to resolve inflammation and tissue injury via prosaposin-GPR37 signaling. *Cell Reports, 42*(7), 112808.
170. Wang, H., et al. (2024). Activating macrophage continual Efferocytosis via microenvironment biomimetic short fibers for reversing inflammation in bone repair. *Advanced Materials, 36*(30), e2402968.
171. Xian, X., et al. (2017). LRP1 integrates murine macrophage cholesterol homeostasis and inflammatory responses in atherosclerosis. *eLife*, 6.
172. Wynn, T. A., & Vannella, K. M. (2016). Macrophages in tissue repair, regeneration, and fibrosis. *Immunity, 44*(3), 450–462.
173. Bekes, M., Langley, D. R., & Crews, C. M. (2022). PROTAC targeted protein degraders: The past is prologue. *Nature Reviews. Drug Discovery, 21*(3), 181–200.
174. Rana, A., et al. (2017). Promoting Drp1-mediated mitochondrial fission in midlife prolongs healthy lifespan of Drosophila melanogaster. *Nature Communications, 8*(1), 448.

Apoptosis and Cell Clearance in Skin Wound Healing

Hannah K. L. De Cleene, Burcu N. Keçeli, and Sophia Maschalidi

Abstract

The skin is our body's largest natural barrier, and daily, it undergoes numerous challenges, such as physical injury, chemical insults, and UV radiation. The significance of repairing breaches of this barrier as quickly as possible is especially highlighted in chronic nonhealing wounds, including those associated with diabetes, which affects one in ten people worldwide, with vascular disease and aging. Dying cells in the wound perpetuate inflammation, contribute to dysregulated tissue repair, and increase the risk of infection. Cutaneous wound healing typically consists of four phases, which are hemostasis, inflammation, growth, re-epithelialization, and remodeling. In this chapter, we will discuss how various cell types, immune and nonimmune cells that reside in the epidermis and dermis, die from apoptosis and how efficient dead cell clearance by professional and nonprofessional phagocytes contributes to effective skin wound healing. We will also consider a range of cell death modalities, such as ferroptosis, necroptosis, pyroptosis, and others, and present our current knowledge about their contribution to skin injury and repair.

Keywords

Apoptosis · Efferocytosis · Cell death · Phagocytosis · Skin injury · Wound healing · Chronic wound

H. K. L. De Cleene · B. N. Keçeli · S. Maschalidi (✉)
Unit for Cell Clearance in Health and Disease, VIB Center for Inflammation Research, Ghent, Belgium

Department of Biomedical Molecular Biology, Ghent University, Ghent, Belgium
e-mail: hannah.decleene@irc.vib-ugent.be; Burcu.Keceli@irc.vib-ugent.be; sophia.maschalidi@irc.vib-ugent.be

C. D. Gregory et al. (eds.), *Apoptosis and Other forms of Cell Death in the Regulation of Inflammation in Health and Disease*, Advances in Experimental Medicine and Biology 1481, https://doi.org/10.1007/978-3-031-92785-0_5

Abbreviations

ADAM9	Metalloprotease 9
AIM2	Interferon inducible protein
ANGPTL4	Angiopoietin-like 4
CCL2	C-C motif chemokine ligand 2
cDCs	Conventional dendritic cells
CCN1	Cellular communication network factor 1
CGRP	Calcitonin-gene-related peptide
CLEC9A	C-type lectin domain containing 9A
CSF1R	Receptor of the colony-stimulating factor-1
CX43	Connexin43
CXCR2	C-X-C motif chemokine receptor 2
CYR61/CCN1	Cysteine-rich protein 61
DAMPs	Damage-associated molecular patterns
dDCs	Dermal dendritic cells
DETCs	Dendritic epidermal T cells
DFO	Deferoxamine
DFUs	Diabetic foot ulcers
FADD	Fas-associated death domain
Fer-1	Ferrostatin-1
FSP1	Ferroptosis suppressor protein 1
GDF15	Growth differentiation factor 15
GPX4	Glutathione peroxidase 4
GSH	Glutathione
HDFs	Human dermal fibroblasts
HDMECs	Human dermal microvascular endothelial cells
HIF-1α	Hypoxia-inducible factor-1 alpha
HMGB1	High mobility group box 1
IFE	Interfollicular epidermis
IFN-γ	Interferon-gamma
IFNAR1	Interferon alpha/beta receptor 1
IL	Interleukin
ILCs	Innate lymphoid cells like
LCs	Langerhans cells
LYVE1	Lymphatic vessel endothelial hyaluronan receptor 1
MALAT1	Metastasis-associated lung adenocarcinoma transcript 1
MERTK	Mer tyrosine kinase
MFG-E8	Milk-fat-globular-epidermal growth factor 8
miRR-126	microRNA-126
MoDCs	Monocyte-derived DCs
MSCs	Mesenchymal stem cells
NCOA4	Nuclear receptor coactivator 4
NETs	Neutrophils extracellular traps
NF-κB	Nuclear factor kappa B

NK cells	Natural killer cells
NPY	Neuropeptide Y
NRF2	Nuclear factor erythroid 2-related factor 2
PGE2	Prostaglandin E2
PLA2	Phospholipase 2
PPARγ	Peroxisome proliferator-activated receptor gamma
PS	Phosphatidylserine
RAGE	Receptor for advanced glycation end products
RIPK3	Receptor-interacting serine/threonine kinase 3
ROS	Reactive oxygen species
S1P	Sphingosine 1-phosphate
SASP	Senescence-associated secretory phenotype
SIRT3	Sirtuin 3
SP	Substance P
SPD	Surfactant polymer dressing
STAB1	Stabilin 1
STEAP3	Six-transmembrane epithelial antigen of prostate 3
TGF-ß1	Transforming growth factor beta 1
TIM-4	T cell membrane protein 4
TLR	Toll-like receptor
TNFR1	Tumor necrosis factor receptor 1
TNFα	Tumor necrosis factor alpha
TSP-1	Thrombospondin-1
UV radiation	Ultraviolet radiation
VEGF	Vascular endothelial growth factor
XCR1	X-C motif chemokine receptor 1
XIAP	X-linked inhibitor of apoptosis protein
ZBP1	Z-DNA-binding protein 1
α-SMA	α-smooth muscle actin

The Anatomy and Cellular Composition of the Skin

The skin is the largest organ of the body and serves as a multifaceted protective barrier essential for maintaining homeostasis. Three layers constitute the skin. The outermost layer is the epidermis and consists of a network of cells limiting water loss, allowing us to live in air without suffering from dehydration and protecting against external threats and shielding from UV radiation [1]. The dermis lies beneath the epidermis and comprises the connective tissue, blood vessels, and nerve endings that play a pivotal role in supporting the epidermal layer and in facilitating sensory perception (e.g., touch, temperature, pain perception, etc.). The hypodermis is composed almost entirely of adipose tissue, it provides insulation and acts as an energy reservoir (Fig. 1). Human and mouse skin exhibit notable differences, although the overall architecture and function of murine skin parallel those of humans [2]. Major relevant differences between mouse and human are highlighted throughout the chapter.

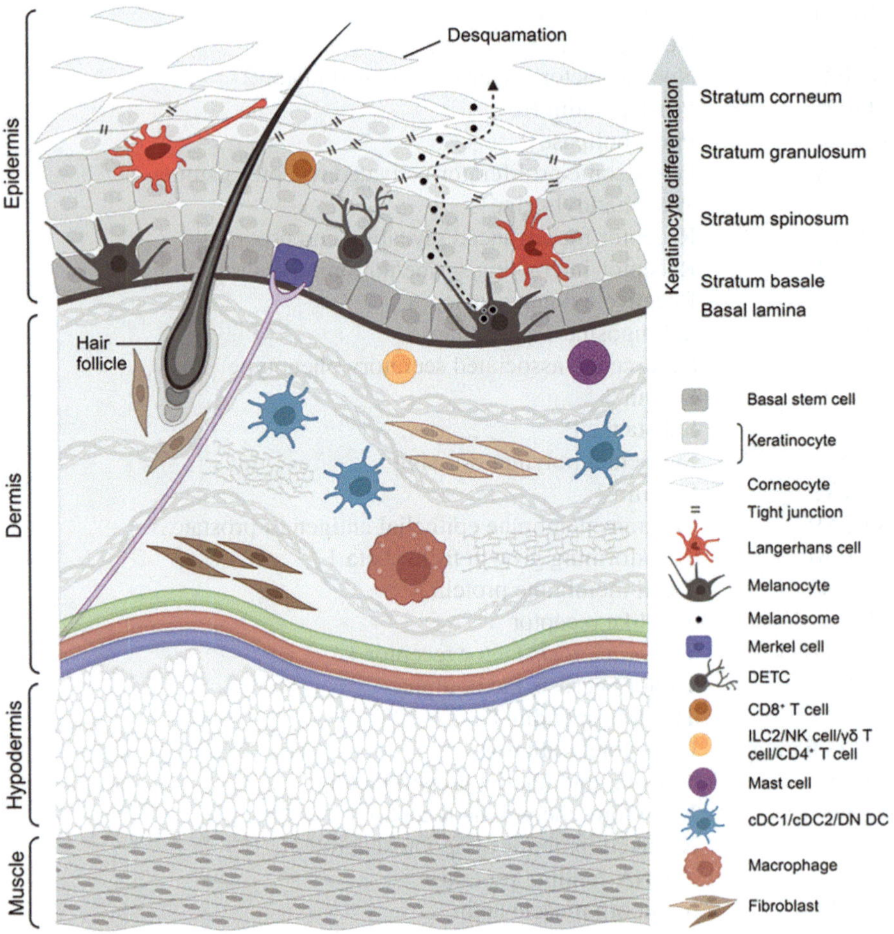

Fig. 1 Anatomy and cellular composition of the skin during homeostasis. The skin is composed of three main layers—the epidermis, dermis, and hypodermis, sitting atop a muscle layer. Together, they serve as a barrier required for protecting the body from external threats, dehydration, and energy loss. The epidermis consists mostly of keratinocytes which originate from basal stem cells that differentiate in a layer-restricted manner and eventually cornify and shed in a process called desquamation. Merkel cells and melanocytes, residing in the basal layer, are tightly associated with nerve endings or produce melanin, respectively. During homeostasis, the epidermis is populated by Langerhans cells, dendritic epidermal T cells (DETCs), and $CD8^+$ T cells, providing rapid protection in the event of infection. The dermis comprises mostly of fibroblasts, endothelial cells, and structural proteins, providing support, elasticity, and strength to the skin. In addition, the healthy dermis is inhabited by myeloid cells, including macrophages, different subsets of dendritic cells, and mast cells, as well as lymphoid cells, such as ILC2s, NK cells, γδ T cells, and $CD4^+$ T cells further providing protection to the skin and body. Skin injury and inflammation induce striking differences in the cellular composition and phenotypic heterogeneity of the skin

Epidermis

Nonimmune Cells

The developing epidermis emerges from the ectoderm. It starts off as a single layer of stem cells on the basal lamina, which separates the epidermis from the underlying dermis. This layer will later on form the layers of the epidermis, hair follicles, sebaceous glands, and the neuroendocrine Merkel cells. The Merkel cells are tightly associated with nerve endings [3–5]. The basal layer of the epidermis is also inhabited by melanocytes, skin pigmentation cells that transfer their melanin-producing and storing melanosomes to keratinocytes, thus protecting against UV radiation [6, 7]. Each layer within the epidermis originates from basal stem cells that differentiate based on layer-restricted expression of transcription factors demarcating commitment points [8–10]. Once in the granular layer of the epidermis, terminally differentiated keratinocytes produce major structural proteins and finally undergo cornification, the process in which they are reduced to a denucleated cell envelope consisting of proteins cross-linked to extruded lipids [11]. These cross-linked, dead "corneocytes," unlike in other tissues, are not cleared through phagocytosis but become an integral part of the tissue and are crucial for its barrier properties [12, 13]. Eventually, they are shed from the surface by a process called desquamation. Collectively, this network of cells contributes to the epidermal architecture, pigmentation, and mechano-transduction in the skin.

Immune Cells

Langerhans cells (LCs) are the sole myeloid cell type in the epidermis, colonizing the tissue as yolk sac macrophages, and later fetal liver monocytes spread throughout the developing embryo [14–17]. Although these CSF1R-dependent self-maintaining cells share many ontogenetic properties with tissue-resident macrophages throughout the body (e.g., microglia in the brain) [18–21], they are functionally considered to resemble conventional dendritic cells (cDCs). LCs play a crucial role in immune surveillance by capturing antigens in the skin and initiating adaptive immune responses through migration and presentation of antigens to skin-draining lymph nodes [22–24]. The phagocytic properties of LCs and their contribution to skin homeostasis and wound healing will be addressed in detail in Section "Langerhans Cells (LCs)". Next to the LCs, lymphoid cells inhabit the healthy epidermis. Infiltrating the skin during infection, antigen-specific effector T cells differentiate into long-lived, resident, memory $CD8^+$ T cells and constitute the majority of resident αβ T cells in the epidermis [25, 26]. In addition, γδ T cells form an interdigitating network within the epidermis and are therefore called dendritic epidermal T cells (DETCs) [27]. Notably, DETCs have not been described in the human epidermis. Together, these lymphoid cells allow rapid intervention in the event of infection by a harmful pathogen and can promote wound closure by promoting proliferation of basal stem cells [28, 29].

Dermis

Nonimmune Cells

Comprising mostly fibroblasts, endothelial cells, and structural proteins, the dermis that is interpolated between the epidermis and subcutaneous tissue is essential for providing support, elasticity, and strength to the skin. Dermal fibroblasts are primarily responsible for the synthesis, deposition, and remodeling of collagens, fibronectin, laminins, elastic fibers, and others, providing structural integrity to the skin [30]. Upon injury, these mesenchymal cells differentiate into α-smooth muscle actin (αSMA)-positive myofibroblasts, which mediate wound contraction and reproduce extracellular matrix (ECM) proteins, thereby exerting crucial roles during wound closure and the healing process [31, 32]. During the remodeling phase of wound healing, these myofibroblasts undergo apoptosis [33]; however, how apoptotic myofibroblasts are eliminated and which phagocyte(s) are involved in this process remains unclear.

Immune Cells

Analogous to the epidermis, the dermis is also inhabited by both myeloid and lymphoid immune cells. The myeloid cells in the healthy dermis primarily consist of macrophages, dermal dendritic cells, and mast cells. Dermal macrophages originating from embryonic seeding or derived from blood monocytes are thought to co-inhabit the dermis [34, 35]. Similarly, the dermal DCs (dDCs) consist of several ontogenetically distinct subsets derived from the bone marrow and exert important phagocytic and immune-regulatory functions in the skin [36]. The different subsets of macrophages and DCs present in the dermis and their phagocytic functions are addressed in more detail below. The mast cells in the skin have been shown to contribute to the homeostatic innate immune defense as well as in early activation of allergic skin response [37]. The lymphoid fraction of dermal immune cells is composed of innate lymphoid cells like innate lymphoid cells (ILCs) and natural killer (NK) cells, being primarily dominated by ILC2s, and adaptive lymphoid cells, mainly by $CD4^+$ T cells and γδ T cells [38, 39]. Because of the constant exposure to harmless or even commensal microbes, it is not surprising that a large population of $CD4^+$ T cells present in the skin are regulatory T cells, and that γδ T cells form a major source of interleukin (IL) 17A, shifting the skin microenvironment into a type 3 immune response targeted toward controlling infections by extracellular bacteria [40–42].

Notably, skin injury and inflammation induce striking differences in the cellular composition of the skin, including expansion of existing populations such as dermal DCs, macrophages, ILCs, and mast cells, as well as recruitment of bone marrow-derived monocytes, neutrophils, and eosinophils [43–45]. Importantly, inflammation can induce phenotypic heterogeneity within ontogenetically identical cell populations due to activation [46, 47].

Cutaneous Phagocytes and Their Role in Tissue Repair: An Appetite for Healing

In this part, we will summarize the characteristics and properties of cells that act as professional and nonprofessional phagocytes to eliminate cellular debris during all phases of cutaneous wound healing (Fig. 2).

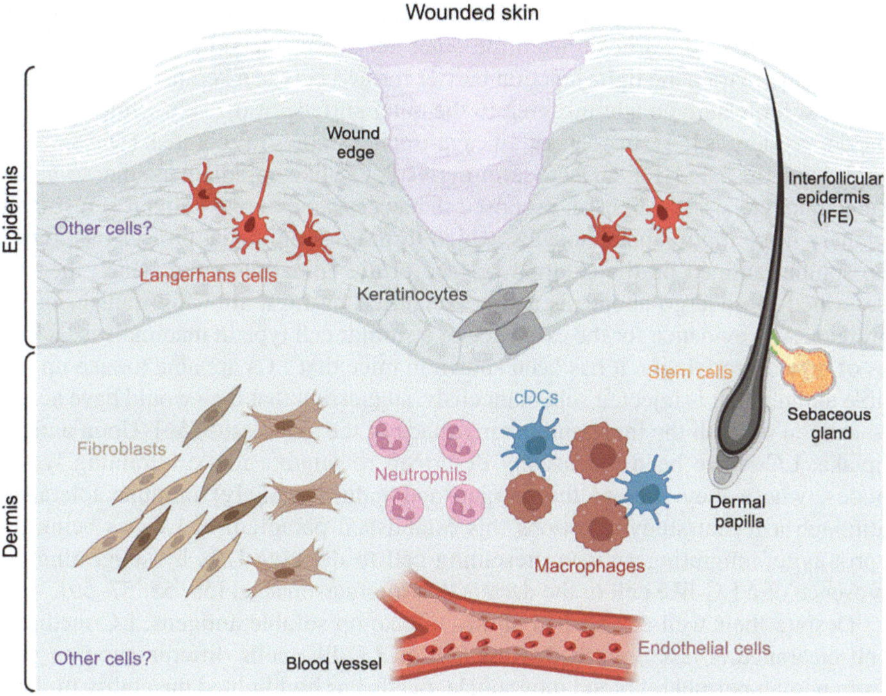

Fig. 2 Phagocytes participate in cell clearance in cutaneous wound healing. Cutaneous wound healing typically consists of the following phases: hemostasis, inflammation, growth, re-epithelialization, and remodeling. During the inflammatory phase, macrophages clear cellular debris and neutrophils that die from apoptosis after having exerted their antimicrobial activity. This causes a boost to tissue remodeling later on. Although the participation of Langerhans cells in apoptotic cell engulfment (e.g., apoptotic keratinocytes) during cutaneous wound healing remains under-studied, the important role of dermal dendritic cells and mainly cDC1 in apoptotic cell clearance during skin injury has recently been addressed. Fibroblasts can be directly involved in cellular clearance, especially in the absence of professional phagocytes, and endothelial cells participate in efferocytosis by increasing macrophage recruitment and apoptotic cell clearance via thrombospondin-1 (TSP-1) expression. However, the contribution of endothelial cells and keratinocytes in the wound healing process as nonprofessional phagocytes merits more in-depth investigation. Lastly, the potential role of the nonmotile stem cells—residing in hair follicles, the sebaceous glands, and the interfollicular epidermis (IFE) or mesenchymal stem cells—in engulfing corpses is recently attracting attention and remains to be fully addressed in skin wound healing

Langerhans Cells (LCs)

LCs, aptly described as "macrophages in DC clothing," constitute the sole myeloid cell that is present in the epidermis [48]. A key role that has been attributed to tissue-resident macrophages is the maintenance of tissue homeostasis through their scavenger function, where they quickly clear cellular and other debris [49]. Given that LCs have long been functionally identified as DCs, several researchers have shown their efficiency in taking up soluble antigens both in mice and humans [49–54]. Consistent with their role in sampling the external environment, upon activation LCs extend their dendrites toward the outer surface of the skin, allowing them to dock and penetrate the tight-junction barrier formed between keratinocytes to endocytose soluble antigen administered to the outer surface of the skin while retaining barrier properties [51]. In contrast, lineage-tracing studies in zebrafish (*Danio rerio)* suggested the need for an ectoderm-derived, LC-like middleman, the so-called "metaphocyte" [55]. Lin et al. proposed that access to external antigen is restricted to these more motile metaphocytes, which form trans-epithelial protrusions to take up soluble antigen on the external surface of the epidermis, after which they are described to undergo apoptosis and are engulfed by LCs, allowing transfer of the antigen. Any evidence for the existence of a similar cell type in mammals is lacking as of now. Surprisingly, it has been shown in mice that LCs are able to take up soluble antigen that is injected subcutaneously, suggesting that they would have access to antigen on both the inner and outer surface of the skin tissue [56]. Upon antigen uptake, LCs have been extensively described to migrate to skin-draining lymph nodes, where they present the antigen and induce (mainly) immune tolerance, although a recent study questions this established paradigm of LCs as being the "prototype" migrating antigen-presenting cell to draining LNs by suggesting the presence of a LC-like cell in the dermis that migrates instead [50, 53, 57–59].

Despite their well-established ability to take up soluble antigens, LC-mediated cell clearance is less described. Research on LC-like cells differentiated ex vivo from human peripheral blood mononuclear cells has highlighted the ability of these cells to engulf dead tumor cells and cross-present their antigens to prime cytotoxic $CD8^+$ T lymphocytes [57]. Close to a decade later, a potential mechanism for recognition and clearance of apoptotic cells was highlighted, showing that transforming growth factor beta 1 (TGF-β1) signaling induces AXL expression and AXL-mediated efferocytosis, and that its (global) expression is crucial for the establishment of a functional LC network and prevention of epidermal inflammation [60]. Recent work in zebrafish has suggested an additional role for LCs in the clearance of degenerated axons upon wounding, during which LCs have been shown to increase their phagocytic capacity [55, 61]. The exact contribution of LC-mediated cell clearance to cutaneous wound healing, however, has not been established.

Nonetheless, multiple studies have shown contradicting roles for LCs during tissue repair, depending on the models used. Highlighting the positive role during wound healing, LC-depletion studies have shown that a lack of LCs is associated with increased presence of apoptotic keratinocytes and exacerbated inflammation, impaired angiogenesis, and overall delayed wound closure [62–67]. In contrast, hypertrophic scarring has been associated with increased presence of LCs during the remodeling phase of wound healing, and depletion of Langerin^{+} cells was shown to enhance keratinocyte proliferation, re-epithelialization, and wound healing [68, 69]. Taken together, although the role of LCs in the uptake of soluble antigen and the presentation thereof has long been established, their ability to engulf cellular debris in homeostatic and injured conditions and their contribution to wound repair remain understudied.

Dermal Dendritic Cells (dDCs)

The bona fide DCs inhabiting the skin are described to reside within the dermis and are therefore referred to as dDCs. As with any other peripheral tissue in the body, the dDCs consist of ontogenetically and functionally distinct subsets that can be distinguished based on surface marker expression. X-C motif chemokine receptor 1 (XCR1)$^{+}$ cDC1s, as in other tissues, are known to excel at cross-presenting exogenous antigen, attributed to their expression of C-type lectin domain containing 9a (CLEC9A/ DNGR1) [36, 70–72]. CD11b^{+} CD172^{+} cDC2s are the most abundant type of cDC located within the dermis and have been shown to be potent inducers of adequate T cell responses in the skin [35, 36]. Similarly, the human dermis is inhabited by cDC1 and cDC2 homologues characterized by their respective expression of CD141 and CD1c [73, 74]. In contrast, the murine dermis has been described to be colonized by an XCR1^{-} CD11b^{lo} population of so-called double negative DCs, that have not been described in human skin or other murine tissues [75]. Finally, the skin is populated by monocyte-derived cells that exert DC-like functions and are often referred to as monocyte-derived DCs (MoDCs) [35].

Despite their differential surface marker expression, dDCs are often overshadowed by the LCs due to their supposed functional similarities. Lack of separation of myeloid cell clusters in single-cell transcriptomics datasets in recent years fails to pick apart these ontogenetically and spatially distinct cell types, thereby masking dDC heterogeneity and function in steady state or injury/inflammation. Nonetheless, several studies have convincingly shown the ability of dermal cDC1s and cDC2s to take up external (soluble) antigens and elicit adequate T cell responses upon migration to the skin-draining lymph nodes [76–78]. Upon injection, dDCs were shown to engulf parasites and beads, differentially affecting their motility and subsequent antigen presentation in the draining lymph nodes [79, 80]. In addition, the combination of recent studies highlighting the ability of dermal cDC1s to take up apoptotic cells ex vivo and earlier research on cross-presentation of keratinocyte-derived antigen support their assumed role as professional phagocytes and potent antigen presenting cells in the dermis [36, 71, 81, 82]. Although the cross-presentation of keratinocyte-derived self-antigen was claimed to be limited to a distinct CD103^{+} CD207^{+} "*subset*" of dermal cDC1s, recent work on splenic cDC1s has convincingly shown the upregulation of both CD103 and CD207 during homeostatic maturation of cDC1s upon uptake of apoptotic cells, suggesting that these efferocytic CD103^{+} CD207^{+} cDC1s likely represent a more advanced maturation state of dermal cDC1s rather than a distinct subset, although skin-specific evidence is lacking [36, 71, 83, 84].

Recent work has proposed an unexpected role for dDC-mediated cell clearance in cutaneous wound healing, showing that local inhibition of a negative regulator of DC-mediated efferocytosis accelerates wound closure in healthy and diabetic wounds [81, 82]. Supporting the notion that dDCs promote wound healing, several earlier studies in mice and rats have indicated a positive correlation between the number of dDCs and the rate of wound closure in different models, enhancing epithelial migration and revascularization [85–87]. Given the possibility to specifically target dDC populations or topically administer them in a hydrogel, as shown recently, this provides an interesting therapeutic target for wound healing, although their precise contribution to tissue repair requires further research [88].

Neutrophils

Other professional phagocytes that have well-orchestrated roles during skin wound healing are neutrophils and macrophages. Macrophages and neutrophils have long been studied in the context of wound healing under sterile and uninterrupted conditions. Early study of wound healing in antibiotic-maintained PU.1 null mice, which lack neutrophils and macrophages, demonstrated reduced programmed cell death in the first 24 h probably due to the lack of apoptotic neutrophil population [89]. However, by day 4 after full thickness punch biopsy, the wound area shows accumulation of dead cells which can be explained by the lack of the functional competence of macrophage population to clear apoptotic cells in PU.1 ablated mice. Under these fetal-like repair and inflammation-free conditions, absence of neither neutrophils nor macrophages impairs the wound healing compared to wildtype wounds.

Neutrophils accumulate in the wound milieu starting from the first inflammatory phase until day 4. By day 2, their increasing accumulation in the wound makes them the biggest community of immune cells in the area. Starting from day 4, their number diminishes, and they are outnumbered by macrophages [90]. Neutrophils mediate the pro-inflammatory environment in the early phases of skin wound healing by secretion of growth factors and cytokines while clearing pathogens. Neutrophils form extracellular traps (NETs) which are commonly found in wound and are important for their antimicrobial activity. However, NETosis is increased in diabetic wounds, and treatment with DNase 1 that disrupts NETs accelerated wound healing in both diabetic and normoglycemic mice [91].

Clearance of apoptotic neutrophils by macrophages is a key event in the transition of this forementioned inflammatory milieu toward the proliferative phase of wound healing [92]. In the case of angiopoietin-like 4 protein (ANGPTL4) deficiency, reduced numbers of monocyte-derived macrophages accompanied by prolonged presence of neutrophils in the wound area perpetuated inflammation that led to delay in proliferative phase and impaired wound healing [93]. Another matricellular protein, cellular communication network factor 1 (CCN1), was shown to be important for apoptotic neutrophil clearance by macrophages. CCN1 knockout mice display impaired wound healing and addition of CCN1 in diabetic wounds improves wound healing. CCN1 serves as a bridging molecule that binds phosphatidylserine (PS) on the apoptotic cells and integrin $\alpha_v\beta_3$ /$\alpha_v\beta_5$ on macrophages [94].

Macrophages

Macrophages play a pivotal role to clear the neutrophils as it has been exemplified earlier. In addition to their role in clearance, they play an important part in tissue vascularization and regenerative growth during wound healing [95–97]. Wound monocytes/macrophages are primarily derived from blood monocytes. Macrophages accumulate in the wound area starting from day 2. Their number increases in the wound milieu by day 7 which marks the end of inflammatory phase, making them the biggest immune cell population in the area [90, 98]. The increased number of macrophages coincides with the reparative phase of the wound healing.

An early study from Leibovich and Ross showed that depletion of macrophages via the application of systemic cortisone and anti-macrophage serum in the adult guinea pig skin resulted in the accumulation of dead cells, fibrin, and debris [99]. Additionally, selective elimination of monocytes and macrophages by diphtheria toxin in the adult skin prior to wounding inferred impaired skin wound healing similarly in mice [96, 97, 100]. In these examples, murine wounds lacking macrophages demonstrated decreased granulation tissue formation, which is a transitional replacement of normal dermis that results in scar tissue in the long run.

Macrophages display a pro-inflammatory phenotype during early phases of wound healing while they shift to a pro-healing phenotype in later stages following the clearance of neutrophils [101, 102]. Pro-inflammatory macrophages die via a Fas-associated death domain (FADD)/receptor-interacting serine/threonine kinase 3 (RIPK3)-dependent manner, and in wounded *Fadd/Ripk3* knockout mice, Ly6C^{high} pro-inflammatory macrophages accumulate in the wounds. The wounds of these show impaired granulation tissue formation as well as retardation in healing [103].

Excessive inflammation, which could be caused by the accumulation of uncleared cell debris, might give rise to impaired wound healing, as it is observed well in cases of defective apoptotic neutrophil removal [93, 94]. A recent observation supporting this notion is the use of valproic acid to accelerate wound healing in murine wounds which makes the wound environment anti-inflammatory [104]. While the application of valproic acid increases the expression of IL-10 and TGF-β, it decreases the secretion of pro-inflammatory cytokines TNFα, IL-6, and IL-1β in the wound area. The same chemical accelerates wound healing and reduces the apoptotic cell burden in the area. Since it increases phagocytic capacity of macrophages in vitro by increasing the expression of phagocytosis receptors, it is likely that the in vivo reduction of apoptotic burden can be a consequence of the increased clearance capacity of the phagocytes in the area.

Macrophage phagocytosis in wounded skin has been studied also from the perspective of phagocytic regulators, receptors, and opsonizing molecules. For instance, peroxisome proliferator-activated receptor gamma (PPARγ) is a ligand-activated transcription factor that coordinates macrophage function, [105] and its levels are increased up till day 7 in full thickness wounds [106]. PPARγ macrophage-specific KO mice showed impaired wound healing and downregulation of phagocytosis-related genes in their peritoneal macrophages. Although the infiltrated neutrophil and macrophage number in the wound milieu has not changed, apoptotic cells have accumulated in the skin wounds of PPARγ KO mice. Addition of PPARγ agonist improved the wound healing in WT mice and decreased the apoptotic cell burden in the wound area [106].

Additionally, wound lymphatic vessel endothelial hyaluronan receptor 1 $(LYVE1)^+$ macrophages express increased levels of efferocytosis receptor *Timd4* and the bridging molecule *Gas6*. T cell membrane protein 4 (TIM-4) neutralization impaired wound healing by increasing apoptotic cell burden and inflammatory gene expression [107]. Finally, the expression of milk-fat-globular-epidermal growth factor 8, MFG-E8, a bridging molecule binding to PS, increases 6 days after wounding, and in the same study, the addition of MFG-E8 accelerated phagocytosis in peritoneal macrophages [108].

Senescent cells can also be cleared from the wound milieu by macrophages via efferocytosis. Senescent fibroblasts and endothelial cells increased in number transiently during full thickness skin wound healing. Murine models depleting senescent cells via p16/p21 double knockout or skin lacking senescent cells demonstrate delayed wound closure [109]. Compromised senescence during wound healing results in impaired granulation tissue formation. In intact human skin, senescent fibroblasts are cleared by macrophages by a stepwise strategy. Macrophages secrete TNFα to induce apoptosis of the senescent cells. Upon induction of apoptosis, they recognize the PS exposed on the surfaces via STAB1 and clear them [110]. Whether transient populations of wound senescent cells are also cleared by macrophages is yet to be investigated.

Fibroblasts

Fibroblasts in the skin are characterized by great heterogeneity, respond to autocrine signals and Wnt-regulated signals from the epidermis, and regulate tissue repair via ECM deposition and secreted factors [111].

In vitro studies modeling the senescence-mediated clearance of human keratinocyte and fibroblast cell lines by human monocyte leukemia cell line THP-1 originating macrophages demonstrated that senescent-cell clearance can be accelerated by carnosine [112]. Carnosine is an endogenous dipeptide composed of L-histidine and β-alanine, which is demonstrated to enhance wound healing without a clear link to improved cell clearance in wounded skin [113, 114]. However, clearance of senescent keratinocytes and fibroblasts in vitro by macrophages improves their clearance capacity through upregulating their phagocytic receptor CD36 and receptor for advanced glycation end products (RAGE) via the AKT2 signaling pathway [112].

In the absence of macrophages, fibroblasts engulf the cellular debris during skin wound healing [89]. Fibroblasts are involved in clearance not only in the absence of professional phagocytes but also during normal cutaneous wound healing in mice and humans. Dermal fibroblasts play a role in clearing the apoptotic endothelial cells generated during vascular regression [115]. In vitro experiments performed by coculturing apoptotic human dermal microvascular endothelial (HDMEC) cells and human dermal fibroblasts (HDF) in 3:1 ratio have shown that ~4.9% HDFs could engulf apoptotic HDMECs after 24 h. The HDFs perform the efferocytosis of HDMECs with an MFG-E8/integrin $\alpha_v\beta_5$-dependent manner. Using microscopy, they showed that in vivo dermal fibroblasts are engulfing apoptotic endothelial cells in unwounded skin. However, further validation of efferocytosis by fibroblasts in wounded skin is lacking.

Fibroblasts in wound beds upregulate efferocytosis-related genes such as *Tyro3*, *Axl*, *Itgav*, *Slc7a11*, *Gas6*, and *Mfge8*. AXL is expressed during the inflammatory phase of wound healing; in naïve skin, it is expressed in a Toll-like receptor 3 (TLR3)-dependent manner, whereas its upregulation requires additional mechanisms in wounded skin [107]. AXL inhibition impaired the wound healing process by affecting revascularization and fibroblast repopulation.

Keratinocytes

Keratinocytes are skin structural cells and are indispensable during the wound repair process for re-epithelialization and for crosstalk with other immune cells via cytokines, chemokines, and extracellular vesicles. For instance, mice defective in C-X-C motif chemokine receptor 2 (CXCR2), expressed by keratinocytes and endothelial cells at sites of epithelialization and neovascularization, exhibit defective neutrophil recruitment and delayed wound healing [116]. C-C motif ligand 2 (CCL2) released by keratinocytes triggers macrophage recruitment and production of epidermal growth factor by macrophages in wounded skin [117]. Apart from their interaction with and recruitment of immune cells like macrophages and neutrophils, keratinocytes interact via LCs with T cells for antigen presentation and T cell activation and kill directly pathogens by secreting antimicrobial peptides [118]. Albeit their versatile roles in cutaneous wound healing, their contribution as nonprofessional phagocytes to cell clearance is not yet thoroughly regarded.

Endothelial Cells

Endothelial cells are traditionally performing angiogenesis that is necessary for nutrients and oxygen delivery promoting cellular proliferation and tissue regeneration during wound healing. Moreover, endothelial cells, in other tissues besides the skin, can induce apoptosis [119], clear apoptotic corpses in the liver [120], or clear myelin debris in neural injury. [121] They also participate in efferocytosis by increasing macrophage recruitment and apoptotic cell clearance via thrombospondin-1 (TSP-1) expression [122]. Further investigation to elucidate their role as nonprofessional phagocytes of dying cells during cutaneous wound repair, especially in nonhealing wounds associated with vascular diseases or when professional phagocytes cannot perform their function to clear excess dead cells, is necessary.

Stem Cells

There are several stem cell subsets in the skin such as unipotent epidermal stem cells in uninjured skin that become multipotent after injury, as well as stem cells residing in hair follicles, the sebaceous glands, and the interfollicular epidermis (IFE). During wound healing, stem cells play a critical role in the damaged tissue regeneration, contributing to re-epithelialization and the epidermal stem cell-fibroblasts interaction to convert the last to myofibroblasts [123]. Apoptosis is crucial in regulating stem-cell-induced regeneration. Mice deficient for the proapoptotic *Sept4*/ARTS gene displayed elevated numbers of apoptosis-resistant hair follicle stem cells, improved wound healing and the inactivation of the caspase inhibitor X-linked inhibitor of apoptosis protein (XIAP) reversed these phenotypes Mice deficient for the proapoptotic *Sept4*/ARTS gene displayed elevated numbers of apoptosis-resistant hair follicle stem cells, improved wound healing, and the inactivation of the caspase inhibitor X-linked inhibitor of apoptosis protein (XIAP) reversed these phenotypes [124]. Mesenchymal stem cells (MSCs) are self-renewing multipotent stem cells that have emerged as a promising therapeutic option for wound healing due to their immunomodulatory function to neutrophils, macrophages' polarization, and T cell, NK cell, and DC activation [125, 126]. The role of mesenchymal stem cells in the context of efferocytosis and their potential role as nonprofessional phagocytes, for instance in the bone marrow microenvironment [127], are recently gaining attention. Stewart et al. have recently investigated how nonmotile stem cells can transiently clear apoptotic corpses during the hair cycle [128]. However, the role of stem cells as nonprofessional phagocytes in apoptotic cell clearance during skin repair upon injury remains to be addressed.

Cell Death Modalities in Skin Wound Healing

Apoptosis

Apoptosis is a highly conserved and thoroughly understood programmed cell death process needed for development and tissue homeostasis. It is distinct from necrosis and is characterized by chromatin condensation, membrane blebbing, and cell disassembly into apoptotic bodies [129]. Apoptosis is initiated via the intrinsic or mitochondrial pathway [130] or through the extrinsic or death receptor pathway [131], while both pathways are utilizing caspases as their downstream effectors of death. Caspase-dependent apoptosis accounts for the majority of the cellular turnover, and the cells committing to apoptosis are dying without releasing deleterious cell contents [132], rendering the process noninflammatory and immunologically silent. Nevertheless, apoptotic corpses need to cleared by phagocytes via efferocytosis in a timely manner without progressing to secondary necrosis [133]. The last involves the release of necrosis-associated DAMPs (damage-associated molecular patterns) that have inflammatory and immunogenic consequences.

Cell death is an integral component of maintaining skin homeostasis. For instance, epidermal keratinocytes undergo a terminal differentiation and cell death program termed cornification [11]. Hair follicle growth is dictated by a repeated apoptosis-driven involution phase (catagen) [134] following the growth stage (anagen), which, when interrupted, can lead to defective hair growth, alopecia, or premature hair greying [135].

In this review, we will discuss apoptosis in the context of acute skin injury and its importance during the wound repair process.

Types of Apoptotic Corpses in Skin Injury and Repair (Apoptotic Meals)

Cutaneous wound healing typically consists of the following phases: hemostasis, inflammation, growth, re-epithelialization, and remodeling [123]. While a vast variety of immune and nonimmune cells are participating temporally and spatially to ensure successful wound closure, they undergo apoptosis all along this process. As early as 1997, Brown et al. studied the apoptotic cell pattern in wounded skin and revealed that inflammatory cells underwent apoptosis already at 12 hours after injury whereas myofibroblasts were dying at later stages, between days 12 and 20 or even at the remodeling stage of wound healing [136]. This study also illustrated that apoptosis in the epithelium and down-regulation of inflammation can be concurrent events [136]. In the context of corneal wound healing, stromal keratocyte apoptosis is an early event propelling the recruitment and proliferation of other keratocytes and myofibroblast generation [137]. During the inflammatory phase of skin wound healing, neutrophils rapidly populate the wounded skin to phagocytose pathogens and generate NETs, they will in turn die by apoptosis and be efferocytosed primarily by macrophages [94, 138, 139]. Prompt clearance of apoptotic neutrophils is necessary for the resolution of inflammation by inhibiting proinflammatory cytokine production and increasing the production of TGF-β1 or prostaglandin E2 (PGE2) that are beneficial for wound closure [140, 141]. In the case of deficiency of the guanine-nucleotide exchange factor Vav3 in macrophages, phagocytosis of apoptotic neutrophils was impaired, resulting in reduced release of TGF-β1, decreased myofibroblast differentiation, and thus delayed wound contraction and closure [142].

Apoptosis of immune cells and their subsequent efferocytosis sets off the proliferative and migratory phase of keratinocytes eliciting re-epithelialization in the healing wound edge to restore the epidermal barrier [118]. During the remodeling phase, apoptosis of myofibroblasts promotes wound closure by triggering the transition of granulation tissue into scar [33]. However, in the case of impaired myofibroblast efferocytosis there is hypertrophic scar formation and fibrosis [143]. At the end of the inflammatory phase and with the release of growth factors from macrophages including vascular endothelial growth factor (VEGF), there is angiogenesis that involves endothelial cell proliferation and the formation of new blood vessels. Neovessel pruning, during the remodeling phase, involves endothelial cell apoptosis, still its contribution to wound repair remains to be elucidated [144, 145]. The significance of apoptosis was emphatically revealed in the case of mice lacking either of the caspases 3 and 7 and presented with delayed cutaneous wound healing and defective liver regeneration [146]. Notably, the authors proposed that signals from dying cells in the wounded tissues, such as the caspase-mediated activation of phospholipase A2 (PLA2) and the subsequent production and release of the lipid signal PGE2, send signals to stimulate the proliferation of stem or progenitor cells, which in turn promote tissue regeneration and wound healing [146].

Other Modalities of Cell Death in Skin Injury and Repair

Ferroptosis

Ferroptosis is a relatively newly defined necrotic, iron-dependent cell death modality [147]. Excessive lipid peroxidation is one of the hallmarks of ferroptosis [148, 149] and glutathione peroxidase 4 (GPX4), the glutamate-cystine antiporter system x_c^-, the ferroptosis suppressor protein 1 (FSP1), the nuclear factor erythroid 2-related factor 2 (NRF2), and p53 are among the molecules that have been implicated to ferroptosis induction or inhibition [150–152]. SLC7A11 is the subunit of system x_c^- that regulates exchange of intracellular glutamate for extracellular cystine which is reduced to cysteine, a rate-limiting precursor for synthesis of the antioxidant glutathione (GSH). When glutathione is reduced, GPX4, which uses glutathione (GSH) as a cofactor, cannot function effectively to reduce lipid peroxides leading to ferroptosis [153]. Notably, mice with keratinocyte-specific deficiency in glutamate cysteine ligase, which catalyzes the rate-limiting step in GSH biosynthesis, developed keratinocyte hyperproliferation and only mild wound healing abnormalities, attributed to strong activation of the cytoprotective transcription factor NRF2 [154]. Recently, ferroptosis has been implicated in the pathogenesis of diabetic wound healing as well as irradiated- and ultraviolet (UV)-driven wounds [155]. In a study from Wilkinson et al., they profiled iron levels which were reduced in diabetic murine wounds [156]. Mechanistically, they found that iron administration promoted both the deposition and remodeling of extracellular matrix proteins via upregulation of the iron-converting metalloreductase six-transmembrane epithelial antigen of prostate 3 (STEAP3), a key mediator of extracellular matrix production [156]. On the other hand, several studies have used the ferroptosis inhibitor deferoxamine (DFO), an iron chelator and FDA-approved pro-angiogenesis molecule. Co-administration of DFO with hydroxysafflor yellow A promoted diabetic wound healing via increased angiogenesis and upregulation of hypoxia-inducible factor-1 alpha (HIF-1α) expression [157]. Similarly, more studies underlined the beneficial effect of DFO administration in diabetic wounds by using various biomaterials such as bioactive injectable hydrogels, bioglass, and nanofibrous scaffolds or a transdermal drug delivery system complexed with the polymer polyvinylpyrrolidone and surfactants [158–160]. Treatment with the ferroptosis inhibitor ferrostatin-1 (Fer-1) reduced the high glucose levels, oxidative stress, inflammation and accelerated healing in a rat diabetic model [161]. The expression of *Slc7a11* was 200-fold higher in the wounds of *db/db* diabetic mice compared with normoglycemic C57BL/6 wounded skin lysates, and the administration of erastin, an SLC7A11 inhibitor, together with a bolus of early-stage apoptotic cells accelerated wound healing in both normoglycemic and diabetic mice [81]. This effect was related to enhanced efferocytosis and augmented apoptotic cell clearance but not to induction of ferroptosis in the skin, as the topical administration of RSL3 [162], a ferroptosis inducer and GPX4 inhibitor, did not improve wound healing [81]. In a streptozotocin-induced type 1 diabetic mouse model, induction of ferroptosis hastened wound healing by sensitizing glucose-induced senescent fibroblasts to ferroptosis and ferritinophagy via nuclear receptor coactivator 4 (NCOA4) overexpression [163].

Necroptosis

Necroptosis is a pro-inflammatory, regulated form of cell death implicated in the pathogenesis of many diseases, but serving as a defense mechanism as well against pathogens and tumors. Necroptosis is induced by ligand-receptor binding including TNF-α/TNF receptor 1 (TNFR1), Fas ligand/FAS, interferon-gamma (IFN-γ)/interferon alpha/beta receptor 1 (IFNAR1), double-stranded RNA/TLR3, and double-stranded DNA/Z-DNA-binding protein 1 (ZBP1), and the molecular pathways underlying necroptosis are nicely summarized in [164, 165]. When apoptosis is blocked, via Caspase-8/10 inhibition, necroptosis can be activated. Necroptosis has been associated with diabetic, chronic nonhealing wounds, where excessive cell death and the inflammatory environment, reactive oxygen species, and DAMPs can lead to necroptosis, especially in keratinocytes [166]. Overall, necroptotic DAMPs can trigger fibrosis and affect the wound healing process [167]. A study showed that keratinocytes treated with a surfactant polymer dressing (SPD) were protected from necroptotic cell death resulting to reduced nuclear depletion of high mobility group box 1 (HMGB1) and mitochondrial fragmentation [168]. Moreover, deficiency of sirtuin 3 (SIRT3), a histone deacetylase, delayed diabetic skin wound healing by suppressing α-SMA expression, delaying fibroblast migration, and increasing necroptosis [169].

Pyroptosis

Pyroptosis is an inflammasome-induced and gasdermin D-mediated lytic form of cell death characterized by the release of inflammatory cytokines such as IL-18, IL-1β, and IL-1α [170, 171]. Signals from TLR ligands lead to nuclear factor-kappa B (NF-κB)-mediated transcriptional upregulation of inflammasome components and pro-inflammatory cytokines while signals from a wide range of triggers (such as ATP, microbial toxins) result to inflammasome assembly, activation of caspase-1, and cell lysis. Bioglass or silicate-based bioactive glass promoted wound healing by inhibiting pyroptosis in endothelial cells through the decrease of connexin43 (CX43) expression and reactive oxygen species (ROS) levels [172]. In nonhealing diabetic foot ulcers (DFUs) associated with recurrent *Staphylococcus aureus* infection, interferon-inducible protein (AIM2) inflammasome activation and pyroptosis contribute to perpetuated inflammation [173]. Also, overexpression of metastasis-associated lung adenocarcinoma transcript 1 (MALAT1) or knockdown of microRNA (miR)-374a-5p in DFUs could inhibit apoptosis and pyroptosis of DFU fibroblasts, suggesting a therapeutic possibility [174]. A recent study revealed that the secretome of macrophages undergoing pyroptosis, independent of IL-1β release, possesses oxylipins, including PGE2, and metabolites, that together promote cutaneous healing when administered topically on wounds [175]. In the same study, blocking IL-1β present at the wound site showed, as anticipated, some beneficial effects; still, the components of pyroptotic supernatants were overwhelmingly more potent at enhancing the rate of wound closure. While the effect of PGE2 that is synthetized and released via the gasdermin D pores during pyroptosis in promoting wound closure is unlikely due to increased efferocytosis [176], the effect of pyroptotic metabolites and the overall secretome links to boosted migration of fibroblasts and macrophage polarization to $CD301^{+}$ macrophages [175].

Efferocytosis-Mediated Regulation

Several immune and nonimmune cells participate in efferocytosis either by providing the dead cell meal(s) or the phagocyte(s) during wound healing. The apoptotic cell metabolite secretome or pyroptotic cell secretome can suppress inflammation, induce cell proliferation and migration in the neighborhood, and accelerate wound healing [175, 177]. Several signals, such as the "find me" signals sphingosine 1-phosphate (S1P) and ATP released by apoptotic cells, can affect keratinocyte proliferation and migration and tissue regeneration [178, 179]. Phagocytes eliminating apoptotic cells produce factors with pro-resolution properties [180] such as TGFβ that is beneficial for skin wound healing [181]. Another member of the TGFβ superfamily, the growth differentiation factor 15 (GDF15), was shown to be released from efferocytic DCs and to facilitate wound closure downstream of SLC7A11 in diabetic wounds [81]. In a model of muscle regeneration, GDF15 was found to be expressed by a subpopulation of tissue repair macrophages and is important for myoblast proliferation and infiltration of myeloid cells in injured muscle [182]. Elucidating further the factors produced by professional and nonprofessional phagocytes upon efferocytosis during the whole course of the wound healing process will undoubtedly provide new therapeutic possibilities for nonhealing wounds.

Skin Wound Healing in Pathology

Diabetes

Chronic nonhealing wounds associated with diabetes, which affects one in ten people worldwide, are a major complication affecting the quality of life, increasing the risk for amputation, and posing an economical burden to those affected. Dying cells that are not cleared in a timely manner by phagocytes, perpetuate inflammation, delay resolution, and impede tissue repair in wounds of diabetic mice [183, 184].

Various cell types dying from distinct cell death modalities in diabetic chronic wounds are illustrated in Fig. 3.

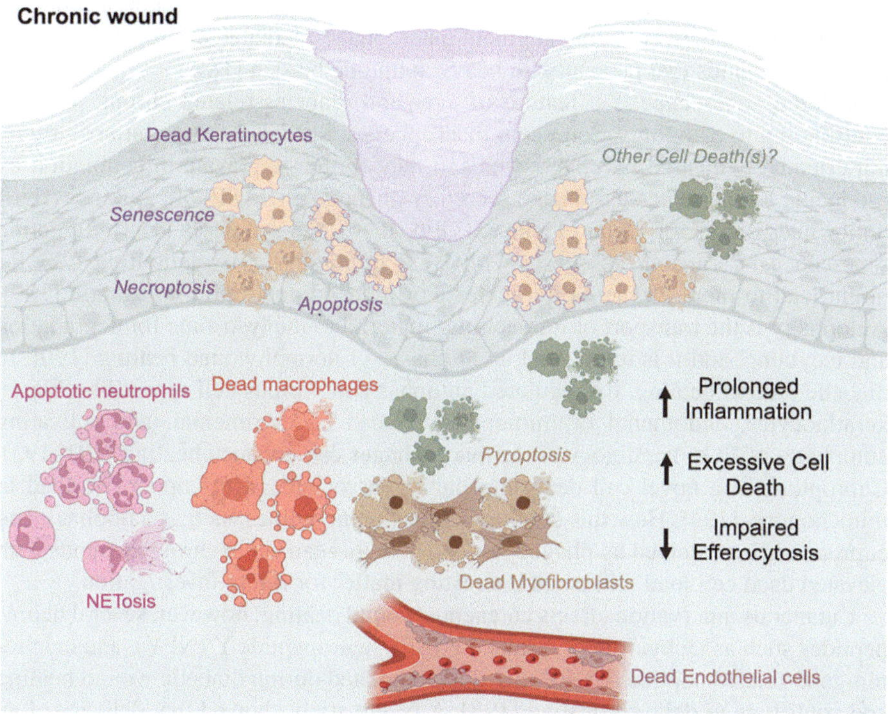

Fig. 3 Excessive cellular demise and impaired efferocytosis are hallmarks of diabetic, nonhealing wounds. Dead cells that are not eliminated in a timely manner by phagocytes, perpetuate inflammation, delay resolution, and tissue repair in wounds of diabetic mice. For instance, uncleared apoptotic neutrophils accumulate and neutrophil extracellular trap (NET) formation is increased in diabetic wounds, while a high glucose environment hinders effective phagocytosis. Other cell death modalities, such as necroptosis in keratinocytes, have been associated with diabetic, chronic nonhealing wounds due to the presence of a hyper-inflammatory environment, high reactive oxygen species, and DAMPs. Also, endothelial cells and (myo)fibroblasts have been shown to die from pyroptosis besides apoptosis and impair re-vascularization and tissue remodeling upon injury. In some cases, induction of ferroptosis promoted wound healing by sensitizing glucose-induced senescent fibroblasts while in other reports treatment with ferroptosis inhibitors (such as deferoxamine and ferrostatin-1) increased wound healing in diabetic models, making it difficult to assess the precise role of ferroptosis in diabetic wound pathology and treatment. Cellular senescence of macrophages, fibroblasts, or keratinocytes aggravates inflammation by producing a senescence-associated secretory phenotype (SASP) of cytokines in chronic wounds. Dysregulated autophagy in various cell types (fibroblasts, keratinocytes, endothelial or immune cells) can also be detrimental in diabetic wounds. Induction or inhibition of these and other yet to be identified in the future cell death modalities and how they affect the appetite of phagocytes (professional and nonprofessional) are attractive protocols to be further investigated for the treatment of chronic wounds

In a model of myocardial injury, macrophages exposed to high glucose decreased miR-126 and metalloprotease 9 (ADAM9) expression, which in turn cleaves Mer tyrosine kinase (MERTK) resulting in decreased efferocytosis of apoptotic cardiomyocytes [185]. The role of microRNAs in the pathogenesis of diabetes and in angiogenesis is gaining increasing attention [186] and over-expression of miR-126 via non-particle delivery could augment macrophage-mediated efferocytosis in diabetic skin wounds that present with heavy apoptotic burden [187].

Cellular senescence is a feature of age- and diabetes-related chronic wounds contributing to pathology. Senescent macrophages, fibroblasts, or keratinocytes are not effectively cleared by phagocytes and they further aggravate inflammation by producing a senescence-associated secretory phenotype (SASP) of cytokines (IL-1, tumor necrosis factor alpha (TNFα), TGFβ, IL-6) and proteases [188]. Targeting senescence by causing apoptosis with BCL inhibitors or by inhibiting CXCR2 might lead to more efficient efferocytosis or reduced neutrophil infiltration [189]. Autophagy is the transport of intracellular material to the lysosome for degradation and recycling, and it is implicated in all phases of normal wound healing [190]. In diabetic wound healing, dysregulated autophagy in various cell types (fibroblasts, keratinocytes, endothelial or immune cells) can be detrimental, and activating autophagy of ECs, keratinocytes, or macrophages can enhance healing [191–193]. Cuproptosis is a novel cell death mechanism associated with copper overload in mitochondria [194]. How this cellular demise by modalities such as autophagy and cuproptosis is perceived by phagocytes and how this can aid in chronic wounds with elevated dead cell load will be an interesting matter for future investigation.

Cutaneous innervation affects cutaneous wound healing; however, several neuropeptides such as tachykinins substance P (SP), neuropeptide Y (NPY), and calcitonin-gene-related peptide (CGRP) are downregulated during diabetic wound healing and contribute to delayed healing [195]. A recent study showed that delivery of an engineered version of CGRP accelerated diabetic wound healing by inhibiting neutrophil recruitment, polarizing macrophages to pro-repair phenotype, accelerating death, and enhancing efferocytosis [196].

Concluding Remarks

Apart from diabetic ulcers, chronic nonhealing can also be associated with vascular diseases and aging. In aging, mitochondria dysfunction, cellular senescence, low-grade chronic inflammation, and defective efferocytosis leading to accumulation of uncleared damaged cells are some of the components of impaired wound healing [197]. Cell death is increasingly being studied in the context of skin homeostasis and injury, and the importance of cell clearance for effective tissue repair by several cellular components receives more appreciation. Boosting efferocytosis by using PPARγ agonists [198, 199], SLC7A11 inhibitors, [81] or other SLCs regulators, or by expression of chimeric efferocytic receptors through virus-like particle delivery, lipid nanoparticles, or other approaches, [200] are interesting therapeutic avenues to be evaluated for chronic wound management.

References

1. Baroni, A., et al. (2012). Structure and function of the epidermis related to barrier properties. *Clinics in Dermatology, 30*, 257–262. https://doi.org/10.1016/j.clindermatol.2011.08.007
2. Zomer, H. D., & Trentin, A. G. (2018). Skin wound healing in humans and mice: Challenges in translational research. *Journal of Dermatological Science, 90*, 3–12. https://doi.org/10.1016/j.jdermsci.2017.12.009
3. Boulais, N., & Misery, L. (2007). Merkel cells. *Journal of the American Academy of Dermatology, 57*, 147–165. https://doi.org/10.1016/j.jaad.2007.02.009
4. Lucarz, A., & Brand, G. (2007). Current considerations about Merkel cells. *European Journal of Cell Biology, 86*, 243–251. https://doi.org/10.1016/j.ejcb.2007.02.001
5. Moll, I., et al. (2005). Human Merkel cells—Aspects of cell biology, distribution and functions. *European Journal of Cell Biology, 84*, 259–271. https://doi.org/10.1016/j.ejcb.2004.12.023
6. Ando, H., et al. (2012). Melanosomes are transferred from melanocytes to keratinocytes through the processes of packaging, release, uptake, and dispersion. *The Journal of Investigative Dermatology, 132*, 1222–1229. https://doi.org/10.1038/jid.2011.413
7. Lerner, A. B., & Fitzpatrick, T. B. (1950). Biochemistry of melanin formation. *Physiological Reviews, 30*, 91–126. https://doi.org/10.1152/physrev.1950.30.1.91
8. Fan, X., et al. (2018). Single cell and open chromatin analysis reveals molecular origin of epidermal cells of the skin. *Developmental Cell, 47*, 21-37 e25. https://doi.org/10.1016/j.devcel.2018.08.010
9. Joost, S., et al. (2016). Single-cell Transcriptomics reveals that differentiation and spatial signatures shape epidermal and hair follicle heterogeneity. *Cell Systems, 3*, 221-237 e229. https://doi.org/10.1016/j.cels.2016.08.010
10. Lin, Z., et al. (2020). Murine interfollicular epidermal differentiation is gradualistic with GRHL3 controlling progression from stem to transition cell states. *Nature Communications, 11*, 5434. https://doi.org/10.1038/s41467-020-19234-6
11. Eckhart, L., Lippens, S., Tschachler, E., & Declercq, W. (1833). Cell death by cornification. *Biochimica et Biophysica Acta, 3471-3480*, 2013. https://doi.org/10.1016/j.bbamcr.2013.06.010
12. Grubauer, G., Feingold, K. R., Harris, R. M., & Elias, P. M. (1989). Lipid content and lipid type as determinants of the epidermal permeability barrier. *Journal of Lipid Research, 30*, 89–96.
13. Murata, T., Honda, T., Mostafa, A., & Kabashima, K. (2022). Stratum corneum as polymer sheet: Concept and cornification processes. *Trends in Molecular Medicine, 28*, 350–359. https://doi.org/10.1016/j.molmed.2022.02.008
14. Ginhoux, F., et al. (2010). Fate mapping analysis reveals that adult microglia derive from primitive macrophages. *Science, 330*, 841–845. https://doi.org/10.1126/science.1194637
15. Ginhoux, F., et al. (2006). Langerhans cells arise from monocytes in vivo. *Nature Immunology, 7*, 265–273. https://doi.org/10.1038/ni1307
16. Hoeffel, G., et al. (2012). Adult Langerhans cells derive predominantly from embryonic fetal liver monocytes with a minor contribution of yolk sac-derived macrophages. *The Journal of Experimental Medicine, 209*, 1167–1181. https://doi.org/10.1084/jem.20120340
17. Kaplan, D. H. (2017). Ontogeny and function of murine epidermal Langerhans cells. *Nature Immunology, 18*, 1068–1075. https://doi.org/10.1038/ni.3815
18. Greter, M., et al. (2012). Stroma-derived interleukin-34 controls the development and maintenance of langerhans cells and the maintenance of microglia. *Immunity, 37*, 1050–1060. https://doi.org/10.1016/j.immuni.2012.11.001
19. Merad, M., et al. (2002). Langerhans cells renew in the skin throughout life under steady-state conditions. *Nature Immunology, 3*, 1135–1141. https://doi.org/10.1038/ni852
20. Wang, Y., & Colonna, M. (2014). Interkeukin-34, a cytokine crucial for the differentiation and maintenance of tissue resident macrophages and Langerhans cells. *European Journal of Immunology, 44*, 1575–1581. https://doi.org/10.1002/eji.201344365

21. Wang, Y., et al. (2012). IL-34 is a tissue-restricted ligand of CSF1R required for the development of Langerhans cells and microglia. *Nature Immunology, 13*, 753–760. https://doi.org/10.1038/ni.2360
22. Merad, M., et al. (2004). Depletion of host Langerhans cells before transplantation of donor alloreactive T cells prevents skin graft-versus-host disease. *Nature Medicine, 10*, 510–517. https://doi.org/10.1038/nm1038
23. Romani, N., et al. (1989). Presentation of exogenous protein antigens by dendritic cells to T cell clones. Intact protein is presented best by immature, epidermal Langerhans cells. *The Journal of Experimental Medicine, 169*, 1169–1178. https://doi.org/10.1084/jem.169.3.1169
24. Schuler, G., & Steinman, R. M. (1985). Murine epidermal Langerhans cells mature into potent immunostimulatory dendritic cells in vitro. *The Journal of Experimental Medicine, 161*, 526–546. https://doi.org/10.1084/jem.161.3.526
25. Clark, R. A. (2015). Resident memory T cells in human health and disease. *Science Translational Medicine, 7*, 269rv261. https://doi.org/10.1126/scitranslmed.3010641
26. Seidel, J. A., et al. (2018). Skin resident memory CD8(+) T cells are phenotypically and functionally distinct from circulating populations and lack immediate cytotoxic function. *Clinical and Experimental Immunology, 194*, 79–92. https://doi.org/10.1111/cei.13189
27. Havran, W. L., & Jameson, J. M. (2010). Epidermal T cells and wound healing. *Journal of Immunology, 184*, 5423–5428. https://doi.org/10.4049/jimmunol.0902733
28. Liu, M., et al. (2022). Dendritic epidermal T cells secreting exosomes promote the proliferation of epidermal stem cells to enhance wound re-epithelialization. *Stem Cell Research & Therapy, 13*, 121. https://doi.org/10.1186/s13287-022-02783-6
29. Wang, Y., et al. (2017). IL-15 enhances activation and IGF-1 production of dendritic epidermal T cells to promote wound healing in diabetic mice. *Frontiers in Immunology, 8*, 1557. https://doi.org/10.3389/fimmu.2017.01557
30. Driskell, R. R., et al. (2013). Distinct fibroblast lineages determine dermal architecture in skin development and repair. *Nature, 504*, 277–281. https://doi.org/10.1038/nature12783
31. Gabbiani, G. (2003). The myofibroblast in wound healing and fibrocontractive diseases. *The Journal of Pathology, 200*, 500–503. https://doi.org/10.1002/path.1427
32. Gabbiani, G., Ryan, G. B., & Majne, G. (1971). Presence of modified fibroblasts in granulation tissue and their possible role in wound contraction. *Experientia, 27*, 549–550. https://doi.org/10.1007/BF02147594
33. Desmouliere, A., Redard, M., Darby, I., & Gabbiani, G. (1995). Apoptosis mediates the decrease in cellularity during the transition between granulation tissue and scar. *The American Journal of Pathology, 146*, 56–66.
34. Jakubzick, C., et al. (2013). Minimal differentiation of classical monocytes as they survey steady-state tissues and transport antigen to lymph nodes. *Immunity, 39*, 599–610. https://doi.org/10.1016/j.immuni.2013.08.007
35. Tamoutounour, S., et al. (2013). Origins and functional specialization of macrophages and of conventional and monocyte-derived dendritic cells in mouse skin. *Immunity, 39*, 925–938. https://doi.org/10.1016/j.immuni.2013.10.004
36. Henri, S., et al. (2010). CD207+ CD103+ dermal dendritic cells cross-present keratinocyte-derived antigens irrespective of the presence of Langerhans cells. *The Journal of Experimental Medicine, 207*, 189–206. https://doi.org/10.1084/jem.20091964
37. Voss, M., Kotrba, J., Gaffal, E., Katsoulis-Dimitriou, K., & Dudeck, A. (2021). Mast cells in the skin: Defenders of integrity or offenders in inflammation? *International Journal of Molecular Sciences, 22*. https://doi.org/10.3390/ijms22094589
38. Boothby, I. C., et al. (2021). Early-life inflammation primes a T helper 2 cell-fibroblast niche in skin. *Nature, 599*, 667–672. https://doi.org/10.1038/s41586-021-04044-7
39. Clark, R. A., et al. (2006). The vast majority of CLA+ T cells are resident in normal skin. *Journal of Immunology, 176*, 4431–4439. https://doi.org/10.4049/jimmunol.176.7.4431
40. Naik, S., et al. (2012). Compartmentalized control of skin immunity by resident commensals. *Science, 337*, 1115–1119. https://doi.org/10.1126/science.1225152

41. Sanchez Rodriguez, R., et al. (2014). Memory regulatory T cells reside in human skin. *The Journal of Clinical Investigation, 124*, 1027–1036. https://doi.org/10.1172/JCI72932
42. Scharschmidt, T. C., et al. (2015). A wave of regulatory T cells into neonatal skin mediates tolerance to commensal microbes. *Immunity, 43*, 1011–1021. https://doi.org/10.1016/j.immuni.2015.10.016
43. Haensel, D., et al. (2020). Defining epidermal basal cell states during skin homeostasis and wound healing using single-cell Transcriptomics. *Cell Reports, 30*, 3932-3947 e3936. https://doi.org/10.1016/j.celrep.2020.02.091
44. Hoste, E., et al. (2021). OTULIN maintains skin homeostasis by controlling keratinocyte death and stem cell identity. *Nature Communications, 12*, 5913. https://doi.org/10.1038/s41467-021-25944-2
45. Kim, B. S., et al. (2013). TSLP elicits IL-33-independent innate lymphoid cell responses to promote skin inflammation. *Science Translational Medicine, 5*, 170ra116. https://doi.org/10.1126/scitranslmed.3005374
46. Dudeck, J., et al. (2017). Mast cells acquire MHCII from dendritic cells during skin inflammation. *The Journal of Experimental Medicine, 214*, 3791–3811. https://doi.org/10.1084/jem.20160783
47. Haniffa, M., Gunawan, M., & Jardine, L. (2015). Human skin dendritic cells in health and disease. *Journal of Dermatological Science, 77*, 85–92. https://doi.org/10.1016/j.jdermsci.2014.08.012
48. Doebel, T., Voisin, B., & Nagao, K. (2017). Langerhans cells – the macrophage in dendritic cell clothing. *Trends in Immunology, 38*, 817–828. https://doi.org/10.1016/j.it.2017.06.008
49. Wang, J., et al. (2023). MiR-23a regulates skin Langerhans cell phagocytosis and inflammation-induced Langerhans cell repopulation. *Biology (Basel), 12*. https://doi.org/10.3390/biology12070925
50. Dioszeghy, V., et al. (2018). Antigen uptake by Langerhans cells is required for the induction of regulatory T cells and the Acquisition of Tolerance during Epicutaneous Immunotherapy in OVA-sensitized mice. *Frontiers in Immunology, 9*, 1951. https://doi.org/10.3389/fimmu.2018.01951
51. Kubo, A., Nagao, K., Yokouchi, M., Sasaki, H., & Amagai, M. (2009). External antigen uptake by Langerhans cells with reorganization of epidermal tight junction barriers. *The Journal of Experimental Medicine, 206*, 2937–2946. https://doi.org/10.1084/jem.20091527
52. Reis e Sousa, C., Stahl, P. D., & Austyn, J. M. (1993). Phagocytosis of antigens by Langerhans cells in vitro. *The Journal of Experimental Medicine, 178*, 509–519. https://doi.org/10.1084/jem.178.2.509
53. Sparber, F., Tripp, C. H., Hermann, M., Romani, N., & Stoitzner, P. (2010). Langerhans cells and dermal dendritic cells capture protein antigens in the skin: Possible targets for vaccination through the skin. *Immunobiology, 215*, 770–779. https://doi.org/10.1016/j.imbio.2010.05.014
54. Wolff, K., & Schreiner, E. (1970). Uptake, intracellular transport and degradation of exogenous protein by Langerhans cells. An electron microscopic-cytochemical study using peroxidase as tracer substance. *The Journal of Investigative Dermatology, 54*, 37–47. https://doi.org/10.1111/1523-1747.ep12551501
55. Lin, X., et al. (2019). An ectoderm-derived myeloid-like cell population functions as antigen transporters for Langerhans cells in zebrafish epidermis. *Developmental Cell, 49*, 605–617 e605. https://doi.org/10.1016/j.devcel.2019.03.028
56. Frugé, R. E., Krout, C., Lu, R., Matsushima, H., & Takashima, A. (2012). Real-time visualization of macromolecule uptake by epidermal Langerhans cells in living animals. *The Journal of Investigative Dermatology, 132*, 609–614. https://doi.org/10.1038/jid.2011.385
57. Ratzinger, G., et al. (2004). Mature human Langerhans cells derived from CD34+ hematopoietic progenitors stimulate greater cytolytic T lymphocyte activity in the absence of bioactive IL-12p70, by either single peptide presentation or cross-priming, than do dermal-interstitial or monocyte-derived dendritic cells. *Journal of Immunology, 173*, 2780–2791. https://doi.org/10.4049/jimmunol.173.4.2780

58. Sheng, J., et al. (2021). Fate mapping analysis reveals a novel murine dermal migratory Langerhans-like cell population. *eLife, 10*. https://doi.org/10.7554/eLife.65412
59. Seneschal, J., Clark, R. A., Gehad, A., Baecher-Allan, C. M., & Kupper, T. S. (2012). Human epidermal Langerhans cells maintain immune homeostasis in skin by activating skin resident regulatory T cells. *Immunity, 36*, 873–884. https://doi.org/10.1016/j.immuni.2012.03.018
60. Bauer, T., et al. (2012). Identification of Axl as a downstream effector of TGF-beta1 during Langerhans cell differentiation and epidermal homeostasis. *The Journal of Experimental Medicine, 209*, 2033–2047. https://doi.org/10.1084/jem.20120493
61. Peterman, E., et al. (2023). Zebrafish cutaneous injury models reveal that Langerhans cells engulf axonal debris in adult epidermis. *Disease Models & Mechanisms, 16*. https://doi.org/10.1242/dmm.049911
62. Fang, Y., Gong, S. J., Xu, Y. H., Hambly, B. D., & Bao, S. (2007). Impaired cutaneous wound healing in granulocyte/macrophage colony-stimulating factor knockout mice. *The British Journal of Dermatology, 157*, 458–465. https://doi.org/10.1111/j.1365-2133.2007.07979.x
63. Hatakeyama, M., et al. (2017). Anti-inflammatory role of Langerhans cells and apoptotic keratinocytes in ultraviolet-B-induced cutaneous inflammation. *Journal of Immunology, 199*, 2937–2947. https://doi.org/10.4049/jimmunol.1601681
64. Joshi, N., et al. (2020). Comprehensive characterization of myeloid cells during wound healing in healthy and healing-impaired diabetic mice. *European Journal of Immunology, 50*, 1335–1349. https://doi.org/10.1002/eji.201948438
65. Kaplan, G., et al. (1992). Novel responses of human skin to intradermal recombinant granulocyte/macrophage-colony-stimulating factor: Langerhans cell recruitment, keratinocyte growth, and enhanced wound healing. *The Journal of Experimental Medicine, 175*, 1717–1728. https://doi.org/10.1084/jem.175.6.1717
66. Li, Z., Lamb, R., Coles, M. C., Bennett, C. L., & Ambler, C. A. (2021). Inducible ablation of CD11c(+) cells to determine their role in skin wound repair. *Immunology, 163*, 105–111. https://doi.org/10.1111/imm.13312
67. Wasko, R., et al. (2022). Langerhans cells are essential components of the angiogenic niche during murine skin repair. *Developmental Cell, 57*, 2699-2713 e2695. https://doi.org/10.1016/j.devcel.2022.11.012
68. Niessen, F. B., Schalkwijk, J., Vos, H., & Timens, W. (2004). Hypertrophic scar formation is associated with an increased number of epidermal Langerhans cells. *The Journal of Pathology, 202*, 121–129. https://doi.org/10.1002/path.1502
69. Rajesh, A., et al. (2020). Depletion of langerin(+) cells enhances cutaneous wound healing. *Immunology, 160*, 366–381. https://doi.org/10.1111/imm.13202
70. Ahrens, S., et al. (2012). F-actin is an evolutionarily conserved damage-associated molecular pattern recognized by DNGR-1, a receptor for dead cells. *Immunity, 36*, 635–645. https://doi.org/10.1016/j.immuni.2012.03.008
71. Bedoui, S., et al. (2009). Cross-presentation of viral and self antigens by skin-derived CD103+ dendritic cells. *Nature Immunology, 10*, 488–495. https://doi.org/10.1038/ni.1724
72. Zhang, J. G., et al. (2012). The dendritic cell receptor Clec9A binds damaged cells via exposed actin filaments. *Immunity, 36*, 646–657. https://doi.org/10.1016/j.immuni.2012.03.009
73. Haniffa, M., et al. (2012). Human tissues contain CD141hi cross-presenting dendritic cells with functional homology to mouse CD103+ nonlymphoid dendritic cells. *Immunity, 37*, 60–73. https://doi.org/10.1016/j.immuni.2012.04.012
74. Robbins, S. H., et al. (2008). Novel insights into the relationships between dendritic cell subsets in human and mouse revealed by genome-wide expression profiling. *Genome Biology, 9*, R17. https://doi.org/10.1186/gb-2008-9-1-r17
75. Malissen, B., Tamoutounour, S., & Henri, S. (2014). The origins and functions of dendritic cells and macrophages in the skin. *Nature Reviews. Immunology, 14*, 417–428. https://doi.org/10.1038/nri3683
76. Flacher, V., et al. (2014). Murine Langerin+ dermal dendritic cells prime CD8+ T cells while Langerhans cells induce cross-tolerance. *EMBO Molecular Medicine, 6*, 1191–1204. https://doi.org/10.15252/emmm.201303283

77. Lee, C. H., et al. (2015). Dermal dendritic cells, but not Langerhans cells, are critical in murine single epicutaneous sensitization. *Experimental Dermatology, 24*, 67–69. https://doi.org/10.1111/exd.12583
78. Tordesillas, L., et al. (2018). PDL2(+) CD11b(+) dermal dendritic cells capture topical antigen through hair follicles to prime LAP(+) Tregs. *Nature Communications, 9*, 5238. https://doi.org/10.1038/s41467-018-07716-7
79. Gerner, M. Y., & Mescher, M. F. (2009). Antigen processing and MHC-II presentation by dermal and tumor-infiltrating dendritic cells. *Journal of Immunology, 182*, 2726–2737. https://doi.org/10.4049/jimmunol.0803479
80. Ng, L. G., et al. (2008). Migratory dermal dendritic cells act as rapid sensors of protozoan parasites. *PLoS Pathogens, 4*, e1000222. https://doi.org/10.1371/journal.ppat.1000222
81. Maschalidi, S., et al. (2022). Targeting SLC7A11 improves efferocytosis by dendritic cells and wound healing in diabetes. *Nature, 606*, 776–784. https://doi.org/10.1038/s41586-022-04754-6
82. Xia, W., Zhu, Z., Xiang, S., & Yang, Y. (2023). Ginsenoside Rg5 promotes wound healing in diabetes by reducing the negative regulation of SLC7A11 on the efferocytosis of dendritic cells. *Journal of Ginseng Research, 47*, 784–794. https://doi.org/10.1016/j.jgr.2023.06.006
83. Bosteels, V., et al. (2023). LXR signaling controls homeostatic dendritic cell maturation. *Science Immunology, 8*, eadd3955. https://doi.org/10.1126/sciimmunol.add3955
84. Edelson, B. T., et al. (2010). Peripheral CD103+ dendritic cells form a unified subset developmentally related to CD8alpha+ conventional dendritic cells. *The Journal of Experimental Medicine, 207*, 823–836. https://doi.org/10.1084/jem.20091627
85. Han, Z., et al. (2017). MiR-21/PTEN Axis promotes skin wound healing by dendritic cells enhancement. *Journal of Cellular Biochemistry, 118*, 3511–3519. https://doi.org/10.1002/jcb.26026
86. Vinish, M., et al. (2016). Dendritic cells modulate burn wound healing by enhancing early proliferation. *Wound Repair and Regeneration, 24*, 6–13. https://doi.org/10.1111/wrr.12388
87. Rajesh, A., et al. (2021). Skin antigen-presenting cells and wound healing: New knowledge gained and challenges encountered using mouse depletion models. *Immunology, 163*, 98–104. https://doi.org/10.1111/imm.13311
88. Henn, D., et al. (2023). Cas9-mediated knockout of Ndrg2 enhances the regenerative potential of dendritic cells for wound healing. *Nature Communications, 14*, 4729. https://doi.org/10.1038/s41467-023-40519-z
89. Martin, P., et al. (2003). Wound healing in the PU.1 null mouse—Tissue repair is not dependent on inflammatory cells. *Current Biology, 13*, 1122–1128. https://doi.org/10.1016/s0960-9822(03)00396-8
90. Kim, M. H., et al. (2008). Dynamics of neutrophil infiltration during cutaneous wound healing and infection using fluorescence imaging. *The Journal of Investigative Dermatology, 128*, 1812–1820. https://doi.org/10.1038/sj.jid.5701223
91. Wong, S. L., et al. (2015). Diabetes primes neutrophils to undergo NETosis, which impairs wound healing. *Nature Medicine, 21*, 815–819. https://doi.org/10.1038/nm.3887
92. Ellis, S., Lin, E. J., & Tartar, D. (2018). Immunology of wound healing. *Curr Dermatol Rep, 7*, 350–358. https://doi.org/10.1007/s13671-018-0234-9
93. Wee, W. K. J., et al. (2022). Single-cell analysis of skin immune cells reveals an Angptl4-ifi20b axis that regulates monocyte differentiation during wound healing. *Cell Death & Disease, 13*, 180. https://doi.org/10.1038/s41419-022-04638-7
94. Jun, J. I., Kim, K. H., & Lau, L. F. (2015). The matricellular protein CCN1 mediates neutrophil efferocytosis in cutaneous wound healing. *Nature Communications, 6*, 7386. https://doi.org/10.1038/ncomms8386
95. Willenborg, S., et al. (2012). CCR2 recruits an inflammatory macrophage subpopulation critical for angiogenesis in tissue repair. *Blood, 120*, 613–625. https://doi.org/10.1182/blood-2012-01-403386
96. Lucas, T., et al. (2010). Differential roles of macrophages in diverse phases of skin repair. *Journal of Immunology, 184*, 3964–3977. https://doi.org/10.4049/jimmunol.0903356

97. Mirza, R., DiPietro, L. A., & Koh, T. J. (2009). Selective and specific macrophage ablation is detrimental to wound healing in mice. *The American Journal of Pathology, 175*, 2454–2462. https://doi.org/10.2353/ajpath.2009.090248
98. Rodero, M. P., et al. (2014). In vivo imaging reveals a pioneer wave of monocyte recruitment into mouse skin wounds. *PLoS One, 9*, e108212. https://doi.org/10.1371/journal.pone.0108212
99. Leibovich, S. J., & Ross, R. (1975). The role of the macrophage in wound repair. A study with hydrocortisone and antimacrophage serum. *The American Journal of Pathology, 78*, 71–100.
100. Goren, I., et al. (2009). A transgenic mouse model of inducible macrophage depletion: Effects of diphtheria toxin-driven lysozyme M-specific cell lineage ablation on wound inflammatory, angiogenic, and contractive processes. *The American Journal of Pathology, 175*, 132–147. https://doi.org/10.2353/ajpath.2009.081002
101. Sanin, D. E., et al. (2022). A common framework of monocyte-derived macrophage activation. *Science Immunology, 7, eabl7482*. https://doi.org/10.1126/sciimmunol.abl7482
102. Willenborg, S., et al. (2021). Mitochondrial metabolism coordinates stage-specific repair processes in macrophages during wound healing. *Cell Metabolism, 33*, 2398-2414 e2399. https://doi.org/10.1016/j.cmet.2021.10.004
103. Injarabian, L., et al. (2024). FADD- and RIPK3-mediated cell death ensures clearance of Ly6C(high) wound macrophages from damaged tissue. *The Journal of Investigative Dermatology, 144*, 152-164 e157. https://doi.org/10.1016/j.jid.2023.06.203
104. Chen, H., Liang, F., Fu, C., Wang, Z., & Zhang, Z. (2023). Valproic acid accelerates skin wound healing in mice via its anti-inflammatory and apoptotic cell clearance-promoting effects. *The Journal of International Medical Research, 51*, 3000605231184038. https://doi.org/10.1177/03000605231184038
105. Majai, G., Sarang, Z., Csomos, K., Zahuczky, G., & Fesus, L. (2007). PPARgamma-dependent regulation of human macrophages in phagocytosis of apoptotic cells. *European Journal of Immunology, 37*, 1343–1354. https://doi.org/10.1002/eji.200636398
106. Chen, H., et al. (2015). Macrophage peroxisome proliferator-activated receptor gamma deficiency delays skin wound healing through impairing apoptotic cell clearance in mice. *Cell Death & Disease, 6*, e1597. https://doi.org/10.1038/cddis.2014.544
107. Justynski, O., et al. (2023). Apoptosis recognition receptors regulate skin tissue repair in mice. *eLife, 12*. https://doi.org/10.7554/eLife.86269
108. Uchiyama, A., et al. (2017). Mesenchymal stem cells-derived MFG-E8 accelerates diabetic cutaneous wound healing. *Journal of Dermatological Science, 86*, 187–197. https://doi.org/10.1016/j.jdermsci.2017.02.285
109. Demaria, M., et al. (2014). An essential role for senescent cells in optimal wound healing through secretion of PDGF-AA. *Developmental Cell, 31*, 722–733. https://doi.org/10.1016/j.devcel.2014.11.012
110. Ogata, Y., et al. (2021). SASP-induced macrophage dysfunction may contribute to accelerated senescent fibroblast accumulation in the dermis. *Experimental Dermatology, 30*, 84–91. https://doi.org/10.1111/exd.14205
111. Driskell, R. R., & Watt, F. M. (2015). Understanding fibroblast heterogeneity in the skin. *Trends in Cell Biology, 25*, 92–99. https://doi.org/10.1016/j.tcb.2014.10.001
112. Li, X., et al. (2020). Carnosine stimulates macrophage-mediated clearance of senescent skin cells through activation of the AKT2 signaling pathway by CD36 and RAGE. *Frontiers in Pharmacology, 11*, 593832. https://doi.org/10.3389/fphar.2020.593832
113. Homaeigohar, S., et al. (2023). Biosynthesis of zinc oxide nanoparticles on l-carnosine biofunctionalized Polyacrylonitrile nanofibers; a biomimetic wound healing material. *ACS Applied Bio Materials, 6*, 4290–4303. https://doi.org/10.1021/acsabm.3c00499
114. Ansurudeen, I., et al. (2012). Carnosine enhances diabetic wound healing in the db/db mouse model of type 2 diabetes. *Amino Acids, 43*, 127–134. https://doi.org/10.1007/s00726-012-1269-z

115. Romana-Souza, B., Chen, L., Leonardo, T. R., Chen, Z., & DiPietro, L. A. (2021). Dermal fibroblast phagocytosis of apoptotic cells: A novel pathway for wound resolution. *The FASEB Journal, 35*, e21443. https://doi.org/10.1096/fj.202002078R
116. Devalaraja, R. M., et al. (2000). Delayed wound healing in CXCR2 knockout mice. *The Journal of Investigative Dermatology, 115*, 234–244. https://doi.org/10.1046/j.1523-1747.2000.00034.x
117. Villarreal-Ponce, A., et al. (2020). Keratinocyte-macrophage crosstalk by the Nrf2/Ccl2/EGF signaling Axis orchestrates tissue repair. *Cell Reports, 33*, 108417. https://doi.org/10.1016/j.celrep.2020.108417
118. Piipponen, M., Li, D., & Landen, N. X. (2020). The immune functions of keratinocytes in skin wound healing. *International Journal of Molecular Sciences, 21*. https://doi.org/10.3390/ijms21228790
119. Motz, G. T., et al. (2014). Tumor endothelium FasL establishes a selective immune barrier promoting tolerance in tumors. *Nature Medicine, 20*, 607–615. https://doi.org/10.1038/nm.3541
120. Dini, L., et al. (1995). Phagocytosis of apoptotic bodies by liver endothelial cells. *Journal of Cell Science, 108*(Pt 3), 967–973. https://doi.org/10.1242/jcs.108.3.967
121. Zhou, T., et al. (2019). Microvascular endothelial cells engulf myelin debris and promote macrophage recruitment and fibrosis after neural injury. *Nature Neuroscience, 22*, 421–435. https://doi.org/10.1038/s41593-018-0324-9
122. Kirsch, T., et al. (2010). Endothelial-derived thrombospondin-1 promotes macrophage recruitment and apoptotic cell clearance. *Journal of Cellular and Molecular Medicine, 14*, 1922–1934. https://doi.org/10.1111/j.1582-4934.2009.00799.x
123. Rodrigues, M., Kosaric, N., Bonham, C. A., & Gurtner, G. C. (2019). Wound healing: A cellular perspective. *Physiological Reviews, 99*, 665–706. https://doi.org/10.1152/physrev.00067.2017
124. Fuchs, Y., et al. (2013). Sept4/ARTS regulates stem cell apoptosis and skin regeneration. *Science, 341*, 286–289. https://doi.org/10.1126/science.1233029
125. Maxson, S., Lopez, E. A., Yoo, D., Danilkovitch-Miagkova, A., & Leroux, M. A. (2012). Concise review: Role of mesenchymal stem cells in wound repair. *Stem Cells Translational Medicine, 1*, 142–149. https://doi.org/10.5966/sctm.2011-0018
126. Gao, M., et al. (2024). Regulation of inflammation during wound healing: The function of mesenchymal stem cells and strategies for therapeutic enhancement. *Frontiers in Pharmacology, 15*, 1345779. https://doi.org/10.3389/fphar.2024.1345779
127. Quarato, E. R., et al. (2023). Efferocytosis by bone marrow mesenchymal stromal cells disrupts osteoblastic differentiation via mitochondrial remodeling. *Cell Death & Disease, 14*, 428. https://doi.org/10.1038/s41419-023-05931-9
128. Stewart, K. S., et al. (2023). Stem cells tightly regulate dead cell clearance to maintain tissue fitness. *bioRxiv*. https://doi.org/10.1101/2023.05.22.541773
129. Kerr, J. F., Wyllie, A. H., & Currie, A. R. (1972). Apoptosis: A basic biological phenomenon with wide-ranging implications in tissue kinetics. *British Journal of Cancer, 26*, 239–257. https://doi.org/10.1038/bjc.1972.33
130. Tait, S. W., & Green, D. R. (2013). Mitochondrial regulation of cell death. *Cold Spring Harbor Perspectives in Biology, 5*. https://doi.org/10.1101/cshperspect.a008706
131. Green, D. R. (2022). The death receptor pathway of apoptosis. *Cold Spring Harbor Perspectives in Biology, 14*. https://doi.org/10.1101/cshperspect.a041053
132. Martin, S. J., Henry, C. M., & Cullen, S. P. (2012). A perspective on mammalian caspases as positive and negative regulators of inflammation. *Molecular Cell, 46*, 387–397. https://doi.org/10.1016/j.molcel.2012.04.026
133. Silva, M. T. (2010). Secondary necrosis: The natural outcome of the complete apoptotic program. *FEBS Letters, 584*, 4491–4499. https://doi.org/10.1016/j.febslet.2010.10.046
134. Botchkareva, N. V., Ahluwalia, G., & Shander, D. (2006). Apoptosis in the hair follicle. *The Journal of Investigative Dermatology, 126*, 258–264. https://doi.org/10.1038/sj.jid.5700007
135. Nishimura, E. K., Granter, S. R., & Fisher, D. E. (2005). Mechanisms of hair graying: Incomplete melanocyte stem cell maintenance in the niche. *Science, 307*, 720–724. https://doi.org/10.1126/science.1099593

136. Brown, D. L., Kao, W. W., & Greenhalgh, D. G. (1997). Apoptosis down-regulates inflammation under the advancing epithelial wound edge: Delayed patterns in diabetes and improvement with topical growth factors. *Surgery, 121*, 372–380. https://doi.org/10.1016/s0039-6060(97)90306-8
137. Wilson, S. E., Chaurasia, S. S., & Medeiros, F. W. (2007). Apoptosis in the initiation, modulation and termination of the corneal wound healing response. *Experimental Eye Research, 85*, 305–311. https://doi.org/10.1016/j.exer.2007.06.009
138. Bratton, D. L., & Henson, P. M. (2011). Neutrophil clearance: When the party is over, clean-up begins. *Trends in Immunology, 32*, 350–357. https://doi.org/10.1016/j.it.2011.04.009
139. Liu, G., et al. (2008). High mobility group protein-1 inhibits phagocytosis of apoptotic neutrophils through binding to phosphatidylserine. *Journal of Immunology, 181*, 4240–4246. https://doi.org/10.4049/jimmunol.181.6.4240
140. Fadok, V. A., et al. (1998). Macrophages that have ingested apoptotic cells in vitro inhibit proinflammatory cytokine production through autocrine/paracrine mechanisms involving TGF-beta, PGE2, and PAF. *The Journal of Clinical Investigation, 101*, 890–898. https://doi.org/10.1172/JCI1112
141. Bouchery, T., & Harris, N. (2019). Neutrophil-macrophage cooperation and its impact on tissue repair. *Immunology and Cell Biology, 97*, 289–298. https://doi.org/10.1111/imcb.12241
142. Sindrilaru, A., et al. (2009). Wound healing defect of Vav3−/− mice due to impaired beta2-integrin-dependent macrophage phagocytosis of apoptotic neutrophils. *Blood, 113*, 5266–5276. https://doi.org/10.1182/blood-2008-07-166702
143. van der Veer, W. M., et al. (2009). Potential cellular and molecular causes of hypertrophic scar formation. *Burns, 35*, 15–29. https://doi.org/10.1016/j.burns.2008.06.020
144. Chavakis, E., & Dimmeler, S. (2002). Regulation of endothelial cell survival and apoptosis during angiogenesis. *Arteriosclerosis, Thrombosis, and Vascular Biology, 22*, 887–893. https://doi.org/10.1161/01.atv.0000017728.55907.a9
145. Dimmeler, S., & Zeiher, A. M. (2000). Endothelial cell apoptosis in angiogenesis and vessel regression. *Circulation Research, 87*, 434–439. https://doi.org/10.1161/01.res.87.6.434
146. Li, F., et al. (2010). Apoptotic cells activate the "phoenix rising" pathway to promote wound healing and tissue regeneration. *Science Signaling, **3**, ra13*. https://doi.org/10.1126/scisignal.2000634
147. Dixon, S. J., et al. (2012). Ferroptosis: An iron-dependent form of nonapoptotic cell death. *Cell, 149*, 1060–1072. https://doi.org/10.1016/j.cell.2012.03.042
148. Yang, W. S., & Stockwell, B. R. (2016). Ferroptosis: Death by lipid peroxidation. *Trends in Cell Biology, 26*, 165–176. https://doi.org/10.1016/j.tcb.2015.10.014
149. Wiernicki, B., et al. (2020). Excessive phospholipid peroxidation distinguishes ferroptosis from other cell death modes including pyroptosis. *Cell Death & Disease, 11*, 922. https://doi.org/10.1038/s41419-020-03118-0
150. Yang, W. S., et al. (2014). Regulation of ferroptotic cancer cell death by GPX4. *Cell, 156*, 317–331. https://doi.org/10.1016/j.cell.2013.12.010
151. Dixon, S. J., et al. (2014). Pharmacological inhibition of cystine-glutamate exchange induces endoplasmic reticulum stress and ferroptosis. *eLife, 3*, e02523. https://doi.org/10.7554/eLife.02523
152. Wiernicki, B., et al. (2022). Cancer cells dying from ferroptosis impede dendritic cell-mediated anti-tumor immunity. *Nature Communications, 13*, 3676. https://doi.org/10.1038/s41467-022-31218-2
153. Feng, S., et al. (2023). The mechanism of ferroptosis and its related diseases. *Molecular Biomedicine, 4*, 33. https://doi.org/10.1186/s43556-023-00142-2
154. Telorack, M., et al. (2016). A glutathione-Nrf2-Thioredoxin cross-talk ensures keratinocyte survival and efficient wound repair. *PLoS Genetics, 12*, e1005800. https://doi.org/10.1371/journal.pgen.1005800
155. Bi, M., Li, D., & Zhang, J. (2023). Research progress and insights on the role of ferroptosis in wound healing. *International Wound Journal, 20*, 2473–2481. https://doi.org/10.1111/iwj.14102

156. Wilkinson, H. N., et al. (2019). Reduced iron in diabetic wounds: An oxidative stress-dependent role for STEAP3 in extracellular matrix deposition and remodeling. *The Journal of Investigative Dermatology, 139*, 2368-2377 e2367. https://doi.org/10.1016/j.jid.2019.05.014
157. Gao, S. Q., et al. (2018). Co-delivery of deferoxamine and hydroxysafflor yellow a to accelerate diabetic wound healing via enhanced angiogenesis. *Drug Delivery, 25*, 1779–1789. https://doi.org/10.1080/10717544.2018.1513608
158. Kong, L., et al. (2018). Bioactive injectable hydrogels containing Desferrioxamine and bioglass for diabetic wound healing. *ACS Applied Materials & Interfaces, 10*, 30103–30114. https://doi.org/10.1021/acsami.8b09191
159. Chen, H., et al. (2016). Upregulating Hif-1alpha by hydrogel Nanofibrous scaffolds for rapidly recruiting angiogenesis relative cells in diabetic wound. *Advanced Healthcare Materials, 5*, 907–918. https://doi.org/10.1002/adhm.201501018
160. Duscher, D., et al. (2015). Transdermal deferoxamine prevents pressure-induced diabetic ulcers. *Proceedings of the National Academy of Sciences of the United States of America, 112*, 94–99. https://doi.org/10.1073/pnas.1413445112
161. Li, S., Li, Y., Wu, Z., Wu, Z., & Fang, H. (2021). Diabetic ferroptosis plays an important role in triggering on inflammation in diabetic wound. *American Journal of Physiology. Endocrinology and Metabolism, 321*, E509–E520. https://doi.org/10.1152/ajpendo.00042.2021
162. Jiang, X., Stockwell, B. R., & Conrad, M. (2021). Ferroptosis: Mechanisms, biology and role in disease. *Nature Reviews. Molecular Cell Biology, 22*, 266–282. https://doi.org/10.1038/s41580-020-00324-8
163. Wei, X., et al. (2023). Defective NCOA4-dependent ferroptosis in senescent fibroblasts retards diabetic wound healing. *Cell Death Discovery, 9*, 138. https://doi.org/10.1038/s41420-023-01437-7
164. Ye, K., Chen, Z., & Xu, Y. (2023). The double-edged functions of necroptosis. *Cell Death & Disease, 14*, 163. https://doi.org/10.1038/s41419-023-05691-6
165. Weinlich, R., Oberst, A., Beere, H. M., & Green, D. R. (2017). Necroptosis in development, inflammation and disease. *Nature Reviews. Molecular Cell Biology, 18*, 127–136. https://doi.org/10.1038/nrm.2016.149
166. Bonnet, M. C., et al. (2011). The adaptor protein FADD protects epidermal keratinocytes from necroptosis in vivo and prevents skin inflammation. *Immunity, 35*, 572–582. https://doi.org/10.1016/j.immuni.2011.08.014
167. Liu, X., Lu, F., & Chen, X. (2022). Examination of the role of necroptotic damage-associated molecular patterns in tissue fibrosis. *Frontiers in Immunology, 13*, 886374. https://doi.org/10.3389/fimmu.2022.886374
168. Khandelwal, P., et al. (2021). A surfactant polymer wound dressing protects human keratinocytes from inducible necroptosis. *Scientific Reports, 11*, 4357. https://doi.org/10.1038/s41598-021-82260-x
169. Yang, S., Xu, M., Meng, G., & Lu, Y. (2020). SIRT3 deficiency delays diabetic skin wound healing via oxidative stress and necroptosis enhancement. *Journal of Cellular and Molecular Medicine, 24*, 4415–4427. https://doi.org/10.1111/jcmm.15100
170. Liu, X., et al. (2016). Inflammasome-activated gasdermin D causes pyroptosis by forming membrane pores. *Nature, 535*, 153–158. https://doi.org/10.1038/nature18629
171. Broz, P., & Dixit, V. M. (2016). Inflammasomes: Mechanism of assembly, regulation and signalling. *Nature Reviews. Immunology, 16*, 407–420. https://doi.org/10.1038/nri.2016.58
172. Zhang, K., et al. (2022). Bioglass promotes wound healing by inhibiting endothelial cell pyroptosis through regulation of the connexin 43/reactive oxygen species (ROS) signaling pathway. *Laboratory Investigation, 102*, 90–101. https://doi.org/10.1038/s41374-021-00675-6
173. Pastar, I., et al. (2021). Intracellular Staphylococcus aureus triggers pyroptosis and contributes to inhibition of healing due to perforin-2 suppression. *The Journal of Clinical Investigation, 131*. https://doi.org/10.1172/JCI133727
174. Chen, C., et al. (2023). MALAT1 participates in the role of platelet-rich plasma exosomes in promoting wound healing of diabetic foot ulcer. *International Journal of Biological Macromolecules, 238*, 124170. https://doi.org/10.1016/j.ijbiomac.2023.124170

175. Mehrotra, P., et al. (2024). Oxylipins and metabolites from pyroptotic cells act as promoters of tissue repair. *Nature, 631*, 207–215. https://doi.org/10.1038/s41586-024-07585-9
176. Meriwether, D., et al. (2022). Macrophage COX2 mediates Efferocytosis, resolution reprogramming, and intestinal epithelial repair. *Cellular and Molecular Gastroenterology and Hepatology, 13*, 1095–1120. https://doi.org/10.1016/j.jcmgh.2022.01.002
177. Medina, C. B., et al. (2020). Metabolites released from apoptotic cells act as tissue messengers. *Nature, 580*, 130–135. https://doi.org/10.1038/s41586-020-2121-3
178. Herzinger, T., Kleuser, B., Schafer-Korting, M., & Korting, H. C. (2007). Sphingosine-1-phosphate signaling and the skin. *American Journal of Clinical Dermatology, 8*, 329–336. https://doi.org/10.2165/00128071-200708060-00002
179. Lucas, C. D., et al. (2022). Pannexin 1 drives efficient epithelial repair after tissue injury. *Science Immunology, 7, eabm4032*. https://doi.org/10.1126/sciimmunol.abm4032
180. Bonnefoy, F., et al. (2018). Factors produced by macrophages eliminating apoptotic cells demonstrate pro-Resolutive properties and terminate ongoing inflammation. *Frontiers in Immunology, 9*, 2586. https://doi.org/10.3389/fimmu.2018.02586
181. Pakyari, M., Farrokhi, A., Maharlooei, M. K., & Ghahary, A. (2013). Critical role of transforming growth factor Beta in different phases of wound healing. *Adv Wound Care (New Rochelle), 2*, 215–224. https://doi.org/10.1089/wound.2012.0406
182. Patsalos, A., et al. (2022). A growth factor-expressing macrophage subpopulation orchestrates regenerative inflammation via GDF-15. *The Journal of Experimental Medicine, 219*. https://doi.org/10.1084/jem.20210420
183. Khanna, S., et al. (2010). Macrophage dysfunction impairs resolution of inflammation in the wounds of diabetic mice. *PLoS One, 5*, e9539. https://doi.org/10.1371/journal.pone.0009539
184. Wetzler, C., Kampfer, H., Stallmeyer, B., Pfeilschifter, J., & Frank, S. (2000). Large and sustained induction of chemokines during impaired wound healing in the genetically diabetic mouse: Prolonged persistence of neutrophils and macrophages during the late phase of repair. *The Journal of Investigative Dermatology, 115*, 245–253. https://doi.org/10.1046/j.1523-1747.2000.00029.x
185. Suresh Babu, S., et al. (2016). MicroRNA-126 overexpression rescues diabetes-induced impairment in efferocytosis of apoptotic cardiomyocytes. *Scientific Reports, 6*, 36207. https://doi.org/10.1038/srep36207
186. Shantikumar, S., Caporali, A., & Emanueli, C. (2012). Role of microRNAs in diabetes and its cardiovascular complications. *Cardiovascular Research, 93*, 583–593. https://doi.org/10.1093/cvr/cvr300
187. Mehrotra, P., & Ravichandran, K. S. (2022). Drugging the efferocytosis process: Concepts and opportunities. *Nature Reviews. Drug Discovery, 21*, 601–620. https://doi.org/10.1038/s41573-022-00470-y
188. Wilkinson, H. N., & Hardman, M. J. (2020). Senescence in wound repair: Emerging strategies to target chronic healing wounds. *Frontiers in Cell and Development Biology, 8*, 773. https://doi.org/10.3389/fcell.2020.00773
189. Wilkinson, H. N., et al. (2019). Elevated local senescence in diabetic wound healing is linked to pathological repair via CXCR2. *The Journal of Investigative Dermatology, 139*, 1171–1181 e1176. https://doi.org/10.1016/j.jid.2019.01.005
190. Ren, H., Zhao, F., Zhang, Q., Huang, X., & Wang, Z. (2022). Autophagy and skin wound healing. *Burns. Trauma, 10*, tkac003. https://doi.org/10.1093/burnst/tkac003
191. Li, L., et al. (2019). High glucose suppresses keratinocyte migration through the inhibition of p38 MAPK/autophagy pathway. *Frontiers in Physiology, 10*, 24. https://doi.org/10.3389/fphys.2019.00024
192. Jin, H., et al. (2018). Melatonin protects endothelial progenitor cells against AGE-induced apoptosis via autophagy flux stimulation and promotes wound healing in diabetic mice. *Experimental & Molecular Medicine, 50*, 1–15. https://doi.org/10.1038/s12276-018-0177-z
193. Xie, X., et al. (2021). The infection characteristics and autophagy defect of dermal macrophages in STZ-induced diabetic rats skin wound Staphylococcus aureus infection model. *Immunity, Inflammation and Disease, 9*, 1428–1438. https://doi.org/10.1002/iid3.492

194. Tsvetkov, P., et al. (2022). Copper induces cell death by targeting lipoylated TCA cycle proteins. *Science, 375*, 1254–1261. https://doi.org/10.1126/science.abf0529
195. Nowak, N. C., Menichella, D. M., Miller, R., & Paller, A. S. (2021). Cutaneous innervation in impaired diabetic wound healing. *Translational Research, 236*, 87–108. https://doi.org/10.1016/j.trsl.2021.05.003
196. Lu, Y. Z., et al. (2024). CGRP sensory neurons promote tissue healing via neutrophils and macrophages. *Nature, 628*, 604–611. https://doi.org/10.1038/s41586-024-07237-y
197. Poon, I. K. H., & Ravichandran, K. S. (2024). Targeting Efferocytosis in Inflammaging. *Annual Review of Pharmacology and Toxicology, 64*, 339–357. https://doi.org/10.1146/annurev-pharmtox-032723-110507
198. Desouza, C. V., & Shivaswamy, V. (2010). Pioglitazone in the treatment of type 2 diabetes: Safety and efficacy review. *Clin Med Insights Endocrinol Diabetes, 3*, 43–51. https://doi.org/10.4137/cmed.s5372
199. Shen, D., et al. (2018). Pioglitazone attenuates aging-related disorders in aged apolipoprotein E deficient mice. *Experimental Gerontology, 102*, 101–108. https://doi.org/10.1016/j.exger.2017.12.002
200. Morioka, S., et al. (2022). Chimeric efferocytic receptors improve apoptotic cell clearance and alleviate inflammation. *Cell, 185*, 4887–4903 e4817. https://doi.org/10.1016/j.cell.2022.11.029

Of Tingible Bodies and Starry Skies: Control of Normal and Malignant Tissues Through Apoptotic Cell-Dependent Communication (ACDC)

Stephen J. Jenkins and Christopher D. Gregory

Abstract

Apoptosis, the original regulated cell death programme, is an established controller of cell populations in all tissues. The hub of the apoptosis machinery, the caspase cascade, mediates the cleavage of hundreds of protein substrates to ultimately effect disabling and dismantling of cells, deleting them in a cell-autonomous fashion that is well understood. Of lesser renown are the intercellular communicative properties of apoptotic cells which, as burgeoning evidence indicates, compound the signalling processes that culminate in the famously speedy, and most commonly non-phlogistic, phagocytic clearance of dying/dead cells and their fragments (efferocytosis) by macrophages and other efferocytes. Here, we discuss more broadly the curious variety of instructive intercellular signals that apoptotic cells can deploy in order to reprogramme their living neighbours, be they close or distant, efferocytes or not. We review the literature, and we consider conceptually the potential of such non-cell-autonomous, apoptotic cell-dependent communication (ACDC) mechanisms to profoundly impact cell populations in health and disease. We illustrate our discourse firstly with the biology of lymphoid germinal centres (GC) and their constituent efferocytic, 'tingible body' macrophages, and secondly with the pathology of malignancies in which counterpart efferocytic, 'starry sky' macrophages feature prominently. In conclusion, we suggest several key avenues in ACDC research.

S. J. Jenkins (✉) · C. D. Gregory (✉)
Centre for Inflammation Research, Institute for Regeneration and Repair, University of Edinburgh, Edinburgh, UK
e-mail: stephen.jenkins@ed.ac.uk; chris.gregory@ed.ac.uk

C. D. Gregory et al. (eds.), *Apoptosis and Other forms of Cell Death in the Regulation of Inflammation in Health and Disease*, Advances in Experimental Medicine and Biology 1481, https://doi.org/10.1007/978-3-031-92785-0_6

Keywords

Apoptosis · B cell · Burkitt lymphoma · Efferocytosis · Germinal centre · Lymphoid follicle · Microenvironment · Non-Hodgkin lymphoma · Tingible body macrophage · Starry sky

Abbreviations

ACDC	Apoptotic cell-dependent communication
AiA	Apoptosis-induced apoptosis
AID	Activation-induced cytidine deaminase
AiP	Apoptosis-induced proliferation
ApoBD	Apoptotic body
ApoEV	Apoptotic cell-derived extracellular vesicle
BCME	B-cell microenvironment
BCR	B-cell receptor
BL	Burkitt lymphoma
BLL	Burkitt-like lymphoma
CSF1(R)	Colony stimulating factor 1 (receptor)
CX3CL1	Fractalkine, neurotactin
CX3CR1	Fractalkine receptor
DZ	Germinal centre dark zone
FDC	Follicular dendritic cell
GC	Germinal centre of lymphoid follicle
iPLA2	Calcium-independent phospholipase A2
LN	Lymph node
LPC	Lysophosphatidylcholine
LPS	Lipopolysaccharide
LZ	Germinal centre light zone
CD91	Lipoprotein receptor-related protein 1, CD91
MCM	Medullary cord macrophage
MFG-E8	Milk fat globule epidermal growth factor 8, lactadherin
MSM	Medullary sinus macrophage
NHL	Non-Hodgkin lymphoma
PANX1	Pannexin 1
PGE2	Prostaglandin E2
PtdSer	Phosphatidylserine
S1P	Sphingosine-1-phosphate
SHM	Somatic hypermutation
SSM	Subcapsular sinus macrophage
SSP	Starry sky pattern
SS-TAM	Starry sky tumour-associated macrophage
TAM	Tumour-associated macrophage
TBM	Tingible body macrophage
TFH	T follicular helper cell

TME	Tumour microenvironment
TUNEL	Terminal deoxynucleotidyl transferase dUTP nick end labelling
TZM	T-cell zone macrophage

Introduction

Apoptosis, the prototypic physiological cell death programme, impacts profoundly on the turnover of cell populations in all tissues, where it acts as an elemental homeostatic regulator [1]. From a cell-autonomous perspective, apoptosis is well renowned for effecting cell deletion through a 'suicidal' process that requires the activation of executioner caspases, resulting in the organised dismantling of affected cells [2, 3]. This initiation of self-destruction may be regarded as 'the end of the beginning' because the cellular impact of apoptosis is not restricted to the deletion of the apoptotic cell itself. Much less appreciated are the non-cell-autonomous effects of apoptosis wherein the caspase-activated, dying cells of tissues act as signalling beacons which rapidly instruct diverse responses in living cells, including phagocytosis, immunomodulation, survival, proliferation, morphogenesis and, in some cases, the initiation of further cell death. We term these communal, intercellular signalling events *Apoptotic Cell-Dependent Communication* (ACDC) and support the growing evidence that, through such processes, apoptosis acts as a critical tissue and organismal regulator, operating well beyond its direct role in autonomous cell deletion. Although virtually any cell type may be capable of responding to apoptosis either in its vicinity or more remotely, currently the most widely studied apoptosis-reactive cells are macrophages, professional phagocytes which efficiently dispose of cell corpses through efferocytosis [4], the coordinated recognition, engulfment and degradation of apoptotic cells.

In this review, we will focus on the mechanisms underlying the communication between apoptotic cells and tissue macrophages and their potential importance in the interactions between apoptotic B lymphocytes and the tingible body macrophages (TBM) found in germinal centres (GC) of lymphoid follicles. GC are anatomically defined, specialised regions of secondary lymphoid tissues that wax and wane during T cell-dependent antibody responses activated, for example, by invading pathogens. GC arise to improve the affinity and effector quality of antibodies and in the production of plasma cells and memory B cells. Key cellular events in GC that underpin the selection of only the highest-affinity antibody-producing cells are the proliferation and subsequent apoptosis of large numbers of low-affinity or self-antigen-reactive B cells that are generated as byproducts of the process of somatic hypermutation, followed by their clearance by TBM. Of particular relevance to the present review is that GC are exemplars of normal tissue microenvironments wherein massive cell proliferation is accompanied by massive apoptosis and communication with macrophages (as well as, potentially, other cell types) in the lymphoid follicle. At the end of the GC reaction, the expanded lymphoid tissue shrinks back to its pre-activation size and architecture. We will consider how ACDC mechanisms between B cells and TBM may be orchestrated in GC and how they may control cell turnover in GC. Furthermore, since it is becoming clear that apoptosis may play decisive roles in tissue expansion—a concept termed the *apoptosis paradox* [5]—we will discuss how the principles of ACDC at play in normal GC may be hijacked in their malignant counterparts, as well as other malignancies in which TBM give the tumours a 'starry-sky' histological appearance. Overall, the purpose of this review is to evaluate the knowns and the unknowns of ACDC in normal and malignant B-cell microenvironments (BCME) and to place them in the perspective of the roles of ACDC in normal and malignant tissue turnover more generally. Our aim is to provide a topical view of concepts and mechanisms and to highlight important knowledge gaps and future research directions.

General Concepts in Apoptotic Cell-Dependent Communication (ACDC)

The Cell-Autonomous Apoptosis Programme

There is little doubt that apoptosis constitutes an efficient, controllable programme for purposeful deletion of cells, diametrically opposing proliferation and providing a critical homeostatic mechanism in cell-population control. Over the past several decades, apoptosis researchers have focused intensely on the molecular machinery of apoptosis and its associated regulatory mechanisms. There are very many routes to apoptosis, ranging from immune cell signalling to growth factor withdrawal/insufficiency, oxidative stress, DNA damage and senescence, among many others; apoptosis occurs within the GC and during the generation of adaptive immune cells in the bone marrow and thymus where it can be considered specialised in being selective based on antigen receptor specificity. All these roads can lead to the cell-autonomous activation of caspases, a family of 'protein-snipping' cysteinyl proteases that cleave their substrates after key aspartate residues. The cleavage preference of individual caspases is determined by an additional trio of amino acid residues located on the N-terminal side of, and contiguous with, the key aspartate residues in the peptide sequences (reviewed in Ref. [6]). Critical to apoptosis are the apical/initiator caspases acting upstream of the effector/executioner caspases to produce, through positive feedback, a caspase cascade that generates all the morphological and functional features of apoptosis and causes the autonomous death of the cell in which the cascade was activated (Fig. 1).

ued) during apoptosis and are functional in ACDC. (Figure adapted from [1]). *APAF1* Apoptotic protease activating factor 1, *ARTS* Apoptosis related protein in TGF-β signalling, *ATP11A/C* Phospholipid flippases, *BAK* BCL2 antagonist killer, *BAX* BCL2 associated X, *BCL2* B cell lymphoma 2, *BID* BH3 interacting domain, *BSG* Basigin, *CASP* Caspase, *CYC* Cytochrome c, *FADD* FAS associated via death domain, *FAS* Death receptor APO-1/CD95, *FASL* FAS ligand, *MOMP* Mitochondrial outer membrane permeabilisation, *MLCK* myelin light-chain kinase, *NOXA* damage (Latin), *OMI* Serine protease high temperature requirement A, *PANX1* Pannexin 1, *PGE2* Prostaglandin E2, *PtdSer* Phosphatidylserine, *PUMA* p53 upregulated modulator of apoptosis, *ROCK1* Rho-associated coiled-coil protein kinase 1, *SMAC* Second mitochondria-derived activator of caspases (Diablo), *tBID* Truncated BID, *XIAP* X chromosome-linked inhibitor of apoptosis protein, *XKR8* XK-related protein 8 phospholipid scramblase

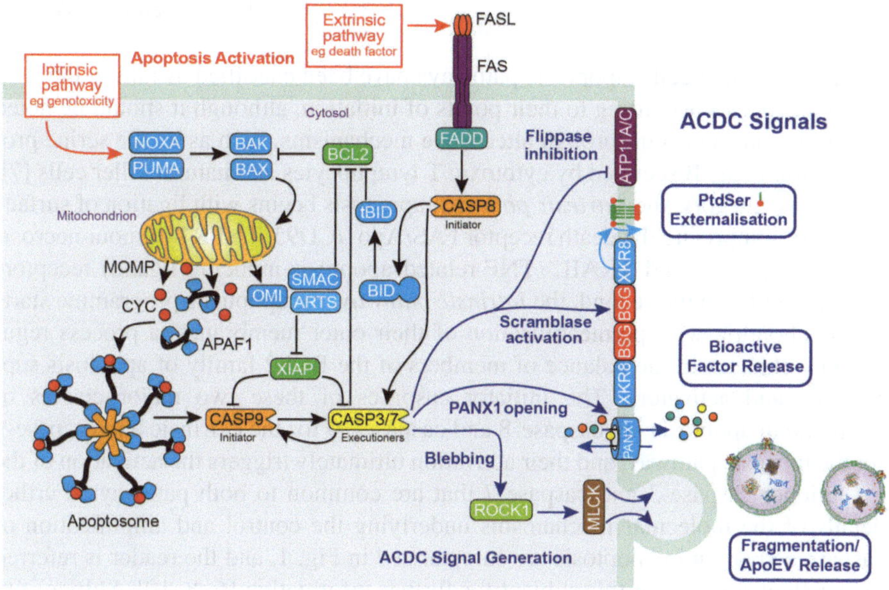

Fig. 1 Molecular cell biology of apoptosis: routes to cell death and to ACDC. Illustration of the coupling of the cell-autonomous apoptosis machinery—in which activation of evolutionarily conserved caspases is of critical importance—to ACDC signal generation. Two major pathways of caspase activation have been detailed in mammals: the extrinsic (death receptor) pathway and the intrinsic (mitochondrial) pathway. Activation of the extrinsic pathway is initiated in the plasma membrane through aggregation of transmembrane proteins known as death receptors, such as FAS, consequent to the binding of their cognate ligands (death factors), such as FASL (top right). This leads to the formation of death-inducing signalling complexes requiring recruitment of adapter proteins such as FADD and causes cleavage and activation of the initiator caspase-8, which ultimately cleaves and activates the executioner caspase-3 and caspase-7. A positive feedback loop with the intrinsic pathway fully activates and amplifies apoptotic signalling. In the intrinsic pathway, wide-ranging cell stresses and toxicity such as genotoxic damage initiate apoptosis by driving mitochondrial outer membrane permeabilisation (MOMP), which is regulated by the BCL2 family of proteins, whose members have the capacity to induce or inhibit apoptosis activation. Inducers such as tBID, PUMA and NOXA act on the mitochondrial pore-forming BCL2 family members BAX and BAK to induce MOMP. This causes the release of proteins into the cytosol, notably cytochrome c (CYC), from the mitochondrial intermembrane space. Pore formation can be tempered by anti-apoptotic BCL2-family members such as BCL2 itself. Cytosolic CYC interacts with APAF1 to form the heptameric apoptosome complex, which cleaves and activates caspase-9, the initiator caspase of the intrinsic pathway. Activated caspase-9 subsequently cleaves and activates the effector caspase-3 and caspase-7. Inhibitors of apoptosis such as XIAP and upstream antagonists OMI, SMAC and ARTS provide further controls on activation of caspase-9, caspase-3 and caspase-7. Among the many hundreds of cellular protein substrates of the activated effector caspases are those that generate ACDC signals (dark blue arrows and labels, right). Three key examples are illustrated: first, coordinated inhibition of the flippases ATP11A and C, along with activation of the phospholipid scramblase XKR8-BSG drive rapid transfer of PtdSer from the inner to the outer plasma membrane leaflet; second, constitutive opening of pannexin 1 (PANX1) channels allows release to the external milieu of a wide range of cellular factors up to approximately 1kDa in size including multiple metabolites, nucleotides and other bioactive molecules during apoptosis; third, blebbing of the plasma membrane via MLCK phosphorylation initiated by effector caspase-mediated cleavage and activation of the Rho kinase, ROCK1 resulting in cell fragmentation and production of apoptotic bodies (ApoBD) as well as smaller apoptotic cell-derived EV (ApoEV) which carry a wide range of cargoes from the apoptotic cell. Apart from these examples, numerous additional processes—such as the generation of internucleosomally cleaved DNA fragments and the synthesis of PGE2—are controlled by effector caspase activation Fig. 1 (contin-

In mammalian cells, apoptosis pathways have been classified as falling into two major categories according to their points of initiation, although it should be noted that initiation can occur through alternative mechanisms, such as by the serine protease, Granzyme B secreted by cytotoxic T lymphocytes and natural killer cells [7]. In the first category, the *extrinsic pathway*, apoptosis begins with ligation of surface receptors comprising the death receptor FAS/Apo1/CD95, TNFR1 (tumour necrosis factor receptor 1) and TRAIL (TNF-related apoptosis-inducing ligand) receptors DR4 and DR5; in the second, the *intrinsic pathway*, the apoptosis programme starts at mitochondria with permeabilisation of their outer membranes, a process regulated by changes in the balance of members of the BCL2 family of apoptosis suppressors and activators. The initiator caspases of these two major classes of mammalian apoptosis are caspase-8 and caspase-10 for the extrinsic and caspase-9 for the intrinsic pathway, and their activation ultimately triggers the activation of the executioner caspase-3 and caspase-7 that are common to both pathways. Further details of the molecular mechanisms underlying the control and amplification of caspase activation in apoptosis are summarised in Fig. 1, and the reader is referred to excellent reviews on this subject for further information [6, 8–10]. Suffice it to say here, the activation of the caspase cascade has the potential to cause the cleavage of many hundreds of cellular proteins during apoptosis. Among these are key substrates whose cleavage is required for the two critical properties of the apoptosis programme: the organised disabling/demolition of the cell [2, 6] and the enabling of ACDC (and beyond).

Potential of Apoptosis to Determine Variable Cell-Population Control in Tissues

In simplest terms of tissue turnover and cell population size, the logical, bare-bones consequences of apoptosis of a particular cell in a tissue depend on whether or not that cell is replaced, either by one or more identical cells or by different cells. We propose that tissue responses to apoptosis-derived effector signals therefore fall into two main categories: **(1) PRODUCTIVE**, representing replacement or expansion of cells, and **(2) REDUCTIVE**, when cells are not replaced, including when apoptosis events may be propagated. Productive and reductive signalling axes from apoptosis have the potential to cooperate both with each other and with 'exogenous', non-apoptosis-driven tissue regulators, resulting in (i) no net change of cell numbers, (ii) net cell gain or (iii) net cell loss (Fig. 2).

Thus, apoptosis-derived signals are required not only for the regulated loss of cells in morphogenesis or tissue involution but also for the regulated replacement and expansion of cell populations. We believe these simplified principles form the basis of novel mechanisms through which cell death actively regulates metazoan tissue populations from morphogenesis to homeostasis, from tissue repair to organ expansion and involution and from normal physiology to disease pathogenesis/immunity. Fundamentally, the basis of ACDC is the capacity of the dying cell to communicate with, and influence, living cells, not least those in its neighbourhood. In the context of

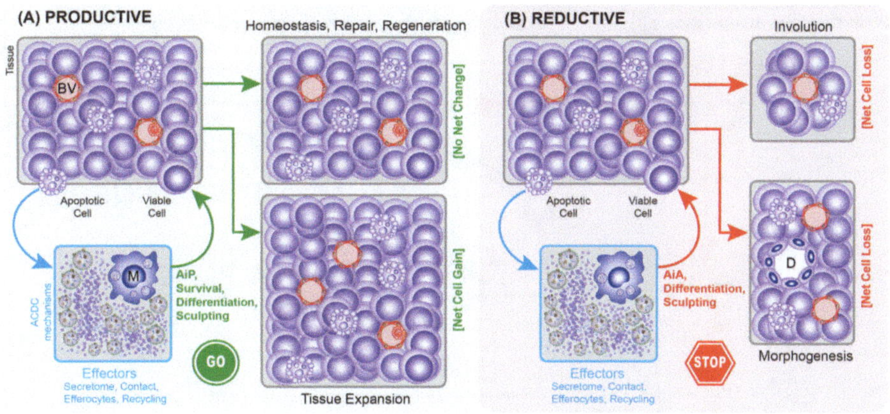

Fig. 2 Schematic representations of classes of tissue responses to apoptosis. Tissue responses to apoptosis can be perceived broadly as falling into two operational categories: (**a**) **Productive**, in which cells are replaced and/or expanded in tissues (represented by grey boxes), and (**b**) **Reductive**, in which cells are lost. (**a**) Examples of productive apoptosis responses are cell turnover, wound repair and tissue regeneration. Apoptotic cells generate secreted and contact-mediated signals (ACDC signals, blue arrow), which, along with recruited efferocytes, comprise ACDC effector mechanisms (blue box). These act on viable cells in the tissue to elicit response programmes such as survival, AiP (apoptosis-induced proliferation), differentiation and tissue remodelling/sculpting (green arrows). (**b**) Examples of reductive tissue responses to apoptosis are involution and atrophy (top right) and developmental morphogenesis (bottom right). Here, the ACDC effectors are proposed to include AiA (apoptosis-induced apoptosis), differentiation and tissue modelling (sculpting) mechanisms (red arrows). Both classes of response may occur simultaneously in individual tissues. *BV* Blood vessel, *D* Ductular structure, *M* Macrophage

tissue, while the apoptotic cell itself is deleted (net cell loss), the non-cell-autonomous instructive molecular effector mechanisms underlying ACDC—which we will review in detail below—can lead to replacement and/or expansion of cells through compensatory proliferation (also known as apoptosis-induced proliferation, AiP). Conversely, apoptosis can be propagated through apoptosis-induced apoptosis (AiA) signalling that compounds cell loss (Fig. 2). It can be appreciated that, by these instructive means, and by interacting with additional control signals unrelated to apoptosis (as well as other cell death routines), ACDC has the capacity to regulate, positively and negatively, the magnitude of cell populations in tissues (Fig. 3). Other consequences of ACDC, such as differentiation, migration and cytostasis, as well as control of inflammation and immunity and other forms of cell death, are also likely to contribute to the broad spectrum of tissue-controlling influences of apoptosis (Fig. 4 and later).

Both AiP and AiA were first defined in the fruit fly, *Drosophila melanogaster* (see also Chapter "Messages from Beyond the Grave: How Can*Drosophila*Help us to Understand the Consequences of Apoptosis?" of this volume). In brief, it was found that apoptosis activation during fly development caused secretion of mitogens by the affected cells, including (i) 'wingless' (Wg), the fly orthologue of the mammalian WNT family, which is induced through JNK activation in the apoptosis-induced cells, (ii) 'decapentaplegic' (Dpp), which is orthologous to mammalian bone morphogenetic proteins and TGF-β, and (iii) the evolutionarily conserved

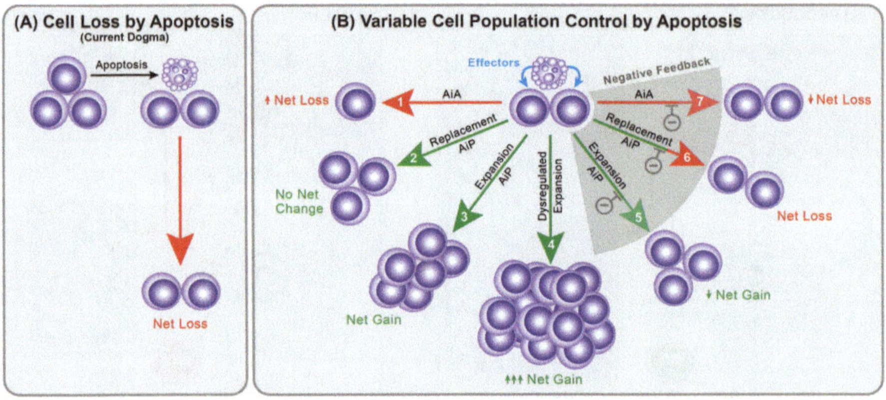

Fig. 3 Control of cell populations by apoptosis. Basic principles of cell population control by apoptosis. These diagrammatic schemes illustrate in simple terms the potential effects of apoptosis initially occurring in a single cell amongst a small group of three cells. (**a**) Cell loss in the absence of apoptosis-induced intercellular signalling. In this scenario—currently the prevailing view in cell-population biology—apoptosis is regarded as agnostic in terms of its ability to affect the tissue in which it occurs and merely reduces cell numbers. (**b**) Based on burgeoning evidence, apoptotic cell-derived communication (ACDC) signals (blue arrows) play instructive roles in regulating cell populations by eliciting productive (green arrows) or reductive (red arrows) responses of healthy cells (in this example immediate neighbours of the dying cell) as described in Fig. 2. Conceptual illustrative pathways (anticlockwise 1–7): 1. Propagation of apoptosis by AiA promotes cell loss; 2. Replacement of the lost, apoptotic cell (through AiP) may reconstitute the population with no net change in cell numbers; 3. Regulated tissue expansion may occur through AiP and other processes induced by apoptosis, such as angiogenesis; 4. Dysregulated expansion of cell numbers leading to massive cell gain (e.g., cancer); 5–7 consequences of productive and reductive responses under the influence of negative feedback signals which may be either dependent upon or independent of apoptosis

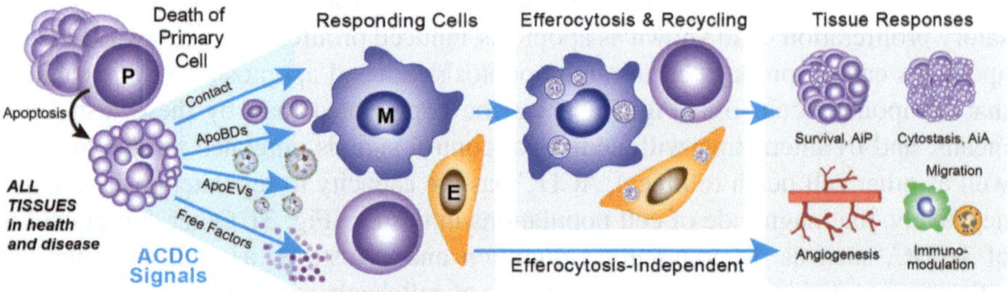

Fig. 4 Effector mechanisms in ACDC. Modes of ACDC are initiated by signals generated by apoptosis together with examples of outcomes for tissues which are realised either through efferocytosis-dependent or -independent routes. Shown schematically is the death of a tissue primary cell (P) with examples of cells that may respond to the ACDC signals generated, such as macrophages (M) and endothelial cells (E) as well as other, living, primary cells. Initial intercellular ACDC signalling is mediated through direct intercellular contact, through the production of cargo-laden ApoBD, smaller ApoEV and via many classes of free bioactive factors released from apoptotic cells. See the text for further details

short-range intracellular signalling protein, Hedgehog (Hh). Intriguingly, Wg and Dpp signal AiP in a proliferative tissue environment such as the wing imaginal discs of fly larvae, and, while dependent on initiator caspase activation, production of these mitogens can occur independently of effector caspase activation. By contrast, Hh signalling for AiP requires effector caspase activation and operates by re-activating differentiated, non-cycling cells (such as those of the developing fly eye) into mitosis. AiA in the fly was also first demonstrated using the wing imaginal disc model, wherein apoptosis induction (and consequent JNK activation) in the posterior aspect of the disc propagated rapid, long-distance apoptosis signalling to its anterior aspect. This ACDC pathway proved to be mediated via the production of Eiger (fly TNF) (reviewed in Ref. [11]). The evolutionary conservation of AiP and AiA are remarkable, both contributing to mammalian hair follicle cycling, for example [12, 13]. Furthermore, WNT signalling in ACDC has proven to be conserved in tissue regeneration from *Hydra* to mammals (reviewed in Ref. [1]), and recent work using zebrafish larvae indicates that relatively large ApoEV (also known as ApoBD, see Section "Apoptotic bodies as intercellular signalling entities" of mean diameter around 2 μm) produced by dying epithelial stem cells transfer WNT8A to neighbouring basal epithelial cells and thereby stimulate their expansion. The same group also reported that such ApoEV also carry macrophage migration inhibitory factor on their surface, which attracts macrophages to the site of epithelial cell death to sustain epithelial AiP [14, 15]

These genetic studies of opposing pathways of ACDC signalling provide a firm biological basis upon which novel concepts of apoptosis-orchestrated variable controls on cell populations can be developed (Fig. 3). Thus, in basal terms, ACDC has the capacity first to keep the numbers of cells in a steady state through the replacement of apoptotic cells via AiP, second to promote net cell loss through AiA, beyond that of cell-autonomous apoptosis alone, and third to increase net cell gain through AiP-mediated population expansion ('over-replacement'). These three principles offer many possible levels of control of cell numbers through ACDC in healthy tissue development and turnover as well as in the pathogenesis of diseases stemming from the inappropriate loss, retention or expansion of cells. Furthermore, as we discuss the subtleties of ACDC in the next sections, it will become clear that the potential consequences of ACDC extend well beyond control of cell numbers in tissues and have profound implications for tissue specialisation and turnover. Here we will illustrate this potential with special reference to the development, function and, ultimately, involution of lymphoid follicles—'mini-organs' in which apoptosis plays fundamental roles in the physiology of B cells—as well as in the dysregulated development of malignant BCME.

Effector Mechanisms in ACDC

As mentioned, the effector mechanisms operating in ACDC are closely coupled to the cell-autonomous apoptosis machinery that culminates in activation of executioner caspases and cleavage of hundreds of protein substrates. Known ACDC mechanisms that are clearly initiated by executioner caspase activation include: (1) changes in plasma membrane architecture that promote or inhibit interactions with neighbouring cells altering their physiological state, (2) active fragmentation of the dying cell to form multiple plasma membrane-bounded entities (ApoBD/large ApoEV), and (3) generation of an active apoptotic cell secretome comprising smaller ApoEV, (with all ApoEV carrying various organellar and molecular cargoes) and free factors, notably lipids, proteins and metabolites (Figs. 1 and 4). While the loss of intercellular contacts of apoptotic cells with their viable neighbours is known to induce profound changes in the latter—for example, survival through transient activation of ERK signalling in the immediate neighbours of dying epithelial cells as they are extruded [16]—the most widely studied plasma membrane alterations of apoptotic cells are those that generate cognate interactions with neighbours, especially macrophages, and promote efferocytosis. In short, ACDC impacts multiple types of neighbouring cells (phagocytes and non-phagocytes) in all tissues through changes in intercellular contact, cell fragmentation and secretory events to engender efferocytosis-dependent and -independent responses, including migration, immunomodulation, survival, proliferation, cytostasis, differentiation and cell death (Fig. 4).

Apoptotic Bodies as Intercellular Signalling Entities

The term ApoBD has been used in different contexts to mean different things. Although mainly applied to the membrane-delimited subcellular fragments of apoptotic cells, the term is also used as a reference to the remains of the shrunken cell body once apoptosis has taken its toll. As subcellular fragments, ApoBD may be regarded as the larger components of a continuum of ApoEV that are released during the apoptosis programme, essentially as part of the secretome of the apoptotic cell. Here, we define ApoBD as plasma membrane-bound fragments of apoptotic cells—arbitrarily in the size range of approximately 1000 to >5000 nm as has been previously proposed [1, 17, 18]—although we note that it remains unclear whether their mode of biogenesis is shared with, or independent from, those ApoEV in the smaller size range (discussed in Section "Secretome of the apoptotic cell: free factors and ApoEV"). Fragmentation of cells is generally viewed as a classical, common feature of apoptosis, and ApoBD have been described *in vivo* and *in vitro* as being produced by various cell types. Not all cell types, however, undergo fragmentation. For example, during apoptosis of granulocytes, the cell body rounds up and production of ApoBD is not overtly evident [19]; thymocytes show similar features [20]. Many questions remain, such as whether the degree of fragmentation is related to how rapidly a cell is efferocytosed (i.e., fragmentation occurs when efferocytosis is delayed), and how thymocytes and neutrophils are inherently resistant to apoptosis-induced fragmentation. As we shall discuss, the degree of fragmentation of apoptotic B cells within the GC is the subject of debate.

What little knowledge of ApoBD biogenesis we have currently has been gleaned largely from *in vitro* studies of apoptosis in purified populations of cells, in the absence of macrophages. The composition and biogenesis of ApoBD have been recently reviewed in detail elsewhere (see [1] and references therein). Briefly,

ApoBD production is in part coupled to caspase-3-activated ROCK1-dependent blebbing along with activation of MLCK, subsequent actomyosin contraction and accompanying increases in hydrostatic pressure. The latter permit flow of intracellular components, including cytoplasmic organelles and nuclear fragments, into the nascent ApoBD, which are pinched off and released as heterogeneous populations in terms of sizes and cargo contents. It is conceivable that ApoBD biogenesis assists in the efficient transfer of safely compartmentalised components of apoptotic cells to closely apposed recipient phagocytes, thereby promoting the efficiency of efferocytosis. In this respect, speedy transfer of ApoBD to phagocytes at close quarters may provide a route for efficient molecular recycling in ACDC. Perhaps free ApoBD are only observed in circumstances of massive apoptosis when efferocytosis capacity is challenged, in order to ensure bioactive and phlogistic intracellular constituents of apoptotic cells are contained in a safe haven? We shall return to this question when we consider ApoBD in the context of the GC (Section "ACDC in the regulation of lymphoid follicles").

Production of ApoBD and smaller ApoEV may also enhance the release of dying cells from the spatial constraints of intercellular contacts in order to facilitate liberation to the external environment or phagocytosis [21, 22]. It is widely assumed that the process of fragmentation is commonly part of the regulated dismantling of apoptotic cells into 'bite-sized' pieces that are more efficiently engulfed by phagocytes than whole cell corpses—although efferocytosis of whole cell bodies undoubtedly occurs [23, 24]. While macrophages are the efferocytes of highest renown—and can cooperate in groups to digest cells that are much larger than themselves through exocytosis of lysosomal enzymes [25]—a diverse array of cell lineages representing perhaps the majority of nucleated cell types, both normal and malignant, have some capacity for efferocytosis [26–30]. Production of ApoBD may be especially important for the effective activity of these so-called non-professional phagocytes, as has been proven in the case of fibroblasts, dendritic cells, alveolar epithelial cells and monoblastic leukaemia cells [31–33].

We conclude that, where it occurs, fragmentation into ApoBD may be relevant to (i) target specific cell types or particular aspects of ACDC, (ii) enhance speed and efficiency of transfer, engulfment, degradation and recycling, and therefore (iii) optimise efferocytosis-dependent signalling in tissue microenvironments. Overall, it seems likely that ApoBD are multipurpose entities that also promote accessibility of apoptotic cells too large for macrophages to eat whole (e.g., hepatocytes or adipocytes). A further mechanism has recently been described to take precedence under these circumstances, at least in the liver. In this case, neutrophils are reported to burrow into apoptotic hepatocytes and ingest them from the inside [34]. Intriguingly, this mechanism, termed 'perforocytosis', may not be redundant since experimental depletion of neutrophils led to an increased detection of apoptotic hepatocytes under steady-state conditions despite greater association of these cells with macrophages. Neutrophil-depleted mice also developed higher levels of autoantibodies, suggesting less effective clearance by macrophages may compromise tolerance to self-antigen [34]. In the context of the removal of overly large cells, extracellular digestion by 'frustrated phagocytosis' may also be important (see Section "Efferocytosis: phagocytic clearance of apoptotic cells"; [25]).

Secretome of the Apoptotic Cell: Free Factors and ApoEV

The secretion of bioactive factors by apoptotic cells is a growing area of knowledge expansion in caspase-dependent cell death processes. The secretome of apoptotic cells includes chemoattractants, growth factors, immunomodulators, angiogenic factors and metabolites (reviewed elsewhere [1, 28, 29, 35]). Furthermore, crude apoptotic secretome preparations, particularly of irradiated human peripheral blood mononuclear cells—which rapidly undergo apoptosis *in vitro* [36]—have been used successfully to promote repair and regeneration *in vivo* in injured tissues such as the spinal cord and skin [37, 38]. In recent years it has become increasingly appreciated that the secretome of apoptotic cells comprises not only factors released freely into the extracellular milieu but also functional modalities carried by extracellular vesicles (ApoEV), which themselves have therapeutic potential. ApoEV cargoes are broad, ranging from intact organelles to constituent molecules of cells, including DNA, RNA, proteins, lipids, carbohydrates, small metabolites and ions, many constituents having the capacity to act as autoantigens. Cargo quantities and qualities (both at the EV surface and in the lumen) are dependent upon ill-understood mechanisms of biogenesis and constraints by ApoEV surface and volumetric dimensions (reviewed in Refs. [1, 17, 18, 39]). Considering the known functions of EV as vehicles of intercellular transfer of multiple classes of biomolecules, it is noteworthy that ApoEV are no exception, having proven capacity to horizontally transfer functional proteins and RNA species to neighbouring cells. Furthermore, apoptotic cells have a known ability to transfer functional genes to neighbouring recipients [40], but it remains unclear whether such horizontal transfer of genomic DNA can be mediated by ApoEV. It should be noted, however, that ApoEV are renowned for their cargoes of double-stranded DNA in chromatin fragments [41]. As discussed at length elsewhere [1, 18], although present knowledge of ApoEV biology remains limited, largely because of technical issues relating to their identification, visualisation and purification, ApoEV have the potential to act significantly in ACDC both by protecting labile cargoes and by activating target cells through contact and cargo transfer.

In considering mechanistic studies that have linked the generation and release of apoptotic secretome moieties to the apoptosis machinery, three cases in point illustrate how different classes of lipids are active in ACDC. First to be reported was the apoptosis-dependent biogenesis of lysophosphatidylcholine (LPC), which can function as a lipid chemoattractant of mononuclear phagocytes. LPC is generated through activation of the calcium-independent phospholipase A2 (iPLA2), which goes on to catalyse the conversion of phosphatidylcholine to LPC [42] and is subsequently secreted via activation of the lipid transporter ABCA1 [43]. A second lipid mediator of ACDC, sphingosine-1-phosphate (S1P), is coupled to the apoptosis programme through caspase-dependent sphingosine kinase activation and acts as a monocyte/macrophage chemoattractant [44]. S1P generated by apoptotic cells can also promote macrophage survival, which requires PI3K, ERK1/2 and Ca^{2+}

signalling [45]. This mode of induction of survival in macrophages is reminiscent of that of epithelial cells adjacent to apoptotic neighbours that are similarly protected via ERK1/2 activation [16]. In addition, S1P promotes reparatory and immunosuppressive ('M2-like') polarisation of macrophages [46]. A third type of lipid mediator of ACDC, the oxylipin prostaglandin E_2 (PGE2), has also been reported to be upregulated through activation of the apoptosis programme [47]. The evidence indicates that, in mouse models of tissue repair in the skin and of regeneration in the liver, effector caspase-mediated cleavage and activation of iPLA2, in addition to driving LPC production, also generates arachidonic acid [48], which is converted to PGH2 by cyclooxygenases 1 and 2. Finally, PGE2 synthase catalyses the production of PGE2 from PGH2 and the subsequent release of PGE2 into the extracellular milieu to transmit the stem cell growth and regenerating properties of caspase-activated iPLA2 [47]. This ACDC pathway—termed the 'phoenix rising' pathway by the authors— also appears to be active in tumour cell proliferation and repopulation post-therapy and in apoptosis-induced tumour angiogenesis [49–52]. As with all signalling circuits, this pathway of PGE2 generation via arachidonic acid production is likely to be context-dependent, especially since iPLA2-induced arachidonic acid has also been reported to trigger apoptosis [53].

As we have mentioned, it has long been known that genetic material is released from apoptotic cells, and several lines of evidence support its involvement in ACDC in multiple contexts, as has been reviewed in detail recently [1, 18]. Topically, recent work has unearthed an unexpected mechanistic link between innervation of tumours, apoptosis and metastasis. Thus, non-metastatic breast cancers have lesser sensory innervation than their metastatic counterparts, innervation being controlled by SLIT2 provided by the tumour vasculature. In the TME, neuron-derived substance P induces apoptosis in a subpopulation of tumour cells that express high levels of its receptor, TACR1. The apoptotic cells consequently release ssRNA, an ACDC mechanism, which drives tumour cell proliferation, invasion and metastasis [54]. This work has important implications for neuronal control, through ACDC, not only of malignant diseases but also of normal tissues. In the context of the release of genetic material, apoptotic cells are also notable sources of cell-free DNA [55, 56], which is found in the circulation of normal and cancer-bearing hosts and has the potential to operate in ACDC in normal as well as malignant tissue microenvironments. Furthermore, this ACDC mechanism may provide an important mode of genetic interplay between tumour cells and normal stromal/immune cells of the TME.

One of the most compelling secretory mechanisms occurring in apoptotic cells critically involves the effector caspase-mediated C-terminal cleavage of pannexin 1 (PANX1) channel subunits, inducing the formation of a constitutively open channel configuration that allows flux of relatively small molecules (but relatively large permeants of physiological membrane channels) across the plasma membrane. Many bioactive molecules are released from apoptotic cells via these open PANX1 channels, and several have well-established ACDC activities, including the nucleotides ATP and UTP, which have mononuclear phagocyte chemoattractant properties [57], and metabolites having a range of immunomodulatory, survival- and growth-promoting effects on tissue targets [58] and also gut bacteria [59] (and see Chapter

"Cell Death and Infection: Implications for Host and Microbe", this volume). Caspase-mediated activation of PANX1 channels in proteoliposomes has shown directly that passage of anionic and cationic molecules up to ~1 kDa, a size which encompasses the aforementioned nucleotides and metabolites, as well as many more potential ACDC effectors, is permitted [60]. Considering some specific examples, Medina and colleagues first described the active release of arrays of metabolites from apoptotic cells of various types, including spermidine, adenosine monophosphate, guanosine 5′-monophosphate, creatine, glycerol-3-phosphate and ATP, selectively secreted through caspase-opened PANX1 channels and capable of eliciting immunomodulatory activity and tissue repair programmes. These groups of metabolites were aptly termed 'goodbye signals' [58]. A further example of PANX1 channel activity generating ACDC circuitry was described recently in dying lung epithelial cells, which were able to engender a conserved tissue repair response orchestrated by macrophages through the release of ATP and other factors. These caspase-activated PANX1-dependent ACDC effectors led to enhanced reparatory programming in macrophages, including the production of amphiregulin, which drives epithelial proliferation [61]. These examples highlight once again the active nature of apoptotic cells as instructive signalling beacons which have fundamental implications for tissue control through intercellular circuitry of variable complexity.

Efferocytosis: Phagocytic Clearance of Apoptotic Cells

Efferocytosis, the phagocytic clearance of apoptotic cells and their fragments is the most widely researched ACDC sector, and its mechanistic details have been extensively reviewed elsewhere in the literature [1, 28, 62–69] (and see Chapter "Efferocytosis in Homeostasis", this volume). Here we will summarise the salient features of efferocytosis in order to provide a backdrop to our discussion of its characteristics in normal and malignant BCME and its contributions to ACDC in these tissue settings. As we have noted already, efferocytosis is undertaken both by macrophages, which are regarded as the professional phagocytes of apoptotic cells, and by so-called non-professional phagocytes belonging to a wide variety of cell lineages. Importantly, because of its efficiency and its relatively narrow window of histological visibility, especially in a steady state, efferocytosis by different classes of phagocytes has not been properly defined and largely remains underestimated [70]. Novel genetic probes that record the fate of apoptotic cells *in vivo* (e.g., ref. (71)) are being developed, and their deployment in the production of 'apoptosis atlases' in various species throughout development and in tissue turnover in healthy and pathological states will impact significantly on the ACDC field.

The phases of efferocytosis may be usefully divided into its '3Rs': ***Recognition***, ***Response*** and ***Removal*** [72]. The *Recognition* stages are the best defined molecularly and encompass (i) plasma membrane changes on apoptotic cells ('eat-me' signals and loss of 'don't-eat-me' signals), (ii) the production of chemoattractants ('find-me' signals) to enable recognition by phagocytes from a distance together with directional migration of motile phagocytes, and (iii) consequent engagement with phagocytes, which may require their recruitment to the site of apoptosis and/or migration within tissues if mechanical constraints permit. Migration may involve wholesale movement of the cell body of the phagocyte or insinuation of its processes to locate dying cells or their fragments. As we have noted, the efficiency of efferocytosis may be enhanced through the physical transfer of ApoBD by neighbouring cells to immobile efferocytes [21], in other words, by bringing the apoptotic target to the phagocyte as opposed to the phagocytes themselves being actively migratory.

Recognition Phase

The most renowned plasma membrane changes on apoptotic cells to generate 'eat-me' signals involve the translocation of phosphatidylserine (PtdSer) from the inner to the outer plasma-membrane leaflet [73], a process requiring (i) effector caspase-mediated inhibition of the ATP-dependent aminophospholipid translocases, ATP11A and ATP11C, irreversibly disabling their ability to flip PtdSer to the inner plasma-membrane leaflet, and (ii) rapid externalisation of PtdSer by the highly conserved XK-related protein, XKR8 (CED-8 in the nematode worm, *Caenorhabditis elegans*), which, in concert with basigin or neuroplastin, becomes activated following caspase cleavage to acquire phospholipid scramblase activity in apoptotic cells [74–76] (Fig. 1). While necessary as an eat-me signal produced by apoptotic cells, mere PtdSer exposure is insufficient, however, to activate downstream efferocytic events [77, 78]. A variety of additional qualitative and quantitative lipid, carbohydrate and protein changes having implications for efferocytosis are known to occur at the surface of apoptotic cells (reviewed recently in Ref. [1]), although definitive apoptosis-coupled mechanisms and their importance for efferocytosis—or indeed other ACDC effects—remain unclear. An attractive partner for the PtdSer eat-me signal is the disruption of 'don't-eat-me' signals that otherwise prevent phagocytic recognition of healthy cells. The most widely studied of these is the immunoglobulin superfamily member CD47, which engages with SIRPα (signal regulatory protein alpha) on phagocytes, preventing the engulfment of viable cells through downstream inhibitory signalling via its ITIM motifs [79, 80]. Additional experimental evidence, however, indicates that this don't-eat-me signalling pathway does not cooperate with PtdSer exposure in efferocytosis [81]. While it is conceivable that other don't-eat-me signalling axes, including CD31-CD31 [82], CD24/Siglec-10 [83], MHC I (β2M)/LILRB1 [84] and PD-L1/PD1 [85], may cooperate with PtdSer, it remains arguable that such pathways are dispensable for efferocytosis, especially since simply masking PtdSer on apoptotic cells blocks their engulfment effectively [81]. It seems that the irreversibility of PtdSer exposure through inhibition of flippase ATP11A/ATP11C combined with activation of scramblase XKR8—but not the scramblase TMEM16F, which can externalise PtdSer transiently in the absence of apoptosis and without engendering phagocytosis—is critical in the route to efferocytosis [78, 81]. The extent to which PtdSer's modification, either chemically (e.g., through oxidation [86]) or positionally (such as through involvement or exclusion of lipid rafts [87]), or its association with additional tethering, eat-me and/or disabled don't-eat-me mechanisms in enabling or optimising efferocytosis in various tissue contexts remains to be determined.

As discussed above, the secretome of the apoptotic cell includes 'find-me' signals such as the lipids LPC and S1P, which engage with phagocyte G protein-coupled receptors G2A and S1P-R, respectively, and the nucleotides ATP and UTP, which trigger chemotaxis through P2Y2, also a G protein-coupled receptor [88]. In addition, apoptotic lymphocytes secrete, in an apparently caspase-dependent manner, a low molecular weight form of the classical chemokine CX3CL1 (fractalkine/neurotactin) that acts as a mononuclear phagocyte chemoattractant, signalling through a fourth G protein-coupled receptor, CX3CR1 [89]. Furthermore, apoptotic human lymphocytes release lactoferrin, which promotes the resolution of inflammation by acting as a 'keep-out' signal to granulocytes and driving the production of anti-inflammatory mediators, including TGF-β1 [90–92]. This property appears to be context-dependent since lactoferrin is a highly pleiotropic protein that can also act as a chemoattractant of monocytes and dendritic cells and promote innate and adaptive immunity [93]. As facets of the *Recognition* phase, certain surface moieties of apoptotic cells appear to interact with tethering receptors of efferocytes, a key example being CD14, a glycosylphosphatidylinositol-anchored pattern recognition receptor whose apoptotic ligand(s) remain unclear, although two points are worthy of note in this regard: (i) apoptotic cells display moieties that cross-react with anti-lipopolysaccharide (LPS) antibodies [94], and since CD14 is a pattern recognition receptor for LPS, such molecular patterns could theoretically act as CD14 ligands, and (ii) glyceraldehyde 3-phosphate dehydrogenase translocated to the surface of apoptotic cells has been suggested to subserve this function [95]. As a tethering receptor, CD14 may not have direct signalling activity during efferocytosis but instead promotes corpse clearance efficiency [96, 97]. In this respect, it is notable that in human M2-like macrophages, antibody crosslinking of CD14 has been reported to activate MERTK signalling in cis to promote clearance of apoptotic neutrophils [98]. As we will see, other efferocyte receptors have tethering properties for apoptotic targets too and may also provide important mechanistic components to tissue context-specific efferocyte activities.

Response Phase

Engagement of apoptotic cell surface 'eat-me' components with phagocyte receptors in combination with tethering interactions and possible loss of function of don't-eat-me receptor/ligand pairings lead to *Response* phase signalling, comprising activation of engulfment and other efferocyte programmes, including survival, proliferation, repair, resolution and immunomodulation. Activation of such responses is thought to require the establishment of 'phagocytic synapses' (perhaps more appropriately termed 'efferocytic synapses' given the range of responses that can be elicited, apart from phagocytosis) that are not well defined but are conceptually akin to the immunological synapses that are required for T-cell activation. The range of efferocyte receptors involved in these responses is broad and includes integrins, scavenger receptors, C-type lectins and others. Of particular note are receptors recognising the exposed PtdSer of apoptotic targets, either directly or via PtdSer-binding opsonins. Key examples of phagocyte receptors capable of binding PtdSer directly are BAI1, TIM-1/3/4, Stabilin-2, RAGE, TLT2 and the class B scavenger receptor CD36; PtdSer also interacts indirectly with CD36 and with the integrin αvβ3, most likely by bridging thrombospondin, though possibly independently of PtdSer too. Integrin αvβ3 also connects with PtdSer via MFG-E8 (milk fat globule epidermal growth factor 8, lactadherin)—as does αvβ5—and via DEL-1. The TYRO3-AXL-MERTK family of receptor tyrosine kinases interacts with PtdSer via GAS6 and/or protein S to mediate efferocytosis (reviewed in Ref. [1]). Interestingly, MERTK remains capable of tethering apoptotic cells to phagocytes when its kinase activity (and consequent phagocytosis signalling capacity) is blocked [99]. MERTK also appears to cooperate with TIM-4 in efferocytosis in a tissue context-dependent fashion. Thus, resident peritoneal macrophages deploy TIM-4 for apoptotic-cell tethering and MERTK for engulfment signalling, whereas bone marrow-derived macrophages rely on MERTK for both processes [99, 100]. Additional opsonins capable of binding PtdSer and bridging to phagocyte receptors include calreticulin, the complement component C1q, and the lung surfactant proteins SP-A and SP-D. C1q interacts with low-density lipoprotein receptor-related protein 1 (CD91) and scavenger receptor SCARF1 on efferocytes, while SP-A and SP-D have been reported to bridge to alveolar macrophage CD91. It is noteworthy that SP-A promotes IL-4-driven reparative macrophage programming and proliferation in the lung, while C1q performs analogous functions in the liver and peritoneal cavity [101]. Therefore, C1q, SP-A and SP-D may also be important in tissue-specific ACDC. Finally, C-type lectin receptors have reported efferocytic activity supporting seminal work proposing that lectins provide significant molecular mechanisms in apoptotic-cell clearance [102]. A key example is MGL1/CD301a, the deficiency of which leads to the persistence of apoptotic cells in irradiated mouse embryos [103]. Furthermore, colonic macrophages of LSECtin/Clec4g-deficient mice have impaired capacity for efferocytosis and for anti-inflammatory and tissue repair programming [104].

Response signalling pathways are driven by these interactions and are also fuelled by metabolites generated by digestion of the apoptotic meal during the *Removal* phase, as we shall discuss. A key question regarding *Response* signalling pathways, such as anti-inflammatory and immunoregulatory programming, is their biochemical basis, including the extent to which they are coupled to engulfment activity. A notable example is MERTK's roles in co-signalling efferocytosis and anti-inflammatory programming. It seems likely that these distinct functions are mediated by distinct and separable signalling pathways [105]. Thus, it has been long established that the engulfment of apoptotic cells is mediated by the GRB2/PI3K p85/PLCγ-binding domain of MERTK and involves a key tyrosine residue, but the same mutant fails to suppress NFκB-mediated inflammatory signalling [106]. The details of its tissue context-dependent anti-inflammatory/immunoregulatory roles remain unknown. By contrast, the C-type lectin receptor Dectin-1, which interacts with apoptotic cells via annexins A1, A5 and A13 that act as bridging molecules, generates tolerogenic SYK- and NADPH oxidase-2-mediated signals in dendritic cells independently of any demonstrable Dectin-1-mediated efferocytic activity [107]. Therefore, diverse ACDC signals may be activated in efferocytes and/or non-efferocytes by direct contact with apoptotic targets—namely apoptotic cells, ApoBD and smaller ApoEV—in the absence of engulfment. This aspect of ACDC has not been studied in detail as yet.

Removal Phase

Signalling for the *Removal* phase of efferocytosis is conserved from nematode worms to mammals, genetic studies have identified the involvement of two pathways in phagocytosis and in cell/membrane migratory processes. The first requires the *C. elegans* genes CED-1, -6 and -7 (with mammalian orthologues MEGF10 (multiple EGF-like domains 10), GULP1 (phosphotyrosine-binding domain-containing protein 1), and ABCA1 and ABCA7 (ATP-binding cassette transporters 1 and 7), respectively. In the second pathway, CED-2, CED-5 and CED-12 (mammalian orthologues being CRKII, DOCK180 and ELMO1, respectively) are required [62]. Both pathways converge on the Rho GTPase, Rac1 (CED-10 orthologue in *C. elegans*), which orchestrates cytoskeletal rearrangement [108]. Rac1 is known to be activated at specific locations on lamellipodia of the efferocyte, coincident with points of F-actin assembly and of entry of apoptotic cells, and subsequent closure of the phagocytic cup necessitates Rac1 deactivation [109]. In mammals, the first (GULP1-dependent) pathway activates Rac1 through ill-defined molecular steps downstream of the PtdSer receptor Stabilin 2 and of MEGF10, a macrophage and astrocyte efferocytic receptor which binds apoptotic cell-exposed PtdSer using C1q as an opsonin. CD91 can also activate Rac1 through this GULP1-dependent pathway following interaction with apoptotic cells via calreticulin, which either bridges to PtdSer directly or uses C1q as an intermediate [62]. ABCA1/7 somehow cooperates in this pathway, but the details remain unclear. In the second pathway, DOCK180 functions as a guanine nucleotide exchange factor which interacts with the adapter CRKII and, in complex with ELMO1, serves to activate Rac1. Upstream receptors involved in efferocytosis in this pathway include the PtdSer receptor, BAI1, MERTK via its PtdSer-binding opsonins GAS6 and Protein S, and the integrins αvβ3 and αvβ5 via the PtdSer-binding opsonin MFG-E8. The MERTK signalling operates—at least in certain contexts—through cooperation in cis with the aforementioned integrins or the PtdSer receptor TIM-4, which in itself is regarded as a tethering receptor but can also signal via MERTK in this pathway, as we have noted. Elevation of intracellular Ca^{2+} is required for efficient efferocytosis (e.g., the recycling of efferocyte receptors to the cell surface), although full details of its various roles await further research. At least three modes of Ca^{2+} elevation operate in efferocytes: (i) MERTK signalling generating IP_3 leading to mobilisation of Ca^{2+} from the endoplasmic reticulum and consequently; (ii) activation of Ca^{2+} influx from the extracellular milieu [67]; and (iii) mitochondrial sequestration of Ca^{2+} inhibited by Drp-1 and mitochondrial fission, which are upregulated in efferocytes in order to facilitate continued clearance of apoptotic cells [110].

The *Removal* phase of efferocytosis includes lysosomal digestion of apoptotic targets and recycling of their component parts, fuelling the efferocytes and thereby contributing to their responses, including their pro-resolution and reparatory reprogramming. Upon completion of Rac1-dependent efferocytosis [109], the engulfed apoptotic target is isolated within the efferocyte in a phagosome which matures, under the influence of recruited GTPases, Rab5 and subsequently Rab7, which enable fusion of early and late endosomes and lysosomes, ultimately creating the acidified environment of the phagolysosome required for degradation of apoptotic cell-derived cargoes [111]. A non-canonical pathway, LC3-associated phagocytosis [112], which utilises components of autophagy, may also function in the removal of dying cells. Molecular mechanisms underlying the association of these pathways with efferocyte programming, which contrasts starkly with the pro-inflammatory responses that accompany phagocytosis of pathogens, have not yet been properly detailed, although they are partly linked to the activities of phagolysosomal breakdown products of apoptotic cells. It is also noteworthy that, when the sheer magnitude of the apoptotic cell is too challenging, as in the case of apoptotic adipocytes, multiple macrophages link up by deploying processes replete with F-actin to surround the dying cell and to create a defined zone into which lysosomal enzymes can be exocytosed in order to carry out digestion extracellularly in a form of 'frustrated phagocytosis' [25]. While this process may lead to pro-inflammatory sequelae, it is known that anti-inflammatory programming in macrophages can be triggered via by-products of digestion of apoptotic adipocytes, such as triacylglycerols, which are taken up by CD36 scavenger receptors and processed to produce fatty acids to fuel oxidative phosphorylation and M2 (reparatory) macrophage polarisation [113]. The metabolic fuelling of efferocytes is not solely based on access to phagolysosome-catabolised AC-derived cargo products, for example, nucleotides, amino acids and fatty acids to power macrophage proliferation and reparatory activities [114–117]; it also originates from the activation of *Response* signalling pathways, such as stimulation of glucose uptake and lactate production to promote further efferocytosis, expansion of phagocytes, production of anti-inflammatory mediators and activation of pro-resolution and repair programmes [118–120]. Therefore, just as there is overlap between the *Recognition* and *Response* phases of efferocytosis, the *Response* and *Removal* phases are similarly interrelated, resulting in the integration of signalling pathways to tune efferocytosis optimally to the apoptotic cell load.

Continuity of efferocytosis is essential under conditions of high apoptosis challenge, such as those faced by TBM, in order for anti-inflammatory, pro-resolution and reparatory responses—and of course, clearance—to be optimised. Metabolic polarisation plays a key role in enabling continual, efficient efferocytosis. For example, efferocytes have been shown to have a propensity to metabolise glucose to lactate. Thus, efferocytosis stimulates glucose uptake (through the solute carrier (SLC) membrane protein SLC2A1) and lactate release (through SLC16A1). Notably, the resultant extracellular lactate operates as a potent ACDC effector that promotes both inflammation resolution by triggering TGF-β1 and IL-10 and induces angiogenesis through stimulation of VEGF [118]. Furthermore, intracellular lactate production appears to be critical for efficient efferocytosis in the face of a high apoptotic cell burden. Recent work has proposed that continual efferocytosis by macrophages is enabled by lactate production arising from short-lived activation of glycolysis through two Akt-dependent pathways that are activated by an apoptotic cell or body: one stabilising the glucose transporter GLUT1 at the plasma membrane to increase influx of glucose to fuel glycolysis, and the second activating the glycolytic enzyme 6-phosphofructo-2-kinase/fructose-2,6-bisphosphatase 2 (PFKB2). Both pathways lead to transient upregulation of lactate production that mediates continuity of efferocytosis, at least in part, through enhancement of Ca^{2+}-dependent recycling of the efferocytic receptors MERTK and CD91 [119]. It has been suggested that, following this transient phase, lactate may be subsequently converted to pyruvate in order to drive the longer-term reparatory phenotype of efferocytic macrophages via the tricarboxylic acid cycle and oxidative phosphorylation [68]. Apart from their recycling capabilities, efferocytic macrophages are able to deploy the conserved RNA polymerase II pause/release mechanism that allows rapid macrophage transcriptional reprogramming, enabling them to respond efficiently to increasing efferocytic challenges [121].

It is becoming clear, therefore, that efferocytes adapt to different tissue settings and conditions. As well as efferocytes needing to be optimised in their individual capacity to clear multiple apoptotic targets when the challenge demands, so too must the size of the population of efferocytes be tailored to the magnitude of the apoptosis challenge. In the case of macrophages, the efferocytic population can be enlarged through (a) recruitment via chemoattraction and/or (b) proliferation *in situ*. We shall discuss these mechanisms further in relation to GC and starry sky lymphomas in Sections "ACDC in the regulation of lymphoid follicles" and "ACDC in starry sky B-cell NHL". The efficiency of efferocytosis and associated response programmes can be regulated by multiple mechanisms, in addition to those already discussed. Compelling evidence obtained in mouse models of tissue damage in the lung (infection with the helminth *Nippostrongylus brasiliensis*) and gut (dextran sodium sulphate-induced colitis) indicates that IL-4- and IL-13-dependent macrophage anti-inflammatory and pro-repair signals require ACDC, operating via AXL and MERTK [122]. Adipose tissue macrophages may utilise similar mechanisms [122, 123]. In the thymus, efferocytosis of apoptotic thymocytes via the TYRO3-AXL-MERTK family of receptor tyrosine kinases during steady-state T-cell selection appears to suppress the production of IL-23 and BMP4 that are required to drive repair and regeneration of the thymus following damage of the organ [124]. Other adaptations to tissue settings may require mechanical sensing. For example, a recent study reported that sensing of microenvironmental stiffness by the mechanosensitive ion channel Piezo1 regulates efferocytosis. Piezo1-deficient macrophages were found to inhibit the resolution of inflammation, promoting liver fibrosis in mice. Piezo1 was required for efficient efferocytosis and lysosomal degradation of phagocytosed apoptotic cell corpses as well as for associated upregulation of macrophage anti-inflammatory genes [125]. These examples serve to illustrate the wide-ranging mechanisms of efferocytosis 'tunability' and suggest that much remains to be discovered through continued investigation of efferocytosis and its impacts in different tissue contexts.

Cellularity of BCME: Opportunities for ACDC

Before detailing the molecular cell biology of ACDC using our chosen BCME examples, let us first outline the cellular elements of normal and malignant BCME. The anatomy of normal lymphoid follicles and GC has been studied in light microscopic detail for almost 150 years. A schematic overview is provided in Fig. 5. As well as the various categories of B lymphocytes—either resting, proliferating or undergoing selection and apoptosis—the other main classes of cells in and around lymphoid follicles are T cells, macrophages, follicular dendritic cells (FDC) and fibroblasts (also known as fibroblastic reticular cells), the latter cell types forming an extensive reticular network. FDC, previously thought to be of myeloid origin, are actually mesenchymal and represent a class of fibroblastic reticular cells that populate GC [126]. Other cells include blood and lymphatic endothelial cells, pericytes and NK cells, as well as additional reticular cell types [127]. Although all cell types have the theoretical potential to receive and respond to ACDC in their vicinity, it is unlikely that small lymphocytes are capable of efferocytosis, at least of relatively large apoptotic targets (whole cells and ApoBD; our unpublished observations). The macrophages of GC, the tingible body macrophages (TBM), remain the most renowned cells of the follicles to engage with apoptotic B cells. The equivalent cell in malignant BCME is the starry sky macrophage, although, like normal GC, BCME in B-cell non-Hodgkin lymphoma (NHL) also contain a range of other immune and stromal cells, which may participate in responses to the apoptotic tumour cells in their neighbourhood. One of the recent papers estimates that TBM account for 90% of efferocytosis, suggesting other cells contribute but are subordinate [128].

Nomenclature: Tingible Body and Starry Sky Macrophages

There is little doubt today that the tingible bodies first described by Flemming in 1884 when he reported on anatomical structures in lymph nodes (LN) and Peyer's patches, which he termed 'Keimzentren'—later to be commonly known as GC—are the stainable remnants of apoptotic cells, mainly B lymphocytes. Elegant multiphoton live imaging provides graphic confirmation of this [128–130]. Interestingly, particularly in the context of our perspective in this chapter, Flemming's original paper was entitled 'Studien uber Regeneration der Gewebe' ('Studies on Tissue Regeneration') [131]. By the 1960s, it was already widely accepted that tingible bodies were largely phagocytosed cell remnants and that the engulfing cells became known as TBM, ranging in size from around 20 μm to 30 μm or more and containing cytoplasmic and/or nuclear inclusions in various numbers from lymphocytes, plasma cells and erythrocytes, as well as whole cells or large fragments thereof in various degrees of degradation [132].

Although the term 'TBM' tends to be used to describe the macrophages of GC that contain overt remnants of apoptotic cells (in other words, macrophages that have obviously engaged in efferocytosis), as we discuss, TBM are not confined to GC, although this is their most common and conspicuous location in normal tissues.

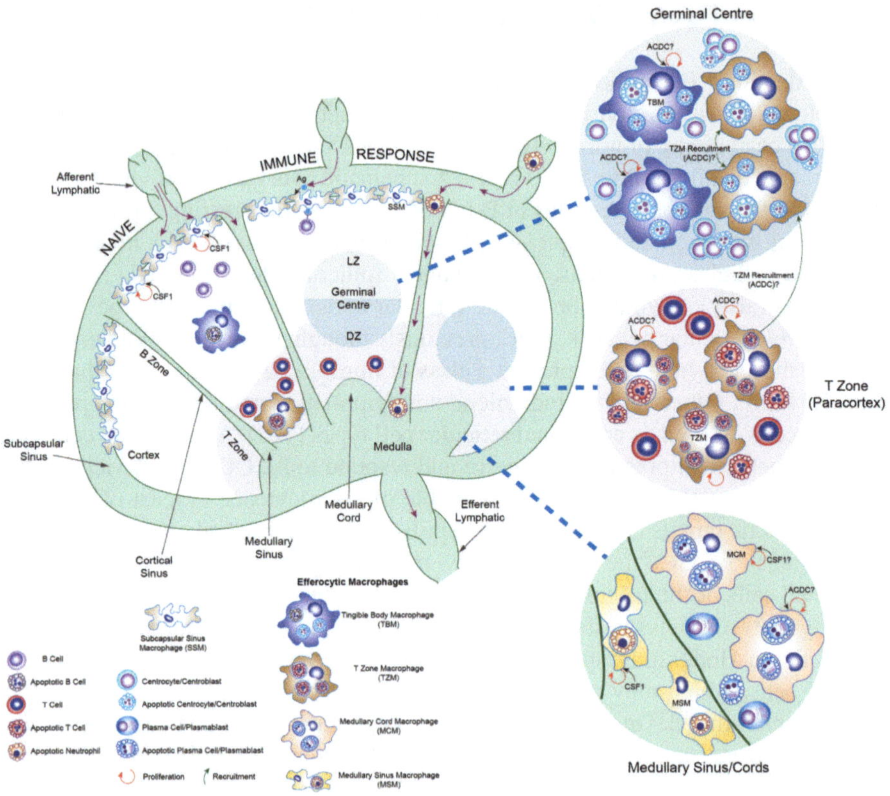

Fig. 5 Schematic anatomy of a lymph node: apoptosis, efferocytosis and potential ACDC. Diagrammatic representation of a lymph node emphasising locations of apoptotic cell types and macrophage subsets under antigen-stimulated ('immune response') and unstimulated ('naïve') conditions. Subcapsular sinus macrophages (SSC) and medullary sinus macrophages (MSM) are maintained in the naïve state through self-renewal and rely upon colony stimulating factor 1 (CSF1) produced by neighbouring endothelial cells for survival. SSM transfer native antigens from draining afferent lymph to the B cell zone, and both populations capture lymph-borne pathogens to prevent dissemination of infection. The efferocytic burden of SSM and MSM is low, although MSM potentially clear apoptotic immune cells from the tissue during an immune response. In contrast, T-cell zone macrophages (TZM), medullary cord macrophages (MCM) and B-cell zone macrophages largely arise during an immune response, with the latter found predominantly within germinal centres and termed tingible-body macrophages (TBM). All exhibit a high degree of efferocytosis, particularly the extensive clearance of dying B cells by TBM. TZM and TBM expand through CSF1-independent proliferation, and TBM potentially arise from TZM. The origin, proliferation and CSF1-dependence of MCM have yet to be determined. Hence, the independence of macrophages from CSF1 in the lymph node corresponds to those micro-anatomical niches that arise upon antigen stimulation and in which the efferocytic burden is high, leaving open the possibility that ACDC may be an integral factor regulating the survival/expansion of these cells. *LZ* Light Zone, *DZ* Dark zone

TBM are also observed in a diversity of cancers in which they are typically depicted as creating a 'starry sky' picture in histological sections due to their relatively bright appearance against a background of darkly stained tumour cells. Hence, they are most commonly called starry sky macrophages in this context. While not uniquely restricted to a specific cancer type, starry sky tumour-associated macrophages (SS-TAM) are most prominent in B-cell NHL, not least those of GC origin. Loose terminology exists too, with starry sky macrophages occasionally being described as TBM and vice versa.

ACDC in the Regulation of Lymphoid Follicles

A Topical Overview of the Germinal Centre (GC) Reaction

LN are complex organs with multiple microanatomical regions and structures that together serve to facilitate and increase the efficacy of the adaptive immune response to infection. They house the GC reaction: the process in which high-affinity antibody-producing B cells are generated and selected during T cell-dependent antibody responses (for detailed reviews, see Refs. [133–135]). GC arise in B-cell follicles following the initial activation and T cell-dependent survival of B cells. Activation of B cells occurs initially within the primary lymphoid follicles of LN (as well as other secondary lymphoid tissue), where the adjacent sub-capsular sinus macrophages (SSM) display intact antigen captured from the afferent lymph for sampling by naïve B cells. The antigen is also captured by follicular dendritic cells (FDC) within the follicle through diffusion or complement receptor-mediated transfer from SSM via B cells, and these cells subsequently become the key reservoir of antigen that supports the GC reaction. Notably, low protease levels specifically in the vicinity of FDC retain intact antigens in these cells [136]. Upon recognition of their specific antigen, activated B cells internalise the antigen, upregulate CCR7, and migrate to the border between the B-cell follicle and T-cell zones. Here, they present antigen-derived peptides in complex with MHC II to T follicular helper cells (TFH), a subset of antigen-specific T helper cells that migrate from the T-cell zone following activation and proliferation in response to the same antigen/pathogen. TFH subsequently provide the B cells with pro-survival and proliferative signals, such as CD40L, IL-21 and IL-4, upon recognition of their cognate antigen. Critically, for an antigen to be presented by B cells efficiently in association with MHC II, it must be captured and internalised through their B-cell receptor (BCR). This so-called linked recognition of antigen, whereby to survive and/or expand, B cells must recognise and present peptides from the same or linked antigen recognised by TFH also forms the subsequent basis for positive selection of high-affinity antibody-producing cells during the GC reaction.

GC arise when some of these B cells return to the primary follicle together with their cognate TFH and undergo rapid proliferation that pushes resting B cells to the periphery of the follicle known as the mantle. Mature GC comprise two microanatomical regions, termed the light zone (LZ) and the dark zone (DZ) (Fig. 5). The intense proliferation of B cells occurs largely within the DZ, wherein they are termed centroblasts [137, 138]. Indeed, the DZ is largely composed of centroblasts and is so named due to the density of nuclear staining of these cells. Notably, the GC centroblast is one of the most rapidly dividing eukaryotic cells, with an estimated cycling time of just 4–6 h [138, 139]. It is during proliferation that centroblasts also undergo activation-induced cytidine deaminase (AID)-driven somatic hypermutation (SHM) of their BCR genes that alters the affinity of their BCR hypervariable regions for antigen. Subsequent selection of daughter clones for those with increased affinity for antigen occurs following their migration to the LZ, a region populated by numerous other cells, including TFH and FDC. The latter give the characteristic histological appearance of light areas scattered among lymphocytes within the LZ, and as discussed above, provide the essential depot of unprocessed antigen for which daughter clones subsequently compete via their BCR affinity. This non-proliferative phase lasts only a few hours [140], and during this period, B cells are often referred to as centrocytes. TFH then provide costimulatory stimuli to those B cells able to capture antigen and bear peptide in MHC II, thereby licensing their survival and re-entry into the cell cycle in the DZ. The intensity of TFH help in the LZ determines both the degree of proliferation and the speed of the cell cycle in the DZ, and this in turn is determined by the amount of antigen presented by the centrocytes [139, 141]. As the capture of antigen by B cells is dependent on the affinity of the BCR [142], this provides the mechanism whereby T cells can sense the affinity of B cells and positively select those of high affinity. The competitive nature of this process is underscored by observations that low-affinity clones can capture and present antigen and gain access to the GC in the absence of other antigen-binding clones [143] and that GC comprised of high- or low-affinity clones grow to similar sizes [144, 145].

GC B cells undergo many iterative rounds of this expansion and selection during affinity maturation, migrating back and forth between the LZ and DZ to do so. Hence, the vast degree of proliferation required during the GC reaction necessitates the loss of defunct, self-reactive, or low-affinity clones on an almost equally extensive scale. Modern advances in the detection of apoptosis have made clear the extent to which apoptosis contributes to this balance and have allowed the identification of the underlying causes. For example, staining for active caspase-3 or the use of apoptosis indicator mice in which a FRET (fluorescence resonance energy transfer) pair is cleaved by active caspase-3 suggests that almost half of all GC B cells undergo apoptosis every 3 h. Hence, the rate of death almost matches that of proliferation [129]. The rate of death is also nearly equivalent in the LZ and DZ compartments, yet the underlying causes differ. Using MYC expression to identify B cells undergoing positive selection, Mayer et al. demonstrated that these were almost completely protected from apoptosis, suggesting that in the LZ, B cells largely die by neglect if not positively selected by TFH [129]. Thus, Vitoria and Nussenzweig neatly summarise affinity maturation as arising from a combination of 'death-limited' selection, whereby only high-affinity clones survive to re-enter the DZ (i.e., affinity-dependent apoptosis), and 'birth-limited' selection, whereby the level of help from TFH determines the resultant degree of expansion in the DZ (affinity-dependent proliferation) [135]. However, low-affinity clones do not necessarily all perish. Rather, the memory B-cell pool is largely comprised of low-affinity cells, suggesting this fate may arise when affinity for antigen fails to support further expansion in the GC [146, 147]. It has been hypothesised that a low-affinity memory pool may provide an expanded group of cells that can anticipate infection with related or subsequently mutated pathogens [147]. Unlike in the LZ, apoptotic cells in the DZ are enriched for clones that do not secrete antibodies despite normal BCR gene expression and apoptosis in this site being reduced by AID deficiency [129]. These findings suggest that apoptosis in the DZ largely arises from AID-induced SHM that gives rise to non-encoding BCR genes or structurally compromised antibodies. These mechanisms account for 75% of all apoptotic cells in the DZ [129]. To facilitate this selective process, GC B cells fully replace their surface BCR following SHM while still in the DZ, and few, if any, that acquire damaging BCR mutations subsequently transition to the LZ [148]. While it remains unclear how defective BCR expression leads to apoptosis, it seems likely related to the requirement for continual signalling through the BCR reported for survival of naïve B cells [149]. Notably, ectopic expression of BCL2 fails to rescue LZ entry by B cells with damaged BCR, suggesting signalling through the BCR may also provide active cues to initiate migration to the LZ [148]. Interestingly, antibody isotype class-switching, which occurs alongside SHM in the GC, also appears to make cells more prone to apoptosis [129]. Another predicted consequence of SHM is the generation of clones with self-reactive BCR, but what happens to these clones is not known. Intriguingly, apoptotic GC B cells do not appear enriched for cells with autoreactivity, suggesting that autoreactive B cells generated by SHM may instead be redeemed by continued SHM or polyreactivity—as polyreactive cells were equally represented in live and apoptotic compartments—or are too transient and/or too rare to detect [129]. Comprehending the mechanisms for maintaining tolerance to self in the GC and defining the role of apoptosis and efferocytosis in these processes remains important for understanding the potential for self-reactive B cells to escape and contribute to the pathogenesis of autoimmune disease.

Origin and Expansion of TBM in the GC Reaction

Overview of LN Macrophages

We will discuss the function of GC macrophages in detail below, but further subpopulations of macrophages are also important to consider when examining the biology of efferocytic macrophages and ACDC in the LN (Fig. 5). As discussed already, SSM line the border between the follicle and subcapsular sinus, where they capture antigen from the tissue-draining lymph fluid for transfer to the underlying B cell follicle [150, 151]. SSM also capture and transfer antigens to B cells when direct infection of the LN occurs following viraemia [152]. Lymph fluid subsequently passes the medullary sinus macrophages (MSM) before exiting through the efferent lymphatics. Along with SSM, MSM capture any pathogens that drain from tissues and prevent their entry into the LN parenchyma and dissemination beyond

[151]. MSM may also efferocytose apoptotic granulocytes that enter the LN during an immune response [153]. Although originally mislabelled as dendritic cells, T-cell zone macrophages (TZM) act as the sole scavengers of the substantial number of apoptotic T cells generated during the contraction phase of an immune response [154]. A small number of TZM are also present in the steady state, which appear to clear rarer apoptotic naïve T cells [154]. Similarly, primary B cell follicles contain a small number of macrophages that undertake low-level efferocytosis of dying naïve B cells and share the phenotype and core transcriptional signature of GC TBM [130]. The number of macrophages in the follicles increases roughly 3-fold upon development of GC and can be divided into GC TBM and non-GC follicle cells [128]. The latter exhibit fewer vacuoles and longer cellular processes, consistent with lower rates of efferocytosis [130]. Lastly, the medullary cords contain a population of macrophages (medullary cord macrophages, MCM) that also largely emerge following immunisation. MCM are found alongside plasmablasts that dominate these transient structures and which derive from activated B cells that migrate from the follicle border and produce the initial wave of antibody secretion that precedes affinity maturation in the GC. Notably, substantial numbers of plasma cells undergo apoptosis following the peak of a primary immune response [155] and can be observed within MCM [156]. However, like TZM, the level of apoptotic bodies within MCM are markedly lower than observed within the TBM of the GC, likely reflecting differences in apoptotic burdens at these sites. It should be noted that monocytes can also be found in the LN, but at least some of these may represent undifferentiated cells that migrate from the tissues to present antigen [157].

Identification of GC TBM

Numerous markers and genetic approaches are now used alongside microanatomical location and/or tingible body morphology to define, identify, and image GC macrophages. Advances in live imaging have also greatly facilitated progress in our understanding of the behaviour and function of TBM. As a brief guide, we describe the common approaches and latest advances in identifying and imaging these cells and in distinguishing these from other LN macrophage subsets. CD68 is probably the most widely used marker to identify TBM and follicle macrophages by immunofluorescence. Mice that report CD68 or CX3CR1 expression (*Cd68*^{GFP} and *Cx3cr1*^{GFP}, respectively) are also often used and have the additional benefit of allowing live imaging of these cells [128, 130]. Notably, GFP expression in *Cx3cr1*^{GFP} also appears to fully overlap with immunofluorescent staining of CD68 expression by GC TBM, suggesting these methods obtain equivalent results [130]. The relatively unrestricted expression of CX3CR1 and CD68 among LN macrophages means these approaches must be combined with methods that identify other macrophage subsets or determine microanatomical location. For example, CD169 is often employed to specifically identify MSM and MCM; pan B-cell markers including B220 and CD21/35 are used to identify follicles, while markers to define the GC include PNA, which selectively binds to GC B cells, Ki67, which identifies proliferating GC centroblasts, or lack of IgD, which is downregulated by activated B cells. For a comprehensive review on antibody markers for SSM, MSM and MCM, see [158]. GC markers can also be used in combination with those for T cells and/or FDC (e.g., CD3 or FDC-M2, respectively) to distinguish LZ and DZ cells. When live-imaging, follicles and GC can be identified by the transfer of fluorescently tagged B cells (e.g., [128, 130]). A recent study found that GC TBM are also largely positive for tdTomato in mice that lineage-trace expression of CD169 (*Cd169*$^{Cre/+}$ *.LSL-tdTomato*) despite being negative for CD169 protein, demonstrating an unexpected utility of this line for live imaging of TBM and follicle macrophages [130]. Light-field imaging suggests that within the GC, tdTomato expression overlaps uniformly with TBM morphology in these mice, yet intriguingly not all GC CD68$^+$ cells appear to be tomato$^+$ [130]. The authors were also able to identify putative TBM from these mice by flow cytometry, gating on tdTomato$^+$ MERTK$^+$ CD11b$^+$ CD169$^-$ cells, of which 70% expressed CX3CR1. However, it is unclear whether this approach would also capture MERTK$^+$ CD11b$^+$ CD169$^-$ CX3CR1$^+$ TZM [154], which imaging suggests may also be labelled in these mice [130]. Along with TZM and SSM, follicle and GC macrophages are also often described as negative for F4/80 by immunostaining, whereas this marker is used together with the endothelial marker Lyve1 to identify and distinguish MCM and MSM (e.g., [154, 158–162]). However, putative TBM identified within single-cell RNA sequencing datasets of CD169-lineage cells from naïve and activated LN appear to express *Adgre1*, the gene encoding F4/80, and protein expression could be detected using CITE-seq (**C**ellular **I**ndexing of **T**ranscriptomes and **E**pitopes by **Seq**uencing) [130]. How well TBM survive the necessary cell-sorting required for this approach and whether this selects for a less fragile subset of TBM (e.g., follicle resident cells) remains to be determined, but the authors point to expression of F4/80 by TBM identified using electron microscopy as confirmation of their findings [163].

Function of GC TBM

Most would agree that the primary function of TBM is clearance of the massive burden of apoptotic centroblasts and centrocytes generated during affinity maturation. Intravital imaging has revealed the extraordinary rate at which GC TBM must eat within the GC to achieve this [128, 130]. Efferocytosis appears almost continual, with each TBM containing on average four apoptotic corpses (or fragments) and undergoing an efferocytic event every 10 min [128]. The number of TBM per GC detected in these studies ranges from 18 to 25 [128, 130], which data modelling suggests is sufficient to remove most dying cells from the GC [128]. For this purpose, TBM adopt an approach that has been termed 'stand-hunting' or 'lazy searching', whereby cellular processes are used to search for and capture apoptotic material while the macrophages remain relatively stationary [128, 130]. Modelling of data from these studies suggests that migration of TBM is only advantageous as a strategy when apoptotic material is at low density, whereas a stationary approach is favoured for the size of TBM and density and motility of apoptotic material observed. Interestingly, the stage of apoptosis at which B cells are eaten by TBM is controversial, with Gurcowicz et al. suggesting that dying B cells are predominantly phagocytosed as intact apoptotic cells at an early stage of apoptosis [128], whereas Grootveld et al. observed fragmentation of B cells followed by capture of motile

ApoBD [130]. Fragmentation of apoptotic GC B cells has also been observed by others [129, 140], and the motility of ApoBD appears to benefit the stand-hunting approach [130]. However, fragmentation of B cells is also evident in close vicinity to TBM, suggesting there may be other advantages to generating ApoBD over uptake of whole cells.

The molecular mechanisms involved in the capture and uptake of GC apoptotic cells are also intricate and suggestive of a tunable and regulatable process. For example, the soluble opsonin MFG-E8 that binds PtdSer induces integrin-mediated clearance by macrophages and is essential for normal clearance of apoptotic GC B cells by TBM. Whereas TBM can bind apoptotic B cells in the absence of MFG-E8, they cannot effectively engulf them [164, 165], and MFG-E8-deficient mice develop expanded GC, splenomegaly, autoantibody production and glomerulonephritis [164, 165]. However, in the LN, MFG-E8 is made by FDC rather than autonomously by TBM [164]. The reason for conferring control of GC efferocytosis to a third party remains unclear, particularly one that has the potential to delete autoreactive B cells should they capture and display autoantigen [166]. TIM-4 is another PtdSer receptor, the absence or blockade of which can lead to elevated production of autoantibodies [167, 168]. TIM-4 is reportedly expressed by GC TBM [169], although this requires confirmation since putative TBM identified by single-cell RNA sequencing appear not to express the gene encoding TIM-4, despite detectable expression by MSM within the same dataset [130]. Accumulation of apoptotic cells within GC and development of autoantibodies have also been observed in mice deficient in MERTK, which otherwise operates mainly via GAS6 that is expressed by most macrophages, including TBM [160, 170, 171].

Overall, these findings suggest the rapid and effective removal of dying B cells by TBM prevents spreading of self-antigen within the GC/LN, particularly to FDC, that may otherwise be recognised by mutated B cells and lead to the development of autoantibody-driven autoimmunity. However, efferocytosis would likely be deficient in numerous other tissue sites in many of the studies described above, and hence, definitive evidence of the importance of GC TBM in the prevention of autoimmune responses is lacking. Impaired uptake of apoptotic cells by TBM has been reported in the GC of patients with systemic lupus erythematosus [172], but whether this is an effect rather than a cause of the disease is unclear, although earlier studies on C1q deficiency support the notion that defective apoptotic cell clearance causes autoimmunity [173]. A non-redundant function for TBM in the rapid clearance of GC B cells is supported by an increase in fragmented DNA detected throughout the GC after timed depletion of TBM in CX3CR1-DTR mice [128]. Unfortunately, the lethal effects of prolonged administration of diphtheria toxin in these mice prevent study of the long-term consequences of TBM depletion on tolerance to self-antigen. Once tolerance is broken to one self-antigen, autoreactive GC generate B cells targeting others [174]. The role, if any, for FDC and TBM in this epitope spreading phenomenon remains unclear. Intriguingly, the effect of single deficiencies in efferocytosis machinery on loss of B-cell tolerance is strain dependent. For example, unlike MFG-E8-null mice on a mixed 129/B6 background [165], significant autoimmune antibody development only arises spontaneously on the C57BL/6 background when mice are doubly deficient in both MFG-E8 and TIM-4 [175]. Study of the underlying reasons for such strain-specific redundancy may identify novel pathways that prevent breakdown of B-cell tolerance. Intriguingly, the breadth of macrophages that express TIM-4 also appears to be highly strain-dependent, with expression seemingly limited to established long-lived resident macrophages in C57BL/6 mice [176–178] but much less restricted in BALB/c mice [179]. As most contemporary studies utilise mice on the BL/6 background, further work is required to determine if the regulation and function of GC TBM are equivalent in other strains.

Origin of GC TBM

The last 15 years have seen a renaissance in our understanding of the diverse mechanisms through which macrophages are maintained in tissues. We now appreciate that in most tissues these cells have much greater longevity and/or capacity for self-renewal than previously thought (reviewed in [180]). For example, SSM largely appear to derive from monocytes that colonise the LN after birth but then largely self-maintain via proliferation in adult mice, both in the steady state and following immunisation [130, 161]. In steady state, TZM also largely derive from monocytes that seed the LN following birth but subsequently persist through a mixture of longevity and slow continual replenishment by monocytes rather than through macrophage proliferation [154, 161]. However, during an immune response, TZM undergo rapid proliferation to populate the growing T-cell zone as the LN swells [154]. These studies highlight the sole importance of proliferation in the expansion of TZM and SSM that occurs following immunisation.

What then of GC macrophages? Although the ontogeny of follicle macrophages prior to an immune response remains unclear, Grootveld et al. recently employed a robust mixture of photo-inducible and tamoxifen-inducible labelling and fate mapping using tissue-protected bone marrow chimaeras to demonstrate that nearly all GC macrophages arise from cells present in the LN prior to immunisation. Although actively dividing TBM have not yet been reported, evidence of the recent proliferation of some TBM was found following daily administration of the proliferation label 5-ethynyl 2′-deoxyuridine post-vaccination [130]. Despite this, the extent of proliferation and whether this occurs prior to the entry of macrophages into the GC is unclear. Indeed, whether mature TBM can proliferate when full of apoptotic cell corpses poses its own physical dilemma. We also do not know whether GC macrophages simply arise from the few follicular macrophages present prior to immunisation or if they migrate from further afield. However, a strong argument can be made that TBM do not arise from MSM or the adjacent SSM despite labelling using the CD169-lineage mapping mouse, since eradication of these cells using either genetic systems ($CD169^{DTR}$) or antibody-mediated blockade of the CSF1R does not affect the number of GC macrophages [128, 130]. Hence, on balance, the two most plausible origins of TBM would seem to be proliferation and differentiation of follicle macrophages and/or proliferation combined with immigration of TZM, which also remain untouched by the strategies that ablate SSM and MSM [128, 130].

ACDC Effector Mechanisms in the Regulation of Lymphoid Follicles

We currently know little about the molecular and cellular interactions that dictate the development, function and survival of GC TBM, but the roles of apoptosis and ACDC have begun to be explored. Infiltration of the follicles by macrophages following immunisation appears to require development of GC rather than general inflammatory cues, since this occurs in parallel with GC development and is absent in mice that are genetically unable to generate GC [128] and also in aged mice which have impaired GC development [181]. Putative TBM from vaccinated mice exhibit a markedly altered transcriptional state compared with the few follicle macrophages present in steady state, with higher expression of genes associated with apoptosis, efferocytosis, and cell motility suggesting they matured to an enhanced efferocytic state that mirrors the greater number of phagocytic vacuoles within them [130]. As ingestion of apoptotic cells drives the ability for continual efferocytosis [121], these changes could arise as a consequence of efferocytosis. In a set of elegant experiments designed to test the role of apoptosis in regulating recruitment and maturation of TBM, Grootveld et al. showed that experimental induction of B-cell apoptosis in the absence of an immune response was sufficient to drive the few macrophages positioned within the follicle to acquire morphological features of TBM, such as greater vacuolation and shortening of their cellular processes [130]. Whether clearance of apoptotic B cells also leads to proliferation of the follicular macrophages or alteration of their transcriptional profile towards that of TBM remains to be evaluated. However, it remains highly possible that the onset of B-cell death could be the trigger of an early phase of macrophage proliferation within the GC, given the potential for efferocytosis to fuel proliferation of macrophages *in vitro* (e.g., see Refs. [116, 120]). Apoptosis of B cells may also trigger recruitment of macrophages into the developing GC. In particular, apoptotic B cells release high levels of the 'find-me' signal CX3CL1 [89], the receptor for which, CX3CR1, is expressed by most macrophages within the LN, including TBM (see above). CX3CL1 appears important for recruitment and/or maintenance of follicle macrophages during an immune response, as fewer of these cells are found in the spleen of immunised mice that lack CX3CR1 [89]. However, some questions over the generality of this requirement across models remain (see below).

As mentioned, an intriguing feature that distinguishes follicle and GC macrophages from SSM and MSM is their lack of dependence on CSF1R signalling for survival and generation [130, 161]. Indeed, these cells appear not to express CSF1R protein [162] despite expressing *Csf1r* mRNA [130]. Furthermore, immunostaining confirms a distinct lack of CSF1-producing cells within follicles [161]. This lack of expression and dependence upon CSF1R demarks TBM as highly unusual compared with macrophages in most tissues, which largely rely on production of CSF1 by a network of 'niche' cells for survival and proliferation (reviewed in [182, 183]). Exactly what signals or processes sustain follicle macrophages following immunisation remains unclear. Notably, CSF1R-independent proliferation of tissue resident macrophages can be triggered by Th2 cells during helminth infection via production of IL-4 [184, 185]. However, whereas IL-4 produced by Th2 cells during worm infection can permeate the entire draining LN in a non-cognate manner [186], IL-4 produced by TFH following immunisation is largely restricted to cognate interactions due to FDC, which sequester free cytokine in order to limit activation of bystander B cells and enable affinity maturation [187]. Hence, TFH-derived IL-4 seems an unlikely candidate to affect TBM dynamics following vaccination. CSF2 can also stimulate survival and proliferation of macrophages in a tissue/inflammation-dependent manner (reviewed in Ref. [182]), and intriguingly, aged CSF2-deficient mice develop a systemic lupus erythematosus-like disorder that may be linked to impaired phagocytosis of apoptotic cells [188]. However, no direct evidence for a role of CSF2 in regulating follicle and GC macrophages has yet been reported. Another factor that can drive context-dependent survival of macrophages is CX3CL1 [189], which, as discussed, is released in abundance by apoptotic B cells [89]. Hence, the massive burden of apoptotic B cells generated within the GC could itself fuel the survival as well as recruitment of the macrophages required for its disposal, meaning the capacity to eat could be intrinsically regulated by the need to eat. Whether CX3CL1 could replace the need for growth factors like CSF1 that drive AKT activation is unclear, but as discussed above, there is substantial evidence that efferocytosis *per se* can support the survival of macrophages and monocytes *in vitro* and *in vivo*, including via AKT and ERK1/2 phosphorylation [190, 191]. However, this idea is challenged by recent findings that CX3CR1-deficient mice have normal numbers of GC macrophages in LN [130]. The reason for the conflicting data between the Truman and Grootveld studies is unclear. Distinct requirements for CX3CR1 between the spleen and the LN or the different background strains of mice that were used in these studies are one possibility. Alternatively, Truman et al. assessed numbers using immunohistochemical detection of $CD68^+$ cells, while Grootveld enumerated *Cx3cr1*$^{GFP+}$ cells by immunofluorescence microscopy, and neither study provided further confirmation of the phenotype or morphology of these cells. As both CD68 and CX3CR1 are expressed by monocytes, it is possible that compensatory and continual recruitment of immature cells could be detected with either method. Interestingly, efficient engulfment of apoptotic corpses also does not appear essential for maintenance of TBM since *Mertk*$^{-/-}$ mice exhibit normal numbers of $CD68^+$ cells within GC [160]. However, these findings do not exclude a role for other ACDC factors in driving survival of GC macrophages or the possibility that compensatory recruitment of monocytes also occurs in these mice.

Like follicle macrophages, TZM also appear independent of CSF1R signalling [130, 161, 192]. However, they also do not require CX3CR1 expression for survival in the steady state despite utilising this receptor to detect and/or respond to naïve apoptotic T cells [154]. Whether CX3CR1 and apoptosis of T cells is required for the expansion and proliferation of TZM during an immune response has not been tested. Similarly, no data yet exists on the dependence of MCM on CSF1 or the role that ACDC may play in their development. However, substantial numbers of $F4/80^+$ macrophages are reported in the medullary regions of CSF1-deficient op/op mice [192, 193], which likely represent MCM given the dependence of MSM on CSF1 [161]. Hence, it is provocative to consider that all regions of the LN in which apoptosis occurs at a high level during an immune response appear populated by CSF1-independent macrophages.

As we have discussed, TBM are not confined to lymphoid follicles. Focussing on the immune system more generally, occasional TBM, defined as macrophages containing stainable cellular debris, have been described in the thymus [194], where it is estimated that 95% of thymocytes undergo apoptosis at some point following somatic recombination of their T-cell receptor genes [195, 196]. This is greatly exaggerated following treatment with immunotoxic xenobiotics such as dexamethasone, which cause substantial synchronised apoptosis of thymocytes and rapid (<3 h) appearance of numerous engorged TBM-like macrophages within the thymic cortex, which give a 'starry sky' appearance to histological sections [197]. Notably, mice that lack both MERTK and AXL appear to have fewer thymic macrophages, suggesting a potential role for ACDC in the maintenance of these cells [198]. We are not the first to point out that TBMs are not unique to GC [158, 199]. Within secondary lymphoid tissue, TZM also undertake relatively continual efferocytosis, with 80% of these cells containing efferocytic vacuoles positive for DNA, active caspase-3 and TUNEL staining in the steady state [154]. How different or overlapping the transcriptional signature of TZM and GC TBM remains to be seen, but notably, TZM do not undertake efferocytosis in sufficient quantities to exhibit more than one or two efferocytic vacuoles, even during the contraction phase of an immune response when the rate of T cell apoptosis rises [154, 200]. Hence, to what degree the high rate of efferocytosis of GC TBM determines their transcriptional identity and the degree to which differences between GC TBM cells and other efferocytic macrophage populations are a product of their distinct rates of efferocytosis versus distinct biochemical signals within their respective microenvironments remains to be determined. However, when addressing these questions, we should also consider the intriguing possibility that the quality of the apoptotic meal (e.g., B cells vs. T cells or thymocytes) may itself also be important, as suggested by recent *in vitro* studies [201]. By extension, we should also consider whether the effects of apoptotic fragments differ from those of intact apoptotic cells and to what extent fragmentation of apoptotic cells differs between microanatomical sites.

In summary, efferocytosis is currently the main established function of GC macrophages, and these cells may arise simply from the need for efferocytosis. Thus, efferocytosis appears to be the key process that both programmes and sustains these cells within the GC. As highlighted by Baudon and Bajenoff [202], another important question is whether the TBM of GC are also involved in the recycling of components of the apoptotic cells, mainly lymphocytes, they engulf and whether this fuels further proliferation within the GC. This was first proposed in 1957 by Oswald Trowell, who also noted earlier observations that GC with the highest numbers of phagocytosed tingible bodies (which subsequently proved to be apoptotic lymphocytes) also had the highest rates of mitosis. Trowell suggested that GC-based lymphopoiesis is actively driven by reutilisation of DNA from phagocytosed dead and degraded lymphocytes [203].

Perhaps of importance to ACDC, a recent study indicates that nociceptive sensory nerve signalling promotes splenic GC reactions and anti-influenza antibody responses in mice. Antigen-induced accumulation of PGE2, a known ACDC mediator, appears to play an initiating role in activation of nociceptors, which then directly influence B cell responses through release of calcitonin gene-related peptide [204]. These findings open up the possibility that TBM could also be directly affected by neurotransmitters during the GC reaction.

ACDC in Starry Sky B-Cell NHL

The Apoptosis Paradox in Cancer and Its Special Importance in Starry Sky Malignancies

Although there is little doubt that apoptosis acts as an effective tumour suppressive mechanism, curiously the rates of apoptosis in aggressive tumours are commonly higher than those seen in more benign cancers. This is true of many different cancer types, including colorectal, cholangiocellular, pancreatic and breast carcinomas, glioblastoma and NHL. Furthermore, at least anecdotally, apoptosis is associated with a propensity for metastasis in malignant cell populations. Apoptosis is a well-known hallmark of highly proliferative cell populations, and accumulating experimental evidence suggests that, in addition to its anti-cancer properties, apoptosis also has pro-oncogenic characteristics which operate through ACDC in the TME (reviewed in [1, 5]). We have already discussed this in relation to the effector caspase-dependent activation of iPLA2 in producing a key ACDC effector lipid, PGE2, which operates in skin repair and liver regeneration. PGE2 is also implicated in ACDC mechanisms driving proliferation and angiogenesis in cancer models [47, 49–52]. Further recent evidence for oncogenic ACDC has been reported in oral squamous cell carcinoma, in which higher apoptosis is associated with more highly proliferative tumours and more aggressive clinical disease. Conditioned medium from cancer-associated fibroblasts co-cultured with apoptotic oral squamous cell carcinoma cells was found to promote proliferation of tumour cells. Although *in vivo* studies have not yet been reported, mechanistically, this pro-oncogenic ACDC activity appears to be associated with decreased STING/IRF3 phosphorylation, reducing production of type I interferons [205]. The potential for ACDC to mediate the apoptosis paradox by contributing to the pathogenesis of starry sky tumours is especially tantalising since, in these cases, the high apoptotic burden of the tumour cell populations and the consequent communication with large numbers of tumour-associated macrophages (TAM) are obvious in standard histological preparations. Next, we consider the evidence for oncogenic ACDC activities, especially in the context of starry sky B-cell malignancies, prototypically Burkitt's lymphoma (BL).

Starry Sky Malignancies of B Cells and Other Lineages

Although in pathology the term 'starry sky' has been used to describe patterns in various lesions, such as those appearing on magnetic resonance images of neurotuberculosis [206] and neurocysticercosis [207], its original and indeed most renowned use is in descriptions of standard histological preparations of aggressive NHL. Here, the bright 'stars' observed in a 'sky' of densely packed and darkly stained highly proliferative lymphoma cells proved to be the TAM that accumulate to efferocytose the large numbers of apoptotic tumour cells that are generated *in situ*. According to Dy-Ledesma and colleagues, the first recorded description of a starry sky cancer appears to have been in the late 1950s by Gall and Rappaport, who used the term to detail the histological picture of BL [208], which had been reported at this time in African children by the missionary surgeon, Denis Burkitt [209, 210]. Burkitt correctly identified cases of this tumour, which commonly presented as large jaw masses with variable abdominal, thoracic and central nervous system dissemination, as a single pathological entity, and these were later found to be extranodal lymphomas showing the starry sky histological picture, also described as the 'water pot' effect [211, 212], most likely referring to the perforated sprinkler patterns of plant watering vessels [213]. Subsequently, Burkitt's tumour, or BL as it came to be known, was described in humans in other parts of the world as well as in dogs, cats and mice (see [214] and references therein). Readers are referred to excellent reviews on BL and its history in which the importance of this tumour in the development of our understanding of the molecular biology and epidemiology of cancer as well as the discovery and role of the oncogenic Epstein-Barr virus in the African form of the disease are discussed in depth [215, 216]. Here we focus particularly on the significance of the starry sky architecture and on the (potential) significance of ACDC in BL pathogenesis with consideration of additional cancers—haematological and others—in order that we may begin to develop general concepts of ACDC mechanisms in cancer biology.

Table 1 illustrates the broad diversity of tumour types amongst human and experimental murine malignancies in which the starry sky picture—minimally comprising frequent apoptotic tumour cells and efferocytic macrophages—features. Examples include B-cell lymphomas of both GC and non-GC origin, T-cell lymphomas, myeloid malignancies and some non-haemopoietic cancers. While B-cell malignancies are the most common and best known, the broad diversity of cancers that can exhibit a starry sky pattern indicates that starry sky TME architecture is neither inextricably linked with B-cell biology nor indeed with lymphomas or haematological cancers more generally. Furthermore, neoplasias with starry sky features often show focal rather than widely prominent starry sky zones and may involve lymph nodes or may appear extranodally [208]. It is noteworthy that diagnosis of rare lymphomatous neoplasms in tissues such as the thyroid and the central nervous system can be aided by starry sky patterning, presumably as a consequence of the predilection of lymphomas to generate such TME zones [217, 218]. A starry sky pattern (SSP) can also aid early diagnosis of epithelioid angiosarcoma, a highly aggressive malignancy [219]. What, then, drives development of the starry sky TME? Association with *c-MYC* deregulation has been suggested [208, 220], but starry sky architecture has been reported in diverse cancers harbouring a range of driver oncogenes, including malignancies without overt *c-MYC* mutations. However, since aberrant MYC expression is common in many tumours, it remains conceivable

Table 1 Starry sky pattern (SSP) in haematological and non-haematological cancers

Cancer	Cell Lineage	Comments	SSP Frequency	References
BL	B cell	Prototypical starry sky malignancy; GC origin	Most	[222]
λ-Myc BLL	B cell (mouse)	BL-like but pre-GC	Most	[223, 224]
LMP2A/λ-Myc BLL	B cell (mouse)	BL-like but pro-B/pre-B	Most	[225]
iMyc$^{E\mu}$ BLL	B cell (mouse)	BL-like but pre-GC	Most	[224]
DLBCL	B cell	GC and non-GC origin	5-10%	[208]
HGBL	B cell	Intermediate between DLBCL and BL; double- and triple-hit lymphomas	Common in double-hit	[208]
MCL	B cell	Aggressive variants	Common in blastoid variants	[208]
PL	B cell (plasma cell)	Post-GC; starry sky often prominent	Common	[208]
ALK$^+$ large B-cell lymphoma	B cell	Immunoblastic or plasmablastic	Common when the proliferation rate is high	[208]
B-LBL	B cell	Can present as ALL; starry sky may be linked to poorer prognosis	Uncommon	[208, 226]
T-LBL	T cell	Focal or prominent starry sky	Uncommon	[208]
PTCL-NOS	T cell	Starry sky indicative of high proliferation rate	Uncommon; may be seen when the proliferation rate is high	[208]
ALCL	T cell (or null cell)	Both ALK$^+$ and ALK$^-$ types can show a starry sky	Observed when tumour cell turnover is high	[208]
EL4	T cell (mouse)	Starry sky induced by drug treatment	Common following FBZ treatment	[227]

(continued)

Table 1 (continued)

Cancer	Cell Lineage	Comments	SSP Frequency	References
Myeloid sarcoma	Myeloid	Starry sky is the most common in the blastic subtype	Uncommon	[208]
AML	Myeloid	NPM1-mutated AML starry sky bone marrow	Uncommon	[228]
Neuroblastoma	Neuron	Undifferentiated subtype	Rare	https://www.iccr-cancer.org/wp-content/uploads/2023/11/ICCR-Neuroblastoma-1st-ed-v1.0-bookmark.pdf
Medulloblastoma	Neuron	Large cell/anaplastic variant	Variant restricted	[229]
Dysgerminoma	Ovary (dog)	Seminal description	Rare	[230]
Angiosarcoma	Dermal endothelium	High proliferation	Rare	[231]

ALCL Anaplastic large-cell lymphoma, *ALK* Anaplastic lymphoma kinase, *AML* Acute myeloid leukaemia, *B-LBL* B-lymphoblastic leukaemia/lymphoma, *BL* Burkitt lymphoma, *BLL* Burkitt-like lymphoma, *DLBCL* Diffuse large B-cell lymphoma, *HGBL* High-grade B-cell lymphoma, *MCL* Mantle cell lymphoma, *PL* Plasmablastic lymphoma, *PTCL-NOS* Peripheral T-cell lymphoma, not otherwise specified, *SSP* Starry sky pattern, *T-LBL* T-lymphoblastic leukaemia/lymphoma

that a correlation exists. It seems that the only consistent feature that correlates with the appearance of the SSP is a high tumour cell turnover rate (Table 1), a scenario that aligns with the normal physiology of the GC. In support of this, it has been demonstrated that inhibition of apoptosis in xenografts of BL abolished their starry sky architecture and reduced the lymphoma cell proliferation rate in parallel [221]. Thus, ACDC is likely to lie at the heart of starry sky patterning, and emerging evidence strongly suggests that such apoptosis-driven intercellular networking has oncogenic potential.

ACDC Effector Mechanisms in Starry Sky B-Cell Tumours

Since (i) macrophage efferocytosis in starry sky B-cell lymphoma is arguably the most prominent example of apoptotic cell clearance in humans, and (ii), as we have discussed, macrophages are actively programmed by apoptotic cells to promote reparatory and regenerative responses, it follows that apoptotic tumour cells interacting with macrophages may provide important pro-tumour signals in the genesis and evolution of starry sky lymphomas. Furthermore, major known functional attributes of TAM in many types of established tumours—including growth-promoting, angiogenic and immunosuppressive activities—are features that are substantially driven in macrophages by apoptotic cells in many different normal tissue contexts. Therefore, the hijacking of ACDC signals otherwise involved in homeostatic tissue control by efferocytes represents a compelling theoretical contributory factor in the pathogenesis of malignant diseases, not least starry sky B-cell NHL. While ample scope remains for adding mechanistic details, some of the underlying principles of ACDC in starry sky B-cell tumours have begun to be established through seminal studies in BL and in BL-like mouse models [221], which may be summarised by two major lines of evidence:

1. **The starry sky architecture of the TME is determined by the availability of large numbers of apoptotic cells.** Intriguingly, this ACDC effect works across species, as demonstrated by the ability of orthotopic starry sky tumours to develop in severe combined immunodeficiency mice after administration of apoptosis-competent human BL cells, indicating that the accumulation of TAM in these tumours is orchestrated by conserved ACDC mechanisms. The evidence also reveals that TAM accumulation includes stimulation of macrophage proliferation as part of this ACDC effect [221]. These causative associations between apoptosis and TAM accumulation reflect the strong correlation between these two parameters observed in cases of human BL.
2. **Lymphoma cell proliferation, angiogenesis and pro-tumour (M2-like) reparatory macrophage polarisation are all promoted by ACDC.** Notably, laser capture microdissection of overtly efferocytic SS-TAM permitted their *in situ* transcriptome to be determined for the first time, and pathway analyses demonstrated enhancement of phagocytosis, wound healing, lipid metabolism and matrix remodelling. Efferocytosis and anti-inflammatory genes such as *Mertk*, *Axl*, *Gas6*, *Cd91*, *Lgals3* (for Galectin-3) and *Tgfb1* were found to be upregulated in SS-TAM through ACDC [221], and later, *Mertk*$^{-/-}$ and *Lgals3*$^{-/-}$ mice proved to be compromised in their capacity to host starry sky BLL (λ-Myc) lymphoma establishment and growth [232, 233]. Apoptotic cell-engaging SS-TAM of BL-like mouse lymphomas bore transcriptional profiles indicative of their potential to produce growth factors (e.g., IGF-1 and PDGF-CC) and to drive angiogenesis and tissue remodelling/expansion via production of CD13, Galectin-3, GPNMB, Heme Oxygenase-1, TIMP-2 and matrix metalloproteinases MMP-2, -3 and -12. Intriguingly, both human BL and mouse λ-Myc lymphoma cells were also themselves activated to express and process MMPs upon apoptosis induction [221], suggesting for the first time that apoptosis may directly endow tumour cell populations with tissue invasive functions.

Additional suggestive evidence of ACDC being proactive in supporting starry sky lymphoma growth is provided by early work showing that rosette formation between starry sky macrophages and lymphoma cells [234, 235] may be important in trophic crosstalk aiding tumour growth and facilitating rapid responses to ACDC, including efferocytosis, macrophage reprogramming and proliferation. Intriguingly, these studies suggested that TBM are not able to phagocytose other particles such as bacteria, latex or carbon particles [235], implying that ACDC dominantly controls macrophage phagocytosis. A further notable example suggestive of ACDC 'in action' in lymphoma growth is provided by a recent report that Fenbendazole, a benzimidazole microtubule-targeting anti-parasitic drug which has anti-cancer potential, can promote the proliferation of murine EL4 lymphoma cells in tumours *in vivo*, despite strongly showing apoptosis-inducing effects against EL4 cells *in vitro* [227]. EL4 (RRID:CVCL_0255) is a long-established mouse precursor T cell lymphoblastic lymphoma/leukaemia line (https://www.cellosaurus.org/CVCL_0255), which produces densely packed lymphomas consisting mainly of large, proliferating immature cells when transplanted into syngeneic (C57BL/6) animals. Upon treatment with fenbendazole, the tumours become replete with M2-like, SS-TAM, grow at least as well as untreated control tumours, and display higher numbers of mitotic figures than the controls [227]. This picture parallels that of BL in which, as we have discussed, apoptosis of B-lymphoma cells drives the tumour-supportive accumulation and polarisation of SS-TAM. Thus, the results of Jung and colleagues [227] are consistent with our view that apoptosis and its effects on TAMs provide key oncogenic factors in the TME through ACDC. Further reports indicate that, although histological SSP becomes more marked with advancement of disease in B-cell NHL and is associated with poorer prognosis [226], at the same time the SSP is linked with favourable responses to R-CHOP (Rituximab, cyclophosphamide, doxorubicin, vincristine and prednisolone) treatment [236]. If this association is in any way causative, one conceivable mechanism is that the increased presence of SS-TAM in SSP lymphomas provides the opportunity for these increased numbers of pro-tumour macrophages to become switched to a tumour-suppressive phenotype by Rituximab, which is known to impart such activity [237].

Conclusions and Future Perspectives

The interactions between apoptotic cells and macrophages in normal and malignant lymphoid microenvironments provide much food for thought concerning the roles of ACDC in tissue turnover in health and disease, raising more questions than answers. To conclude our perspective, we highlight the following as important interrelated questions and topics for future research into ACDC and its impacts on tissues:

1. Do TBM have special properties in ACDC? As we have discussed, in essence, TBM are macrophages that overtly contain the remnants of apoptotic cells and, as such provide a sign of an efferocytically challenging environment. Other macrophages are engaged less obviously in efferocytosis in locations of lesser apoptotic cell load, such as the paracortex (T zone) of lymph nodes [154] and in otherwise starry sky lymphomas that have been altered by radical reduction of the levels of apoptosis through BCL2 overexpression [221]. Thus, it remains

conceivable that TBM are programmed similarly to those efferocytic macrophages that handle a lower burden of corpse clearance, including in steady state.

2. How extensive is the range of responses in ACDC? It is intriguing that apoptotic cells sustain certain metabolic activities and are capable of transferring biologically active metabolites to influence cells in their neighbourhood and beyond. Not least are metabolites generated by the evolutionarily conserved polyamine pathway that controls cell proliferation, differentiation, survival and death. Of particular note is the seminal finding that spermidine, the most common polyamine metabolite of mammalian cells, is substantially secreted from apoptotic cells of various lineages, its release during apoptosis being established as the first physiological process known to produce extracellular spermidine [58]. Since spermidine has fundamental and pleiotropic properties, including the capacity to orchestrate either pro- or anti-tumour effects [238], the consequences of its release during apoptosis in normal and malignant tissues merit extensive investigation. The potential of horizontal transfer of nucleic acids in ACDC holds similar intrigue.
3. Why do apoptotic cells engender so many recognition pathways? Even considering PtdSer exposure alone, a panoply of PtdSer-binding receptors and opsonins are deployed in ACDC. Amongst alternative possibilities, this seemingly broad requirement for recognition of PtdSer in ACDC may be illustrative of failsafe mechanisms, of tissue context-dependent regulatory processes, or of responses to EV as well as apoptotic cells. MERTK appears to be involved in multiple ACDC roles as a specific receptor that can orchestrate phagocytic and immunoregulatory responses through engagement with PtdSer-binding proteins. It is important to appreciate that much of the published literature on MERTK functions *in vivo* utilising the original *Mertk* knockout mice—not least the apparently co-regulated roles of MERTK in suppressing inflammation and promoting phagocytosis of apoptotic cells and cell debris—needs to be reevaluated because of the recent realisation that these animals carry additional changes in their genome affecting the expression of other genes in addition to *Mertk*. To make these mice, 129 strain embryonic stem cells (as were state-of-the-art at the time) were used for *Mertk* targeting, and these were injected into C57BL/6 (B6) blastocysts, the resulting 129/B6 chimaeras being extensively backcrossed to B6. The animals were widely used in *in vivo* studies, including retinal degeneration, autoimmunity and cancer immunoregulation, which showed them to display remarkable coupling of downregulated efferocytic and debris-removal phenotypes with upregulated pro-inflammatory signalling. In recent years, however, it has been established that new *Mertk*$^{-/-}$ lines created purely using B6 embryonic stem cells failed to corroborate these findings due to the effects of tissue-dependent modifier genes of 129 origin that contaminated the original 129-targeted *Mertk* knockouts [239]. Therefore, it is recommended that *in vivo* studies of TAM receptor biology be reevaluated using the pure B6 knockouts of *Mertk*, along with those of its paralogues *Tyro3* and *Axl* [105]. This includes the role of MERTK in ACDC signalling in normal lymphoid tissue and in starry sky lymphomas and other malignancies.

4. How similar are different apoptotic cell types in ACDC? Most studies to date have assumed that 'all apoptotic cell types are equal', at least in terms of their effects on efferocytic macrophages. Recent work has begun to challenge this assumption, pointing to different apoptotic cell types asserting diverse responses in recipient macrophages [201]. The possibility that different lineages of apoptotic cells, especially in dissimilar tissue contexts, may differentially control a spectrum of ACDC signalling outcomes is tantalising.
5. Which cells in different tissue settings respond to ACDC, and what are the consequences? This fundamental and challenging question includes (i) the relative importance of so-called professional and non-professional phagocytes and (ii) the functional effects of ACDC that are unrelated to efferocytosis or phagocytes. Of special relevance to this question is a recent report on the mammalian hair cycle in which the authors propose a finely balanced tuning of efferocytic hair follicle stem cells (non-professional phagocytes) with their regenerative functional activities [12].
6. To what extent does ACDC occur during sublethal activation of components of the apoptosis machinery? It is known that caspases, including effector caspases, have non-apoptotic functions in addition to their roles in apoptosis. Furthermore, apoptotic stress without cell death can occur following sublethal mitochondrial outer membrane permeabilisation (reviewed recently by [240] and Chapter "Molecular Cell Biology of Apoptosis in Health and Disease", this volume), and such cells can promote survival of neighbours through release of fibroblast growth factor-2 [241]. While this example of 'ACDC' following sublethal apoptotic stress has been described in the context of cancer, it seems probable that 'ACDC' in the absence of cell death may also occur in normal tissue physiology, especially since apoptotic programme components can regulate fundamental vital processes in normal tissues [242]. Of particular relevance to the present perspective is the knowledge that effector caspase-3 regulates B-cell cycling [243].
7. How important are intercellular signals generated by non-apoptotic cell death modalities? Several alternative regulated cell death routines have been described, including autophagy (which also functions in cell survival), necroptosis, pyroptosis and others [244], which can impact tissue dynamics and outcomes. For example, the PGE2-driven 'phoenix rising' ACDC pathway [47] has a counterpart in pyroptosis, recent work showing that oxylipins, notably PGE2 and metabolites released during pyroptosis, promote wound healing and tissue regeneration [245]. Similarly, necroptosis can create an immunosuppressive TME through secretion of IL-1α by tumour cells, resulting in local reprogramming of macrophages, monocytes and neutrophils and consequent suppression of T-cell recruitment and effector function [246]. Even necrosis caused by tissue damage has reported tissue repair-inducing properties [247] and can signal through apoptosis induction and ACDC: Thus, in Drosophila, long distance, JNK-independent signalling from necrotic cells in the imaginal disc causes apoptosis that is required for subsequent regeneration [248].

A quarter of a century ago, John Savill and Valerie Fadok made the persuasive argument that 'corpse clearance defines the meaning of cell death' [249]. Here we propose that 'ACDC defines the meaning of apoptosis' and its implications and consequences for the organism. Much work is urgently needed to define the molecular mechanisms underlying ACDC and how they impact tissue physiology and pathology. Furthermore, the future holds much promise in unravelling the communicative mechanisms of other cell-death modalities and how they relate to ACDC.

Acknowledgements The authors' research is supported by grants from the Medical Research Council (UKRI), Cancer Research UK and the Engineering and Physical Sciences Research Council (UKRI): MR/W019264/1; EDDCPJT\100003. We are grateful to colleagues in the Edinburgh Cell Death Collective (ECDC) for discussion, especially to Will Wood for constructive suggestions for Fig. 3. For the purpose of open access, the authors have applied a Creative Commons Attribution (CC BY) licence to any Author Accepted Manuscript version arising from this submission.

References

1. Gregory, C. D. (2023). Hijacking homeostasis: Regulation of the tumor microenvironment by apoptosis. *Immunological Reviews, 319*(1), 100–127.
2. Taylor, R. C., Cullen, S. P., & Martin, S. J. (2008). Apoptosis: controlled demolition at the cellular level. *Nature Reviews. Molecular Cell Biology, 9*(3), 231–241.
3. Gregory, C. D. (2014). Cell biology: The disassembly of death. *Nature, 507*(7492), 312–313.
4. deCathelineau, A. M., & Henson, P. M. (2003). The final step in programmed cell death: phagocytes carry apoptotic cells to the grave. *Essays in Biochemistry, 39*, 105–117.
5. Morana, O., Wood, W., & Gregory, C. D. (2022). The Apoptosis paradox in cancer. *International Journal of Molecular Sciences, 23*(3).
6. Green, D. R. (2018). *Cell Death. Apoptosis and other means to an end.* Cold Spring Harbor Laboratory Press.
7. Goping, I. S., Barry, M., Liston, P., Sawchuk, T., Constantinescu, G., Michalak, K. M., et al. (2003). Granzyme B-induced apoptosis requires both direct caspase activation and relief of caspase inhibition. *Immunity, 18*(3), 355–365.
8. Nagata, S. (2018). Apoptosis and clearance of apoptotic cells. *Annual Review of Immunology, 36*, 489–517.
9. Singh, R., Letai, A., & Sarosiek, K. (2019). Regulation of apoptosis in health and disease: the balancing act of BCL-2 family proteins. *Nature Reviews. Molecular Cell Biology, 20*(3), 175–193.
10. Vitale, I., Pietrocola, F., Guilbaud, E., Aaronson, S. A., Abrams, J. M., Adam, D., et al. (2023). Apoptotic cell death in disease-Current understanding of the NCCD 2023. *Cell Death and Differentiation, 30*(5), 1097–1154.
11. Fuchs, Y., & Steller, H. (2015). Live to die another way: modes of programmed cell death and the signals emanating from dying cells. *Nature Reviews. Molecular Cell Biology, 16*(6), 329–344.
12. Stewart, K. S., Abdusselamoglu, M. D., Tierney, M. T., Gola, A., Hur, Y. H., Gonzales, K. A. U., et al. (2024). Stem cells tightly regulate dead cell clearance to maintain tissue fitness. *Nature, 633*(8029), 407–416.
13. Ankawa, R., Goldberger, N., Yosefzon, Y., Koren, E., Yusupova, M., Rosner, D., et al. (2021). Apoptotic cells represent a dynamic stem cell niche governing proliferation and tissue regeneration. *Developmental Cell, 56*(13), 1900–16 e5.

14. Brock, C. K., Wallin, S. T., Ruiz, O. E., Samms, K. M., Mandal, A., Sumner, E. A., et al. (2019). Stem cell proliferation is induced by apoptotic bodies from dying cells during epithelial tissue maintenance. *Nature Communications, 10*(1), 1044.
15. Essien, S. A., Ahuja, I., & Eisenhoffer, G. T. (2024). Apoptotic extracellular vesicles carrying Mif regulate macrophage recruitment and compensatory proliferation in neighboring epithelial stem cells during tissue maintenance. *PLoS Biology, 22*(11), e3002194.
16. Valon, L., Davidovic, A., Levillayer, F., Villars, A., Chouly, M., Cerqueira-Campos, F., et al. (2021). Robustness of epithelial sealing is an emerging property of local ERK feedback driven by cell elimination. *Developmental Cell, 56*(12), 1700–11 e8.
17. Poon, I. K. H., Parkes, M. A. F., Jiang, L., Atkin-Smith, G. K., Tixeira, R., Gregory, C. D., et al. (2019). Moving beyond size and phosphatidylserine exposure: Evidence for a diversity of apoptotic cell-derived extracellular vesicles in vitro. *Journal of Extracellular Vesicles, 8*(1), 1608786.
18. Gregory, C. D., & Rimmer, M. P. (2023). Extracellular vesicles arising from apoptosis: Forms, functions, and applications. *The Journal of Pathology, 260*(5), 592–608.
19. Savill, J. S., Wyllie, A. H., Henson, J. E., Walport, M. J., Henson, P. M., & Haslett, C. (1989). Macrophage phagocytosis of aging neutrophils in inflammation – Programmed cell-death in the neutrophil leads to its recognition by macrophages. *The Journal of Clinical Investigation, 83*(3), 865–875.
20. Wyllie, A. H. (1981). Cell death: A new classification separating apoptosis from necrosis. In I. D. Bowen & R. A. Lockshin (Eds.), *Cell death in biology and pathology* (pp. 9–34). Chapman and Hall.
21. Hoijman, E., Hakkinen, H. M., Tolosa-Ramon, Q., Jimenez-Delgado, S., Wyatt, C., Miret-Cuesta, M., et al. (2021). Cooperative epithelial phagocytosis enables error correction in the early embryo. *Nature, 590*(7847), 618–623.
22. Kira, A., Tatsutomi, I., Saito, K., Murata, M., Hattori, I., Kajita, H., et al. (2023). Apoptotic extracellular vesicle formation via local phosphatidylserine exposure drives efficient cell extrusion. *Developmental Cell, 58*(14), 1282–98 e7.
23. Savill, J. (1995). The innate immune system: Recognition of apoptotic cells. In C. D. Gregory (Ed.), *Apoptosis and the immune response* (pp. 341–369). Wiley-Liss.
24. van den Eijnde, S. M., Boshart, L., Baehrecke, E. H., De Zeeuw, C. I., Reutelingsperger, C. P., & Vermeij-Keers, C. (1998). Cell surface exposure of phosphatidylserine during apoptosis is phylogenetically conserved. *Apoptosis, 3*(1), 9–16.
25. Haka, A. S., Barbosa-Lorenzi, V. C., Lee, H. J., Falcone, D. J., Hudis, C. A., Dannenberg, A. J., et al. (2016). Exocytosis of macrophage lysosomes leads to digestion of apoptotic adipocytes and foam cell formation. *Journal of Lipid Research, 57*(6), 980–992.
26. Kerr, J. F., Wyllie, A. H., & Currie, A. R. (1972). Apoptosis: A basic biological phenomenon with wide-ranging implications in tissue kinetics. *British Journal of Cancer, 26*(4), 239–257.
27. Gregory, C. D., & Devitt, A. (2004). The macrophage and the apoptotic cell: An innate immune interaction viewed simplistically? *Immunology, 113*(1), 1–14.
28. Arandjelovic, S., & Ravichandran, K. S. (2015). Phagocytosis of apoptotic cells in homeostasis. *Nature Immunology, 16*(9), 907–917.
29. Boada-Romero, E., Martinez, J., Heckmann, B. L., & Green, D. R. (2020). The clearance of dead cells by efferocytosis. *Nature Reviews. Molecular Cell Biology, 21*(7), 398–414.
30. Shankman, L. S., Fleury, S. T., Evans, W. B., Penberthy, K. K., Arandjelovic, S., Blumberg, R. S., et al. (2021). Efferocytosis by Paneth cells within the intestine. *Current Biology, 31*(11), 2469–76 e5.
31. Orlando, K. A., Stone, N. L., & Pittman, R. N. (2006). Rho kinase regulates fragmentation and phagocytosis of apoptotic cells. *Experimental Cell Research, 312*(1), 5–15.
32. Atkin-Smith, G. K., Miles, M. A., Tixeira, R., Lay, F. T., Duan, M., Hawkins, C. J., et al. (2019). Plexin B2 Is a regulator of monocyte apoptotic cell disassembly. *Cell Reports, 29*(7), 1821–31 e3.

33. Tixeira, R., Phan, T. K., Caruso, S., Shi, B., Atkin-Smith, G. K., Nedeva, C., et al. (2020). ROCK1 but not LIMK1 or PAK2 is a key regulator of apoptotic membrane blebbing and cell disassembly. *Cell Death and Differentiation, 27*(1), 102–116.
34. Cao, L., Ma, L., Zhao, J., Wang, X., Fang, X., Li, W., et al. (2023). An unexpected role of neutrophils in clearing apoptotic hepatocytes in vivo. *eLife*, 12.
35. Gregory, C. D., Ford, C. A., & Voss, J. J. (2016). Microenvironmental effects of cell death in malignant disease. *Advances in Experimental Medicine and Biology, 930*, 51–88.
36. Ankersmit, H. J., Hoetzenecker, K., Dietl, W., Soleiman, A., Horvat, R., Wolfsberger, M., et al. (2009). Irradiated cultured apoptotic peripheral blood mononuclear cells regenerate infarcted myocardium. *European Journal of Clinical Investigation, 39*(6), 445–456.
37. Haider, T., Hoftberger, R., Ruger, B., Mildner, M., Blumer, R., Mitterbauer, A., et al. (2015). The secretome of apoptotic human peripheral blood mononuclear cells attenuates secondary damage following spinal cord injury in rats. *Experimental Neurology, 267*, 230–242.
38. Vorstandlechner, V., Copic, D., Klas, K., Direder, M., Golabi, B., Radtke, C., et al. (2023). The secretome of irradiated peripheral mononuclear cells attenuates hypertrophic skin scarring. *Pharmaceutics., 15*(4).
39. Gregory, C. D., & Dransfield, I. (2018). Apoptotic tumor cell-derived extracellular vesicles as important regulators of the onco-regenerative niche. *Frontiers in Immunology, 9*, 1111.
40. Holmgren, L. (2010). Horizontal gene transfer: You are what you eat. *Biochemical and Biophysical Research Communications, 396*(1), 147–151.
41. Tsering, T., Nadeau, A., Wu, T., Dickinson, K., & Burnier, J. V. (2024). Extracellular vesicle-associated DNA: Ten years since its discovery in human blood. *Cell Death & Disease, 15*(9), 668.
42. Lauber, K., Bohn, E., Krober, S. M., Xiao, Y. J., Blumenthal, S. G., Lindemann, R. K., et al. (2003). Apoptotic cells induce migration of phagocytes via caspase-3- mediated release of a lipid attraction signal. *Cell, 113*(6), 717–730.
43. Peter, C., Waibel, M., Keppeler, H., Lehmann, R., Xu, G., Halama, A., et al. (2012). Release of lysophospholipid 'find-me' signals during apoptosis requires the ATP-binding cassette transporter A1. *Autoimmunity, 45*(8), 568–573.
44. Gude, D. R., Alvarez, S. E., Paugh, S. W., Mitra, P., Yu, J., Griffiths, R., et al. (2008). Apoptosis induces expression of sphingosine kinase 1 to release sphingosine-1-phosphate as a "come-and-get-me" signal. *The FASEB Journal, 22*(8), 2629–2638.
45. Weigert, A., Johann, A. M., von Knethen, A., Schmidt, H., Geisslinger, G., & Brune, B. (2006). Apoptotic cells promote macrophage survival by releasing the antiapoptotic mediator sphingosine-1-phosphate. *Blood, 108*(5), 1635–1642.
46. Weigert, A., Mora, J., Sekar, D., Syed, S., & Brune, B. (2016). Killing is not enough: how apoptosis hijacks tumor-associated macrophages to promote cancer progression. *Advances in Experimental Medicine and Biology, 930*, 205–239.
47. Li, F., Huang, Q., Chen, J., Peng, Y., Roop, D. R., Bedford, J. S., et al. (2010). Apoptotic cells activate the "phoenix rising" pathway to promote wound healing and tissue regeneration. *Science Signaling, 3*(110), ra13.
48. Zhao, X., Wang, D., Zhao, Z., Xiao, Y., Sengupta, S., Xiao, Y., et al. (2006). Caspase-3-dependent activation of calcium-independent phospholipase A2 enhances cell migration in non-apoptotic ovarian cancer cells. *The Journal of Biological Chemistry, 281*(39), 29357–29368.
49. Huang, Q., Li, F., Liu, X., Li, W., Shi, W., Liu, F. F., et al. (2011). Caspase 3-mediated stimulation of tumor cell repopulation during cancer radiotherapy. *Nature Medicine, 17*(7), 860–866.
50. Donato, A. L., Huang, Q., Liu, X., Li, F., Zimmerman, M. A., & Li, C. Y. (2014). Caspase 3 promotes surviving melanoma tumor cell growth after cytotoxic therapy. *The Journal of Investigative Dermatology, 134*(6), 1686–1692.
51. Kurtova, A. V., Xiao, J., Mo, Q., Pazhanisamy, S., Krasnow, R., Lerner, S. P., et al. (2015). Blocking PGE2-induced tumour repopulation abrogates bladder cancer chemoresistance. *Nature, 517*(7533), 209–213.

52. Feng, X., Yu, Y., He, S., Cheng, J., Gong, Y., Zhang, Z., et al. (2017). Dying glioma cells establish a proangiogenic microenvironment through a caspase 3 dependent mechanism. *Cancer Letters, 385*, 12–20.
53. Penzo, D., Petronilli, V., Angelin, A., Cusan, C., Colonna, R., Scorrano, L., et al. (2004). Arachidonic acid released by phospholipase A(2) activation triggers Ca(2+)-dependent apoptosis through the mitochondrial pathway. *The Journal of Biological Chemistry, 279*(24), 25219–25225.
54. Padmanaban, V., Keller, I., Seltzer, E. S., Ostendorf, B. N., Kerner, Z., & Tavazoie, S. F. (2024). Neuronal substance P drives metastasis through an extracellular RNA-TLR7 axis. *Nature, 633*(8028), 207–215.
55. Lo, Y. M. D., Han, D. S. C., Jiang, P., & Chiu, R. W. K. (2021). Epigenetics, fragmentomics, and topology of cell-free DNA in liquid biopsies. *Science, 372*(6538).
56. Wang, Y., Wang, M., Djekidel, M. N., Chen, H., Liu, D., Alt, F. W., et al. (2021). eccDNAs are apoptotic products with high innate immunostimulatory activity. *Nature, 599*(7884), 308–314.
57. Chekeni, F. B., Elliott, M. R., Sandilos, J. K., Walk, S. F., Kinchen, J. M., Lazarowski, E. R., et al. (2010). Pannexin 1 channels mediate 'find-me' signal release and membrane permeability during apoptosis. *Nature, 467*(7317), 863–867.
58. Medina, C. B., Mehrotra, P., Arandjelovic, S., Perry, J. S. A., Guo, Y., Morioka, S., et al. (2020). Metabolites released from apoptotic cells act as tissue messengers. *Nature, 580*(7801), 130–135.
59. Anderson, C. J., Medina, C. B., Barron, B. J., Karvelyte, L., Aaes, T. L., Lambertz, I., et al. (2021). Microbes exploit death-induced nutrient release by gut epithelial cells. *Nature, 596*(7871), 262–267.
60. Narahari, A. K., Kreutzberger, A. J., Gaete, P. S., Chiu, Y. H., Leonhardt, S. A., Medina, C. B., et al. (2021). ATP and large signaling metabolites flux through caspase-activated Pannexin 1 channels. *eLife*, 10.
61. Lucas, C. D., Medina, C. B., Bruton, F. A., Dorward, D. A., Raymond, M. H., Tufan, T., et al. (2022). Pannexin 1 drives efficient epithelial repair after tissue injury. *Science Immunology, 7*(71), eabm4032.
62. Park, S. Y., & Kim, I. S. (2017). Engulfment signals and the phagocytic machinery for apoptotic cell clearance. *Experimental & Molecular Medicine, 49*(5), e331.
63. Morioka, S., Maueroder, C., & Ravichandran, K. S. (2019). Living on the edge: Efferocytosis at the Interface of Homeostasis and Pathology. *Immunity, 50*(5), 1149–1162.
64. Doran, A. C., Yurdagul, A., Jr., & Tabas, I. (2020). Efferocytosis in health and disease. *Nature Reviews. Immunology, 20*(4), 254–267.
65. Trzeciak, A., Wang, Y. T., & Perry, J. S. A. (2021). First we eat, then we do everything else: The dynamic metabolic regulation of efferocytosis. *Cell Metabolism, 33*(11), 2126–2141.
66. Mehrotra, P., & Ravichandran, K. S. (2022). Drugging the efferocytosis process: Concepts and opportunities. *Nature Reviews. Drug Discovery, 21*(8), 601–620.
67. Moon, B., Yang, S., Moon, H., Lee, J., & Park, D. (2023). After cell death: The molecular machinery of efferocytosis. *Experimental & Molecular Medicine, 55*(8), 1644–1651.
68. Schilperoort, M., Ngai, D., Sukka, S. R., Avrampou, K., Shi, H., & Tabas, I. (2023). The role of efferocytosis-fueled macrophage metabolism in the resolution of inflammation. *Immunological Reviews, 319*(1), 65–80.
69. Adkar, S. S., & Leeper, N. J. (2024). Efferocytosis in atherosclerosis. *Nature Reviews. Cardiology, 21*(11), 762–779.
70. Gregory, C. D., & Pound, J. D. (2011). Cell death in the neighbourhood: Direct microenvironmental effects of apoptosis in normal and neoplastic tissues. *The Journal of Pathology, 223*(2), 177–194.
71. Raymond, M. H., Davidson, A. J., Shen, Y., Tudor, D. R., Lucas, C. D., Morioka, S., et al. (2022). Live cell tracking of macrophage efferocytosis during Drosophila embryo development in vivo. *Science, 375*(6585), 1182–1187.

72. Gregory, C. D., & Pound, J. D. (2010). Microenvironmental influences of apoptosis in vivo and in vitro. *Apoptosis, 15*(9), 1029–1049.
73. Fadok, V. A., Voelker, D. R., Campbell, P. A., Cohen, J. J., Bratton, D. L., & Henson, P. M. (1992). Exposure of phosphatidylserine on the surface of apoptotic lymphocytes triggers specific recognition and removal by macrophages. *Journal of Immunology, 148*(7), 2207–2216.
74. Suzuki, J., Denning, D. P., Imanishi, E., Horvitz, H. R., & Nagata, S. (2013). Xk-related protein 8 and CED-8 promote phosphatidylserine exposure in apoptotic cells. *Science, 341*(6144), 403–406.
75. Suzuki, J., Imanishi, E., & Nagata, S. (2016). Xkr8 phospholipid scrambling complex in apoptotic phosphatidylserine exposure. *Proceedings of the National Academy of Sciences of the United States of America, 113*(34), 9509–9514.
76. Nagata, S., Sakuragi, T., & Segawa, K. (2020). Flippase and scramblase for phosphatidylserine exposure. *Current Opinion in Immunology, 62*, 31–38.
77. Fadok, V. A., de Cathelineau, A., Daleke, D. L., Henson, P. M., & Bratton, D. L. (2001). Loss of phospholipid asymmetry and surface exposure of phosphatidylserine is required for phagocytosis of apoptotic cells by macrophages and fibroblasts. *Journal of Biological Chemistry, 276*(2), 1071–1077.
78. Segawa, K., Suzuki, J., & Nagata, S. (2011). Constitutive exposure of phosphatidylserine on viable cells. *Proceedings of the National Academy of Sciences of the United States of America, 108*(48), 19246–19251.
79. Lv, Z., Bian, Z., Shi, L., Niu, S., Ha, B., Tremblay, A., et al. (2015). Loss of cell surface CD47 clustering formation and binding avidity to SIRPalpha facilitate apoptotic cell clearance by macrophages. *Journal of Immunology, 195*(2), 661–671.
80. Kelley, S. M., & Ravichandran, K. S. (2021). Putting the brakes on phagocytosis: "Don't-eat-me" signaling in physiology and disease. *EMBO Reports, 22*(6), e52564.
81. Nagata, S., & Segawa, K. (2021). Sensing and clearance of apoptotic cells. *Current Opinion in Immunology, 68*, 1–8.
82. Brown, S., Heinisch, I., Ross, E., Shaw, K., Buckley, C. D., & Savill, J. (2002). Apoptosis disables CD31-mediated cell detachment from phagocytes promoting binding and engulfment. *Nature, 418*(6894), 200–203.
83. Barkal, A. A., Brewer, R. E., Markovic, M., Kowarsky, M., Barkal, S. A., Zaro, B. W., et al. (2019). CD24 signalling through macrophage Siglec-10 is a target for cancer immunotherapy. *Nature, 572*(7769), 392–396.
84. Barkal, A. A., Weiskopf, K., Kao, K. S., Gordon, S. R., Rosental, B., Yiu, Y. Y., et al. (2018). Engagement of MHC class I by the inhibitory receptor LILRB1 suppresses macrophages and is a target of cancer immunotherapy. *Nature Immunology, 19*(1), 76–84.
85. Gordon, S. R., Maute, R. L., Dulken, B. W., Hutter, G., George, B. M., McCracken, M. N., et al. (2017). PD-1 expression by tumour-associated macrophages inhibits phagocytosis and tumour immunity. *Nature, 545*(7655), 495–499.
86. Greenberg, M. E., Sun, M., Zhang, R., Febbraio, M., Silverstein, R., & Hazen, S. L. (2006). Oxidized phosphatidylserine-CD36 interactions play an essential role in macrophage-dependent phagocytosis of apoptotic cells. *The Journal of Experimental Medicine, 203*(12), 2613–2625.
87. Ishii, H., Mori, T., Shiratsuchi, A., Nakai, Y., Shimada, Y., Ohno-Iwashita, Y., et al. (2005). Distinct localization of lipid rafts and externalized phosphatidylserine at the surface of apoptotic cells. *Biochemical and Biophysical Research Communications, 327*(1), 94–99.
88. Gregory, C. (2009). Cell biology: Sent by the scent of death. *Nature, 461*(7261), 181–182.
89. Truman, L. A., Ford, C. A., Pasikowska, M., Pound, J. D., Wilkinson, S. J., Dumitriu, I. E., et al. (2008). CX3CL1/fractalkine is released from apoptotic lymphocytes to stimulate macrophage chemotaxis. *Blood, 112*(13), 5026–5036.
90. Bournazou, I., Pound, J. D., Duffin, R., Bournazos, S., Melville, L. A., Brown, S. B., et al. (2009). Apoptotic human cells inhibit migration of granulocytes via release of lactoferrin. *The Journal of Clinical Investigation, 119*(1), 20–32.

91. Bournazou, I., Mackenzie, K. J., Duffin, R., Rossi, A. G., & Gregory, C. D. (2010). Inhibition of eosinophil migration by lactoferrin. *Immunology and Cell Biology, 88*(2), 220–223.
92. Willems, J. J., Arnold, B. P., & Gregory, C. D. (2014). Sinister self-sacrifice: The contribution of apoptosis to malignancy. *Frontiers in Immunology, 5*, 299.
93. de la Rosa, G., Yang, D., Tewary, P., Varadhachary, A., & Oppenheim, J. J. (2008). Lactoferrin acts as an alarmin to promote the recruitment and activation of APCs and antigen-specific immune responses. *Journal of Immunology, 180*(10), 6868–6876.
94. Tennant, I., Pound, J. D., Marr, L. A., Willems, J. J., Petrova, S., Ford, C. A., et al. (2013). Innate recognition of apoptotic cells: Novel apoptotic cell-associated molecular patterns revealed by crossreactivity of anti-LPS antibodies. *Cell Death and Differentiation, 20*(5), 698–708.
95. Chaudhary, S., Patidar, A., Dhiman, A., Chaubey, G. K., Dilawari, R., Talukdar, S., et al. (2021). Exposure of a specific pleioform of multifunctional glyceraldehyde 3-phosphate dehydrogenase initiates CD14-dependent clearance of apoptotic cells. *Cell Death & Disease, 12*(10), 892.
96. Devitt, A., Moffatt, O. D., Raykundalia, C., Capra, J. D., Simmons, D. L., & Gregory, C. D. (1998). Human CD14 mediates recognition and phagocytosis of apoptotic cells. *Nature, 392*(6675), 505–509.
97. Devitt, A., Parker, K. G., Ogden, C. A., Oldreive, C., Clay, M. F., Melville, L. A., et al. (2004). Persistence of apoptotic cells without autoimmune disease or inflammation in CD14-/- mice. *The Journal of Cell Biology, 167*(6), 1161–1170.
98. Zizzo, G., & Cohen, P. L. (2018). Antibody cross-linking of CD14 activates MerTK and promotes human macrophage clearance of apoptotic neutrophils: The dual role of CD14 at the crossroads between M1 and M2c polarization. *Inflammation, 41*(6), 2206–2221.
99. Dransfield, I., Zagorska, A., Lew, E. D., Michail, K., & Lemke, G. (2015). Mer receptor tyrosine kinase mediates both tethering and phagocytosis of apoptotic cells. *Cell Death & Disease, 6*(2), e1646.
100. Nishi, C., Toda, S., Segawa, K., & Nagata, S. (2014). Tim4- and MerTK-mediated engulfment of apoptotic cells by mouse resident peritoneal macrophages. *Molecular and Cellular Biology, 34*(8), 1512–1520.
101. Minutti, C. M., Jackson-Jones, L. H., Garcia-Fojeda, B., Knipper, J. A., Sutherland, T. E., Logan, N., et al. (2017). Local amplifiers of IL-4Ralpha-mediated macrophage activation promote repair in lung and liver. *Science, 356*(6342), 1076–1080.
102. Duvall, E., Wyllie, A. H., & Morris, R. G. (1985). Macrophage recognition of cells undergoing programmed cell death (apoptosis). *Immunology, 56*(2), 351–358.
103. Yuita, H., Tsuiji, M., Tajika, Y., Matsumoto, Y., Hirano, K., Suzuki, N., et al. (2005). Retardation of removal of radiation-induced apoptotic cells in developing neural tubes in macrophage galactose-type C-type lectin-1-deficient mouse embryos. *Glycobiology, 15*(12), 1368–1375.
104. Yang, Z., Li, Q., Wang, X., Jiang, X., Zhao, D., Lin, X., et al. (2018). C-type lectin receptor LSECtin-mediated apoptotic cell clearance by macrophages directs intestinal repair in experimental colitis. *Proceedings of the National Academy of Sciences of the United States of America, 115*(43), 11054–11059.
105. Ghosh, S., Finnemann, S. C., Vollrath, D., & Rothlin, C. V. (2024). In the eyes of the beholder-new mertk knockout mouse and re-evaluation of phagocytosis versus anti-inflammatory functions of MERTK. *International Journal of Molecular Sciences, 25*(10).
106. Tibrewal, N., Wu, Y., D'Mello, V., Akakura, R., George, T. C., Varnum, B., et al. (2008). Autophosphorylation docking site Tyr-867 in Mer receptor tyrosine kinase allows for dissociation of multiple signaling pathways for phagocytosis of apoptotic cells and down-modulation of lipopolysaccharide-inducible NF-kappaB transcriptional activation. *The Journal of Biological Chemistry, 283*(6), 3618–3627.
107. Bode, K., Bujupi, F., Link, C., Hein, T., Zimmermann, S., Peiris, D., et al. (2019). Dectin-1 binding to annexins on apoptotic cells induces peripheral immune tolerance via NADPH Oxidase-2. *Cell Reports, 29*(13), 4435–46 e9.

108. Kinchen, J. M., Cabello, J., Klingele, D., Wong, K., Feichtinger, R., Schnabel, H., et al. (2005). Two pathways converge at CED-10 to mediate actin rearrangement and corpse removal in C. elegans. *Nature, 434*(7029), 93–99.
109. Nakaya, M., Kitano, M., Matsuda, M., & Nagata, S. (2008). Spatiotemporal activation of Rac1 for engulfment of apoptotic cells. *Proceedings of the National Academy of Sciences of the United States of America, 105*(27), 9198–9203.
110. Wang, Y., Subramanian, M., Yurdagul, A., Jr., Barbosa-Lorenzi, V. C., Cai, B., de Juan-Sanz, J., et al. (2017). Mitochondrial fission promotes the continued clearance of apoptotic cells by macrophages. *Cell, 171*(2), 331–45 e22.
111. Yin, C., & Heit, B. (2021). Cellular responses to the efferocytosis of apoptotic cells. *Frontiers in Immunology, 12*, 631714.
112. Heckmann, B. L., & Green, D. R. (2019). LC3-associated phagocytosis at a glance. *Journal of Cell Science, 132*(5).
113. Huang, S. C., Everts, B., Ivanova, Y., O'Sullivan, D., Nascimento, M., Smith, A. M., et al. (2014). Cell-intrinsic lysosomal lipolysis is essential for alternative activation of macrophages. *Nature Immunology, 15*(9), 846–855.
114. Zhang, S., Weinberg, S., DeBerge, M., Gainullina, A., Schipma, M., Kinchen, J. M., et al. (2019). Efferocytosis fuels requirements of fatty acid oxidation and the electron transport chain to polarize macrophages for tissue repair. *Cell Metabolism, 29*(2), 443–456.e5.
115. Yurdagul, A., Jr., Subramanian, M., Wang, X., Crown, S. B., Ilkayeva, O. R., Darville, L., et al. (2020). Macrophage metabolism of apoptotic cell-derived arginine promotes continual efferocytosis and resolution of injury. *Cell Metabolism, 31*(3), 518–33.e10.
116. Gerlach, B. D., Ampomah, P. B., Yurdagul, A., Jr., Liu, C., Lauring, M. C., Wang, X., et al. (2021). Efferocytosis induces macrophage proliferation to help resolve tissue injury. *Cell Metabolism, 33*(12), 2445–63.e8.
117. Ampomah, P. B., Cai, B., Sukka, S. R., Gerlach, B. D., Yurdagul, A., Jr., Wang, X., et al. (2022). Macrophages use apoptotic cell-derived methionine and DNMT3A during efferocytosis to promote tissue resolution. *Nature Metabolism, 4*(4), 444–457.
118. Morioka, S., Perry, J. S. A., Raymond, M. H., Medina, C. B., Zhu, Y., Zhao, L., et al. (2018). Efferocytosis induces a novel SLC program to promote glucose uptake and lactate release. *Nature, 563*(7733), 714–718.
119. Schilperoort, M., Ngai, D., Katerelos, M., Power, D. A., & Tabas, I. (2023). PFKFB2-mediated glycolysis promotes lactate-driven continual efferocytosis by macrophages. *Nature Metabolism, 5*(3), 431–444.
120. Ngai, D., Schilperoort, M., & Tabas, I. (2023). Efferocytosis-induced lactate enables the proliferation of pro-resolving macrophages to mediate tissue repair. *Nature Metabolism, 5*(12), 2206–2219.
121. Tufan, T., Comertpay, G., Villani, A., Nelson, G. M., Terekhova, M., Kelley, S., et al. (2024). Rapid unleashing of macrophage efferocytic capacity via transcriptional pause release. *Nature*.
122. Bosurgi, L., Cao, Y. G., Cabeza-Cabrerizo, M., Tucci, A., Hughes, L. D., Kong, Y., et al. (2017). Macrophage function in tissue repair and remodeling requires IL-4 or IL-13 with apoptotic cells. *Science, 356*(6342), 1072–1076.
123. Wu, D., Molofsky, A. B., Liang, H. E., Ricardo-Gonzalez, R. R., Jouihan, H. A., Bando, J. K., et al. (2011). Eosinophils sustain adipose alternatively activated macrophages associated with glucose homeostasis. *Science, 332*(6026), 243–247.
124. Kinsella, S., Evandy, C. A., Cooper, K., Iovino, L., deRoos, P. C., Hopwo, K. S., et al. (2021). Attenuation of apoptotic cell detection triggers thymic regeneration after damage. *Cell Reports, 37*(1), 109789.
125. Wang, Y., Wang, J., Zhang, J., Wang, Y., Wang, Y., Kang, H., et al. (2024). Stiffness sensing via Piezo1 enhances macrophage efferocytosis and promotes the resolution of liver fibrosis. *Science Advances, 10*(23), eadj3289.
126. Onder, L., Cheng, H. W., & Ludewig, B. (2022). Visualization and functional characterization of lymphoid organ fibroblasts. *Immunological Reviews, 306*(1), 108–122.

127. Rahman, S. A., Billingsley, J. M., Sharma, A. A., Styles, T. M., Govindaraj, S., Shanmugasundaram, U., et al. (2022). Lymph node CXCR5+ NK cells associate with control of chronic SHIV infection. *JCI Insight, 7*(8).
128. Gurwicz, N., Stoler-Barak, L., Schwan, N., Bandyopadhyay, A., Meyer-Hermann, M., & Shulman, Z. (2023). Tingible body macrophages arise from lymph node-resident precursors and uptake B cells by dendrites. *The Journal of Experimental Medicine, 220*(4).
129. Mayer, C. T., Gazumyan, A., Kara, E. E., Gitlin, A. D., Golijanin, J., Viant, C., et al. (2017). The microanatomic segregation of selection by apoptosis in the germinal center. *Science, 358*(6360).
130. Grootveld, A. K., Kyaw, W., Panova, V., Lau, A. W. Y., Ashwin, E., Seuzaret, G., et al. (2023). Apoptotic cell fragments locally activate tingible body macrophages in the germinal center. *Cell, 186*(6), 1144–61 e18.
131. Flemming, W. (1884). Studien über regeneration der Gewebe. *Archive for Microsk Anatomy, 24*, 50–91.
132. Swartzendruber, D. C., & Congdon, C. C. (1963). Electron microscope observations on tingible body macrophages in mouse spleen. *The Journal of Cell Biology, 19*(3), 641–646.
133. Mesin, L., Ersching, J., & Victora, G. D. (2016). Germinal center B cell dynamics. *Immunity, 45*(3), 471–482.
134. Young, C., & Brink, R. (2021). The unique biology of germinal center B cells. *Immunity, 54*(8), 1652–1664.
135. Victora, G. D., & Nussenzweig, M. C. (2022). Germinal centers. *Annual Review of Immunology, 40*, 413–442.
136. Aung, A., Cui, A., Maiorino, L., Amini, A. P., Gregory, J. R., Bukenya, M., et al. (2023). Low protease activity in B cell follicles promotes retention of intact antigens after immunization. *Science, 379*(6630), eabn8934.
137. Allen, C. D., Okada, T., & Cyster, J. G. (2007). Germinal-center organization and cellular dynamics. *Immunity, 27*(2), 190–202.
138. Victora, G. D., Schwickert, T. A., Fooksman, D. R., Kamphorst, A. O., Meyer-Hermann, M., Dustin, M. L., et al. (2010). Germinal center dynamics revealed by multiphoton microscopy with a photoactivatable fluorescent reporter. *Cell, 143*(4), 592–605.
139. Gitlin, A. D., Shulman, Z., & Nussenzweig, M. C. (2014). Clonal selection in the germinal centre by regulated proliferation and hypermutation. *Nature, 509*(7502), 637–640.
140. Allen, C. D., Okada, T., Tang, H. L., & Cyster, J. G. (2007). Imaging of germinal center selection events during affinity maturation. *Science, 315*(5811), 528–531.
141. Gitlin, A. D., Mayer, C. T., Oliveira, T. Y., Shulman, Z., Jones, M. J., Koren, A., et al. (2015). Humoral Immunity. T cell help controls the speed of the cell cycle in germinal center B cells. *Science, 349*(6248), 643–646.
142. Batista, F. D., & Neuberger, M. S. (2000). B cells extract and present immobilized antigen: implications for affinity discrimination. *The EMBO Journal, 19*(4), 513–520.
143. Schwickert, T. A., Victora, G. D., Fooksman, D. R., Kamphorst, A. O., Mugnier, M. R., Gitlin, A. D., et al. (2011). A dynamic T cell-limited checkpoint regulates affinity-dependent B cell entry into the germinal center. *The Journal of Experimental Medicine, 208*(6), 1243–1252.
144. Stoler-Barak, L., Biram, A., Davidzohn, N., Addadi, Y., Golani, O., & Shulman, Z. (2019). B cell dissemination patterns during the germinal center reaction revealed by whole-organ imaging. *The Journal of Experimental Medicine, 216*(11), 2515–2530.
145. Shih, T. A., Meffre, E., Roederer, M., & Nussenzweig, M. C. (2002). Role of BCR affinity in T cell dependent antibody responses in vivo. *Nature Immunology, 3*(6), 570–575.
146. Taylor, J. J., Pape, K. A., Steach, H. R., & Jenkins, M. K. (2015). Humoral immunity. Apoptosis and antigen affinity limit effector cell differentiation of a single naive B cell. *Science, 347*(6223), 784–787.
147. Viant, C., Weymar, G. H. J., Escolano, A., Chen, S., Hartweger, H., Cipolla, M., et al. (2020). Antibody affinity shapes the choice between memory and germinal center B cell fates. *Cell, 183*(5), 1298–311 e11.

148. Stewart, I., Radtke, D., Phillips, B., McGowan, S. J., & Bannard, O. (2018). Germinal center B cells replace their antigen receptors in dark zones and fail light zone entry when immunoglobulin gene mutations are damaging. *Immunity, 49*(3), 477–89 e7.
149. Kraus, M., Alimzhanov, M. B., Rajewsky, N., & Rajewsky, K. (2004). Survival of resting mature B lymphocytes depends on BCR signaling via the Igalpha/beta heterodimer. *Cell, 117*(6), 787–800.
150. Phan, T. G., Grigorova, I., Okada, T., & Cyster, J. G. (2007). Subcapsular encounter and complement-dependent transport of immune complexes by lymph node B cells. *Nature Immunology, 8*(9), 992–1000.
151. Junt, T., Moseman, E. A., Iannacone, M., Massberg, S., Lang, P. A., Boes, M., et al. (2007). Subcapsular sinus macrophages in lymph nodes clear lymph-borne viruses and present them to antiviral B cells. *Nature, 450*(7166), 110–114.
152. Aguilar, C. C., Kalia, A., Brisse, M. E., Dowd, K. A., Wise-Dent, O., Burgomaster, K. E., et al. (2024). Subcapsular sinus macrophages maximize germinal center development in non-draining lymph nodes during blood-borne viral infection. *Science Immunology, 9*(93), eadi4926.
153. Steer, H. W., & Foot, R. A. (1987). Changes in the medulla of the parathymic lymph nodes of the rat during acute gastro-intestinal inflammation. *Journal of Anatomy, 152*, 23–36.
154. Baratin, M., Simon, L., Jorquera, A., Ghigo, C., Dembele, D., Nowak, J., et al. (2017). T cell zone resident macrophages silently dispose of apoptotic cells in the lymph node. *Immunity, 47*(2), 349–62 e5.
155. Smith, K. G., Hewitson, T. D., Nossal, G. J., & Tarlinton, D. M. (1996). The phenotype and fate of the antibody-forming cells of the splenic foci. *European Journal of Immunology, 26*(2), 444–448.
156. Nossal, G. J., Abbot, A., & Mitchell, J. (1968). Antigens in immunity. XIV. Electron microscopic radioautographic studies of antigen capture in the lymph node medulla. *The Journal of Experimental Medicine, 127*(2), 263–276.
157. Jakubzick, C., Gautier, E. L., Gibbings, S. L., Sojka, D. K., Schlitzer, A., Johnson, T. E., et al. (2013). Minimal differentiation of classical monocytes as they survey steady-state tissues and transport antigen to lymph nodes. *Immunity, 39*(3), 599–610.
158. Gray, E. E., & Cyster, J. G. (2012). Lymph node macrophages. *Journal of Innate Immunity, 4*(5-6), 424–436.
159. Hume, D. A., Perry, V. H., & Gordon, S. (1983). Immunohistochemical localization of a macrophage-specific antigen in developing mouse retina: Phagocytosis of dying neurons and differentiation of microglial cells to form a regular array in the plexiform layers. *The Journal of Cell Biology, 97*(1), 253–257.
160. Rahman, Z. S., Shao, W. H., Khan, T. N., Zhen, Y., & Cohen, P. L. (2010). Impaired apoptotic cell clearance in the germinal center by Mer-deficient tingible body macrophages leads to enhanced antibody-forming cell and germinal center responses. *Journal of Immunology, 185*(10), 5859–5868.
161. Mondor, I., Baratin, M., Lagueyrie, M., Saro, L., Henri, S., Gentek, R., et al. (2019). Lymphatic endothelial cells are essential components of the subcapsular sinus macrophage niche. *Immunity, 50*(6), 1453–66 e4.
162. Grabert, K., Sehgal, A., Irvine, K. M., Wollscheid-Lengeling, E., Ozdemir, D. D., Stables, J., et al. (2020). A transgenic line that reports CSF1R protein expression provides a definitive marker for the mouse mononuclear phagocyte system. *Journal of Immunology, 205*(11), 3154–3166.
163. Smith, J. P., Kosco, M. H., Tew, J. G., & Szakal, A. K. (1988). Thy-1 positive tingible body macrophages (TBM) in mouse lymph nodes. *The Anatomical Record, 222*(4), 380–390.
164. Kranich, J., Krautler, N. J., Heinen, E., Polymenidou, M., Bridel, C., Schildknecht, A., et al. (2008). Follicular dendritic cells control engulfment of apoptotic bodies by secreting Mfge8. *The Journal of Experimental Medicine, 205*(6), 1293–1302.

165. Hanayama, R., Tanaka, M., Miyasaka, K., Aozasa, K., Koike, M., Uchiyama, Y., et al. (2004). Autoimmune disease and impaired uptake of apoptotic cells in MFG-E8-deficient mice. *Science, 304*(5674), 1147–1150.
166. Yau, I. W., Cato, M. H., Jellusova, J., Hurtado de Mendoza, T., Brink, R., & Rickert, R. C. (2013). Censoring of self-reactive B cells by follicular dendritic cell-displayed self-antigen. *Journal of Immunology, 191*(3), 1082–1090.
167. Miyanishi, M., Tada, K., Koike, M., Uchiyama, Y., Kitamura, T., & Nagata, S. (2007). Identification of Tim4 as a phosphatidylserine receptor. *Nature, 450*(7168), 435–439.
168. Rodriguez-Manzanet, R., Sanjuan, M. A., Wu, H. Y., Quintana, F. J., Xiao, S., Anderson, A. C., et al. (2010). T and B cell hyperactivity and autoimmunity associated with niche-specific defects in apoptotic body clearance in TIM-4-deficient mice. *Proceedings of the National Academy of Sciences of the United States of America, 107*(19), 8706–8711.
169. Wong, K., Valdez, P. A., Tan, C., Yeh, S., Hongo, J. A., & Ouyang, W. (2010). Phosphatidylserine receptor Tim-4 is essential for the maintenance of the homeostatic state of resident peritoneal macrophages. *Proceedings of the National Academy of Sciences of the United States of America, 107*(19), 8712–8717.
170. Scott, R. S., McMahon, E. J., Pop, S. M., Reap, E. A., Caricchio, R., Cohen, P. L., et al. (2001). Phagocytosis and clearance of apoptotic cells is mediated by MER. *Nature, 411*(6834), 207–211.
171. Khan, T. N., Wong, E. B., Soni, C., & Rahman, Z. S. (2013). Prolonged apoptotic cell accumulation in germinal centers of Mer-deficient mice causes elevated B cell and CD4+ Th cell responses leading to autoantibody production. *Journal of Immunology, 190*(4), 1433–1446.
172. Baumann, I., Kolowos, W., Voll, R. E., Manger, B., Gaipl, U., Neuhuber, W. L., et al. (2002). Impaired uptake of apoptotic cells into tingible body macrophages in germinal centers of patients with systemic lupus erythematosus. *Arthritis and Rheumatism, 46*(1), 191–201.
173. Botto, M., Dell'Agnola, C., Bygrave, A. E., Thompson, E. M., Cook, H. T., Petry, F., et al. (1998). Homozygous C1q deficiency causes glomerulonephritis associated with multiple apoptotic bodies. *Nature Genetics, 19*(1), 56–59.
174. Degn, S. E., van der Poel, C. E., Firl, D. J., Ayoglu, B., Al Qureshah, F. A., Bajic, G., et al. (2017). Clonal evolution of autoreactive germinal centers. *Cell, 170*(5), 913–26 e19.
175. Miyanishi, M., Segawa, K., & Nagata, S. (2012). Synergistic effect of Tim4 and MFG-E8 null mutations on the development of autoimmunity. *International Immunology, 24*(9), 551–559.
176. Scott, C. L., Zheng, F., De Baetselier, P., Martens, L., Saeys, Y., De Prijck, S., et al. (2016). Bone marrow-derived monocytes give rise to self-renewing and fully differentiated Kupffer cells. *Nature Communications, 7*, 10321.
177. Bain, C. C., Hawley, C. A., Garner, H., Scott, C. L., Schridde, A., Steers, N. J., et al. (2016). Long-lived self-renewing bone marrow-derived macrophages displace embryo-derived cells to inhabit adult serous cavities. *Nature Communications, 7*, ncomms11852.
178. Louwe, P. A., Badiola Gomez, L., Webster, H., Perona-Wright, G., Bain, C. C., Forbes, S. J., et al. (2021). Recruited macrophages that colonize the post-inflammatory peritoneal niche convert into functionally divergent resident cells. *Nature Communications, 12*(1), 1770.
179. Finlay, C. M., Parkinson, J. E., Zhang, L., Chan, B. H. K., Ajendra, J., Chenery, A., et al. (2023). T helper 2 cells control monocyte to tissue-resident macrophage differentiation during nematode infection of the pleural cavity. *Immunity, 56*(5), 1064–81 e10.
180. Jenkins, S. J., & Allen, J. E. (2021). The expanding world of tissue-resident macrophages. *European Journal of Immunology, 51*(8), 1882–1896.
181. Smith, J. P., Lister, A. M., Tew, J. G., & Szakal, A. K. (1991). Kinetics of the tingible body macrophage response in mouse germinal center development and its depression with age. *The Anatomical Record, 229*(4), 511–520.
182. Jenkins, S. J., & Hume, D. A. (2014). Homeostasis in the mononuclear phagocyte system. *Trends in Immunology, 35*(8), 358–367.
183. Guilliams, M., & Scott, C. L. (2017). Does niche competition determine the origin of tissue-resident macrophages? *Nature Reviews. Immunology, 17*(7), 451–460.

184. Jenkins, S. J., Ruckerl, D., Cook, P. C., Jones, L. H., Finkelman, F. D., van Rooijen, N., et al. (2011). Local macrophage proliferation, rather than recruitment from the blood, Is a signature of TH2 inflammation. *Science*.
185. Jenkins, S. J., Ruckerl, D., Thomas, G. D., Hewitson, J. P., Duncan, S., Brombacher, F., et al. (2013). IL-4 directly signals tissue-resident macrophages to proliferate beyond homeostatic levels controlled by CSF-1. *The Journal of Experimental Medicine, 210*(11), 2477–2491.
186. Perona-Wright, G., Mohrs, K., & Mohrs, M. (2010). Sustained signaling by canonical helper T cell cytokines throughout the reactive lymph node. *Nature Immunology, 11*(6), 520–526.
187. Duan, L., Liu, D., Chen, H., Mintz, M. A., Chou, M. Y., Kotov, D. I., et al. (2021). Follicular dendritic cells restrict interleukin-4 availability in germinal centers and foster memory B cell generation. *Immunity, 54*(10), 2256–72 e6.
188. Enzler, T., Gillessen, S., Manis, J. P., Ferguson, D., Fleming, J., Alt, F. W., et al. (2003). Deficiencies of GM-CSF and interferon gamma link inflammation and cancer. *The Journal of Experimental Medicine, 197*(9), 1213–1219.
189. Lionakis, M. S., Swamydas, M., Fischer, B. G., Plantinga, T. S., Johnson, M. D., Jaeger, M., et al. (2013). CX3CR1-dependent renal macrophage survival promotes Candida control and host survival. *The Journal of Clinical Investigation, 123*(12), 5035–5051.
190. Landsman, L., Bar-On, L., Zernecke, A., Kim, K. W., Krauthgamer, R., Shagdarsuren, E., et al. (2009). CX3CR1 is required for monocyte homeostasis and atherogenesis by promoting cell survival. *Blood, 113*(4), 963–972.
191. Nishi, C., Yanagihashi, Y., Segawa, K., & Nagata, S. (2019). MERTK tyrosine kinase receptor together with TIM4 phosphatidylserine receptor mediates distinct signal transduction pathways for efferocytosis and cell proliferation. *The Journal of Biological Chemistry, 294*(18), 7221–7230.
192. Witmer-Pack, M. D., Hughes, D. A., Schuler, G., Lawson, L., McWilliam, A., Inaba, K., et al. (1993). Identification of macrophages and dendritic cells in the osteopetrotic (op/op) mouse. *Journal of Cell Science, 104*(Pt 4), 1021–1029.
193. Cecchini, M. G., Dominguez, M. G., Mocci, S., Wetterwald, A., Felix, R., Fleisch, H., et al. (1994). Role of colony stimulating factor-1 in the establishment and regulation of tissue macrophages during postnatal development of the mouse. *Development, 120*(6), 1357–1372.
194. Elmore, S. (2007). Apoptosis: A review of programmed cell death. *Toxicologic Pathology, 35*(4), 495–516.
195. Chao, D. T., & Korsmeyer, S. J. (1998). BCL-2 family: regulators of cell death. *Annual Review of Immunology, 16*, 395–419.
196. Zhang, N., Hartig, H., Dzhagalov, I., Draper, D., & He, Y. W. (2005). The role of apoptosis in the development and function of T lymphocytes. *Cell Research, 15*(10), 749–769.
197. Elmore, S. A. (2006). Enhanced histopathology of the thymus. *Toxicologic Pathology, 34*(5), 656–665.
198. Jimenez-Garcia, L., Mayer, C., Burrola, P. G., Huang, Y., Shokhirev, M. N., & Lemke, G. (2022). The TAM receptor tyrosine kinases Axl and Mer drive the maintenance of highly phagocytic macrophages. *Frontiers in Immunology, 13*, 960401.
199. Nieuwenhuis, P., & Opstelten, D. (1984). Functional anatomy of germinal centers. *The American Journal of Anatomy, 170*(3), 421–435.
200. McKinstry, K. K., Strutt, T. M., & Swain, S. L. (2010). Regulation of CD4+ T-cell contraction during pathogen challenge. *Immunological Reviews, 236*, 110–124.
201. Liebold, I., Al Jawazneh, A., Casar, C., Lanzloth, C., Leyk, S., Hamley, M., et al. (2024). Apoptotic cell identity induces distinct functional responses to IL-4 in efferocytic macrophages. *Science, 384*(6691), eabo7027.
202. Baudon, E. M., & Bajenoff, M. (2023). Tingible body macrophages: Gargantuan chameleons of the germinal center. *The Journal of Experimental Medicine, 220*(4).
203. Trowell, O. A. (1957). Re-utilization of lymphocytes in lymphopoiesis. *The Journal of Biophysical and Biochemical Cytology, 3*(2), 317–318.

204. Wu, M., Song, G., Li, J., Song, Z., Zhao, B., Liang, L., et al. (2024). Innervation of nociceptor neurons in the spleen promotes germinal center responses and humoral immunity. *Cell, 187*(12), 2935–51 e19.
205. Yu, Q., Huang, X., Zhang, F., Jin, W., Li, K., Xiao, T., et al. (2024). Interaction between CAFs and apoptotic cancer cells promotes OSCC proliferation via STING signaling. *Oral Diseases*.
206. Garg, M., Kulkarni, S. D., & Sayed, R. (2016). Starry-sky appearance in neurotuberculosis. *Pediatric Neurology, 64*, 101–102.
207. Ahmad, R. (2020). Neurocysticercosis and starry sky appearance on brain imaging. *QJM, 113*(4), 283.
208. Dy-Ledesma, J. L., Khoury, J. D., Agbay, R. L., Garcia, M., Miranda, R. N., & Medeiros, L. J. (2016). Starry sky pattern in hematopoietic neoplasms: A review of pathophysiology and differential diagnosis. *Advances in Anatomic Pathology, 23*(6), 343–355.
209. Burkitt, D. (1958). A sarcoma involving the jaws in African children. *The British Journal of Surgery, 46*(197), 218–223.
210. Lichtman, M. A. (2000). Commentary on and reprint of Burkitt D, A sarcoma involving the jaws of African children, in British Journal of Surgery (1958) 46:218–223. In M. A. Lichtman, J. L. Spivak, L. A. Boxer, S. J. Shattil, & E. S. Henderson (Eds.), *Hematology* (pp. 351–358). Academic Press.
211. O'Conor, G. T., & Davies, J. N. (1960). Malignant tumors in African children. With special reference to malignant lymphoma. *The Journal of Pediatrics, 56*, 526–535.
212. O'Conor, G. T. (1961). Malignant lymphoma in African children. II. A pathological entity. *Cancer, 14*, 270–283.
213. Zhang, D. D., Medeiros, C. G., Liang, M., & Medeiros, L. J. (2024). Origin of the terms 'water pot' and 'starry sky' in Burkitt lymphoma. *Pathology, 56*(1), 118–120.
214. Butler, J. J., Szakacs, A., & Sinkovics, J. G. (1967). Virus-induced murine lymphoma resembling Burkitt's tumor. *The American Journal of Pathology, 51*(4), 629–637.
215. Magrath, I. (2009). Denis Burkitt and the African lymphoma. *Ecancermedicalscience, 3*, 159.
216. Robertson, E. S. (2013). *Burkitt's lymphoma* (ed. El-Deiry W., 332 p.). Springer.
217. Bongiovanni, M., Mazzucchelli, L., Martin, V., Crippa, S., Bolli, M., Suriano, S., et al. (2012). Images in endocrine pathology: A starry-sky in the thyroid. *Endocrine Pathology, 23*(1), 79–81.
218. McCorkle, C., & Graffeo, V. (2020). A rare case of dual expression of CD5 and CD10 in primary central nervous system diffuse large B-cell lymphoma with prominent starry sky appearance. *Clinical Case Reports, 8*(12), 2973–2975.
219. Smith, K. J., Lupton, G. P., & Skelton, H. G. (1997). Cutaneous angiosarcomas with a starry-sky pattern. *Archives of Pathology & Laboratory Medicine, 121*(9), 980–984.
220. Sung, H. J., Kim, D., Yoon, D. H., Cho, H., Huh, J., Suh, C. W., et al. (2023). Clinicopathologic and genetic features of the starry-sky pattern in double-expressor diffuse large B-cell lymphoma. *Human Pathology, 139*, 106–116.
221. Ford, C. A., Petrova, S., Pound, J. D., Voss, J. J., Melville, L., Paterson, M., et al. (2015). Oncogenic properties of apoptotic tumor cells in aggressive B cell lymphoma. *Current Biology, 25*(5), 577–588.
222. Leoncini, L., & Stein, H. (2013). Definition of Burkitt's lymphoma. In E. Robertson (Ed.), *Burkitt's Lymphoma* (pp. 81–93). Springer.
223. Kovalchuk, A. L., Qi, C. F., Torrey, T. A., Taddesse-Heath, L., Feigenbaum, L., Park, S. S., et al. (2000). Burkitt lymphoma in the mouse. *Journal of Experimental Medicine, 192*(8), 1183–1190.
224. Zhu, D., Qi, C. F., Morse, H. C., 3rd, Janz, S., & Stevenson, F. K. (2005). Deregulated expression of the Myc cellular oncogene drives development of mouse "Burkitt-like" lymphomas from naive B cells. *Blood, 105*(5), 2135–2137.
225. Vrazo, A., Chauchard, M., Osman, C., & Longnecker, R. (2013). Animal models of Burkitt's lymphoma. In E. Robertson (Ed.), *Burkitt's lymphoma* (pp. 269–299). Springer.

226. Oels, H. C., Harrison, E. G., Jr., & Kiely, J. M. (1968). Lymphoblastic lymphoma with histiocytic phagocytosis ("starry sky" appearance) in adults. Guide to prognosis. *Cancer, 21*(3), 368–375.
227. Jung, H., Kim, S. Y., & Joo, H. G. (2023). Fenbendazole exhibits differential anticancer effects in vitro and in vivo in models of mouse lymphoma. *Current Issues in Molecular Biology, 45*(11), 8925–8938.
228. Falini, B., Cardinali, V., Brunetti, L., & Martelli, M. P. (2023). NPM1-mutated AML with starry sky pattern. *British Journal of Haematology, 202*(1), 8.
229. McLendon, R. E., Adekunle, A., Rajaram, V., Kocak, M., & Blaney, S. M. (2011). Embryonal central nervous system neoplasms arising in infants and young children: A pediatric brain tumor consortium study. *Archives of Pathology & Laboratory Medicine, 135*(8), 984–993.
230. Flores, A. R., Lobo, J., Nunes, F., Rema, A., Lopes, P., Carvalho, L., et al. (2019). Canine ovarian gonadoblastoma with dysgerminoma overgrowth: A case study and literature review. *Journal of Ovarian Research, 12*(1), 89.
231. North, J. P., & McCalmont, T. H. (2011). Angiosarcoma with tingible body macrophages. *Journal of Cutaneous Pathology, 38*(9), 683. discussion 4–6.
232. Farnworth-McHugh, S., Barth, N., Melville, L., Paterson, M., Lynch, C., Holland, P., et al. (2020). Potential oncogenic effect of the MERTK-dependent apoptotic-cell clearance pathway in starry-sky B-cell lymphoma. *Frontiers in Immunology, 11*, 1759.
233. Voss, J. J. L. P., Ford, C. A., Petrova, S., Melville, L., Paterson, M., Pound, J. D., et al. (2017). Modulation of macrophage antitumor potential by apoptotic lymphoma cells. *Cell Death and Differentiation, 24*(6), 971–983.
234. Gough, J. (1967). Significance of a "starry sky" in lymphosarcomata in Britain. *Journal of Clinical Pathology, 20*(4), 578–580.
235. Bedoya, V., Grimley, P. M., & Rabson, A. S. (1969). Ultrastructural evidence to in vitro interaction among Burkitt lymphoma cells: possible relevance of the "phagocytic" activity of starry sky histiocytes in vivo. *Cancer Research, 29*(4), 753–762.
236. Bouroumeau, A., Bussot, L., Bonnefoix, T., Fournier, C., Chapusot, C., Casasnovas, O., et al. (2021). c-MYC and p53 expression highlight starry-sky pattern as a favourable prognostic feature in R-CHOP-treated diffuse large B-cell lymphoma. *The Journal of Pathology. Clinical Research, 7*(6), 604–615.
237. Mantovani, A., Allavena, P., Marchesi, F., & Garlanda, C. (2022). Macrophages as tools and targets in cancer therapy. *Nature Reviews. Drug Discovery, 21*(11), 799–820.
238. Fan, J., Feng, Z., & Chen, N. (2020). Spermidine as a target for cancer therapy. *Pharmacological Research, 159*, 104943.
239. Akalu, Y. T., Mercau, M. E., Ansems, M., Hughes, L. D., Nevin, J., Alberto, E. J., et al. (2022). Tissue-specific modifier alleles determine Mertk loss-of-function traits. *eLife*, 11.
240. Killarney, S. T., Tait, S. W. G., Green, D. R., & Wood, K. C. (2024). Sublethal engagement of apoptotic pathways in residual cancer. *Trends in Cell Biology, 34*(3), 225–238.
241. Bock, F. J., Sedov, E., Koren, E., Koessinger, A. L., Cloix, C., Zerbst, D., et al. (2021). Apoptotic stress-induced FGF signalling promotes non-cell autonomous resistance to cell death. *Nature Communications, 12*(1), 6572.
242. Galluzzi, L., Joza, N., Tasdemir, E., Maiuri, M. C., Hengartner, M., Abrams, J. M., et al. (2008). No death without life: Vital functions of apoptotic effectors. *Cell Death and Differentiation, 15*(7), 1113–1123.
243. Woo, M., Hakem, R., Furlonger, C., Hakem, A., Duncan, G. S., Sasaki, T., et al. (2003). Caspase-3 regulates cell cycle in B cells: A consequence of substrate specificity. *Nature Immunology, 4*(10), 1016–1022.
244. Hadian, K., & Stockwell, B. R. (2023). The therapeutic potential of targeting regulated non-apoptotic cell death. *Nature Reviews. Drug Discovery, 22*(9), 723–742.
245. Mehrotra, P., Maschalidi, S., Boeckaerts, L., Maueroder, C., Tixeira, R., Pinney, J., et al. (2024). Oxylipins and metabolites from pyroptotic cells act as promoters of tissue repair. *Nature*.

246. Hanggi, K., Li, J., Gangadharan, A., Liu, X., Celias, D. P., Osunmakinde, O., et al. (2024). Interleukin-1alpha release during necrotic-like cell death generates myeloid-driven immunosuppression that restricts anti-tumor immunity. *Cancer Cell.*
247. Li, M., Carpio, D. F., Zheng, Y., Bruzzo, P., Singh, V., Ouaaz, F., et al. (2001). An essential role of the NF-kappa B/Toll-like receptor pathway in induction of inflammatory and tissue-repair gene expression by necrotic cells. *Journal of Immunology, 166*(12), 7128–7135.
248. Klemm, J., Stinchfield, M. J., & Harris, R. E. (2021). Necrosis-induced apoptosis promotes regeneration in Drosophila wing imaginal discs. *Genetics, 219*(3).
249. Savill, J., & Fadok, V. (2000). Corpse clearance defines the meaning of cell death. *Nature, 407*(6805), 784–788.

Cell Death and Infection: Implications for Host and Microbe

Anna Davey, Thibaut Sanchez, Christopher D. Lucas, and Christopher J. Anderson

Abstract

Cell death is a part of life. Every day, billions of cells undergo programmed cell death within the human body as part of normal tissue turnover and homeostasis. The factors that initiate programmed cell death, the resulting signalling pathways that occur within the dying cell, the means to dispose of the dying cell, and the response of the neighbouring tissue to the dying cell are all highly evolutionarily conserved and tightly regulated processes. Decades of outstanding research have identified critical components of programmed cell death initiation, corpse clearance, and tissue response, particularly in the context of homeostatic cell death or sterile injury-induced cell death; however, multiple tissues are not sterile under homeostasis. In particular, the intestinal and respiratory tracts are 'external environment-facing' tissues that are in constant contact with commensal microorganisms and are also frequently exposed to invading pathogenic microbes. Indeed, many pathogenic microbes are capable of inducing or inhibiting various forms of cell death during infection, and some go as far as to utilise the metabolites released to the environment during apoptosis for their own growth. The primary emphasis of the field has been on the identification of what form of programmed cell death is initiated during infection and what the immunological consequences are. In contrast, less is known about how the critical stages of cell death and cell clearance directly or indirectly impact the microbes (both commensal and pathogenic) themselves. In this review, we will pay particular attention to the host cell-cell communication events occurring downstream of programmed cell death

A. Davey · T. Sanchez · C. D. Lucas (✉) · C. J. Anderson (✉)
Centre for Inflammation Research, Institute for Regeneration & Repair, University of Edinburgh, Edinburgh, UK
e-mail: a.davey-3@sms.ed.ac.uk; thibaut.sanchez@ed.ac.uk; christopher.lucas@ed.ac.uk; cj.anderson@ed.ac.uk

C. D. Gregory et al. (eds.), *Apoptosis and Other forms of Cell Death in the Regulation of Inflammation in Health and Disease*, Advances in Experimental Medicine and Biology 1481, https://doi.org/10.1007/978-3-031-92785-0_7

and highlight new research and new questions surrounding how these fundamental host processes interact with the bacterial communities, with a particular focus on the intestine and the lung.

Keywords

Apoptosis · Bacteria · Host-microbe interactions · Metabolism

Abbreviations

DISC	Death-inducing signalling complex
GI	Gastrointestinal
TLR	Toll-like receptor
IEC	Intestinal epithelial cell
LPS	Lipopolysaccharide
Tregs	Regulatory T cells
IBD	Inflammatory bowel disease
WHO	World Health Organisation
COPD	Chronic obstructive pulmonary disease
MOMP	Mitochondrial outer membrane permeabilisation
TNF	Tumour necrosis factor
NF-κB	Nuclear factor kappa B
T3SS	Type 3 secretion system
LLO	Listeriolysin O
LEC	Lung epithelial cell
CF	Cystic fibrosis
AMPK	AMP-activated protein kinase
SEB	Staphylococcal enterotoxin B
Mtb	*Mycobacterium tuberculosis*
Panx-1	Pannexin 1
EV	Extracellular vesicle
PtdSer	Phosphatidylserine
MerTK	Mer receptor tyrosine kinase
Th17	T helper 17 cells
ROS	Reactive oxygen species
BMDM	Bone marrow-derived macrophage
Mcl-1	Myeloid cell leukaemia 1
CDK	Cyclin-dependent kinase
PICD	Phagocytosis-induced cell death
RvE1	Resolvin E1
ASM	Acid sphingomyelinase

Introduction

The process of programmed cell death is a critical event to remove damaged or diseased cells and maintain tissue homeostasis. Indeed, dysregulated or aberrant cell death can have disastrous consequences such as neurodegenerative diseases and cancer [1]. There are a wide variety of programmed cell death pathways that can be distinguished on the basis of the effector molecules involved and the morphological changes the cell undergoes [2, 3]. This review will focus on one particular mode of programmed cell death: apoptosis. The molecular pathways leading to apoptosis will be defined briefly below; however, insights into the detailed molecular mechanisms involved can be found elsewhere [1, 4–6]. At their simplified core, programmed cell death pathways involve (i) the induction or triggering of the death-inducing signalling cascade, (ii) the release of secreted mediators from the dying corpse into the surrounding environment, and (iii) the clearance of the corpse. Each of these three general steps of the programmed cell death process can have direct and indirect effects on the ability of the host to cope with infection. Whether to classify apoptosis as a host defence mechanism or pathogen-induced adaptation strategy has been a long-standing back and forth within the field of host-microbe interactions. While the data suggesting one or the other may be context- or organism-dependent, is it not possible for both to be true?

Mucosal Tissues

The gastrointestinal (GI) tract and the respiratory tract have 3-tiered defence systems against infecting microbes: intrinsic, innate, and adaptive [7, 8]. Intrinsically, intestinal and airway anatomy promotes the constant flow of luminal content that can limit the exposure time to pathogens. Additionally, further stimulation can increase the expulsion of luminal content (i.e., diarrhoea or coughing/sneezing) in an attempt to expel pathogens. Following this, the innate defence uses pattern recognition receptors such as Toll Like Receptors (TLRs) to engage with pathogen-associated molecular patterns and activate signalling cascades to trap pathogens in mucus for removal, recruit first responder innate cells, and initiate the adaptive immune response. The adaptive response combines resident and recruited immune cells to help remove the pathogen and provide antigen-specific memory for future exposure [7, 8].

The GI tract is composed of cells that work together to provide a barrier against microbes, allow absorption of water and nutrients, and help regulate immune responses [9]. The outermost tissue layer, the mucosa, is comprised of multiple cell types and, broadly speaking, is largely composed of intestinal epithelial cells (IECs) that are in direct contact with the external environment. The epithelial barrier includes stem cells, which are responsible for replenishing the cells in the tissue every 4–5 days; specialised secretory cells, including Paneth and goblet cells; enteroendocrine cells that facilitate local and systemic hormone-based communication; and enterocytes that facilitate nutrient absorption and are the most abundant cell type within the epithelium [10, 11]. Each specialised cell within the epithelium performs critical functions to protect against invading pathogens. For example, Paneth cells secrete potent antimicrobial peptides to help protect the all-important stem cells within the crypts [12]. Fellow secretory cells, the goblet cells, secrete a protective mucus barrier and deliver antigens to the immune system [11], and a functioning mucus barrier is critical to protect against bacterial infection [13]. Enteroendocrine cells recognise microbial-produced products, including lipopolysaccharide (LPS), and prompt hormone release [14, 15]. LPS engages with TLRs on enteroendocrine cells following epithelial injury, which subsequently induces the release of intestinal repair-promoting hormones [14]. Finally, enterocytes help form a selective physical barrier to ensure absorption of various compounds and can secrete immunoglobulins and present antigen fragments to T cells as part of the immune response [10, 16].

The mucosal space is further occupied by myeloid and lymphoid cells that support the epithelial barrier and serve as a critical connection between intestinal and systemic immune responses. Macrophages occupy multiple tissue spaces within the intestinal tract and contribute to tissue homeostasis via the recognition, uptake (phagocytosis), and killing of microbes that penetrate the epithelial barrier [17]. Further, macrophages help maintain the epithelial barrier by assisting in the removal of dying cells and producing cytokines to support IEC renewal [17]. Single-cell transcriptomic approaches have revealed an incredible diversity of macrophages within various intestinal niches, though the functional implications of this heterogeneity remain an ongoing area of research [18]. Dendritic cells are found within specialised lymphoid tissue, such as Peyer's patches and draining lymph nodes, as well as within the lamina propria underlying the epithelial barrier. Dendritic cells sample luminal antigens across the epithelium or receive antigens from microfold cells [19] and subsequently help dictate T cell effector status and intestinal tropism [20]. Intestinal T cell effector populations, such as Th1, Th2, and Th17 cells, have significant impacts on inflammation and infection [21], and a fine balance between clearing the infection and minimising off-target tissue damage is a crucial component of regulatory cells (Tregs) [22]. Innate lymphoid cells are similarly found within lymphoid structures as well as barrier tissues such as the GI tract and help defend against extracellular and intracellular pathogens [23].

Although functioning as a continuous mucosal tissue, the lung is arranged to maximise surface area and can be anatomically divided into proximal branching airways and distal alveoli where gas exchange occurs. Although originally thought to be comprised primarily of ciliated cells, goblet cells, and basal cells, single-cell RNA sequencing has shed light on a vast and heterogeneous cell population within the proximal airway epithelium [24]. As a few examples of cell types and their functions: Ciliated epithelial cells aid mucociliary clearance of foreign particles and microbes, propelling goblet cell-produced mucus out of the airway [7]. Chemosensory brush cells sense bacteria and stress signals and release molecules to promote either innate immunity and/or inflammation [25, 26]. Basal cells are proximal airway resident stem cells involved in epithelial healing following injury and are capable of repopulating multiple epithelial lineages [24]. Alveolar type 1 and 2 cells permit gas exchange and secrete pulmonary surfactant, respectively, with type 2 cells additionally acting as progenitor cells for type 1 cells as well as antigen presentation [27]. Unlike the epithelium of the GI tract, the lung epithelium is extremely quiescent with low rates of proliferation in the absence of injury [28]. Lungs are also a major site of immune cell presence and surveillance, with immune cells present within the lung lumen and interstitium and tightly adherent to the vascular endothelium [29–31]. Changes to the local microenvironment within the lung, such as changes in oxygen abundance, can have profound impacts on immune cells [32, 33].

The cellular heterogeneity and complexity of both of these mucosal tissue compartments underlie the critical need for cell-cell communication during homeostasis and infection. Given the significant overlap in cellularity and essential functions of the GI and respiratory tracts, it is likely that lessons learnt from one tissue may be applied to the other. The ways in which apoptotic cell death influences these communication events will be discussed below.

Exposure to Commensal Bacteria

Aside from potentially harmful pathogen exposure, the GI tract maintains a beneficial level of commensal bacteria that live in symbiosis with the host and actually aid in the defence against invading pathogens [34]. The respiratory tract differs somewhat in this respect, with a much lower and more fluid biomass present [35].

The normal intestinal microbiota is predominantly composed of the bacterial phyla Firmicutes (Bacillota) and Bacteroidetes (Bacteroidota), with smaller compositions from Actinobacteria (Actinomycetota), Proteobacteria (Pseudomonadota), Fusobacteriota, and Verrucomicrobiota, although this will vary between individuals and the part of the gut sampled [36]. The intestinal microbiome is often altered during disease states, a process generally referred to as dysbiosis, and is closely tied to overall health [36, 37]. For instance, patients with inflammatory bowel disease (IBD) or colorectal cancer undergo dysbiosis, with a bloom of the Enterobacteriaceae family being one of the most commonly reported changes in these disease settings [38–41]. Although whether or not dysbiosis is merely a symptom of tissue disease or a direct contributor to disease has been debated for many years [42–47], the Enterobacteriaceae bloom itself has a causal relationship with intestinal tissue inflammation [48]; thus, we will consider the relationship between programmed host apoptosis and the microbiota in the same vein as exogenously introduced infections. The healthy gut microbiota defends against the invasion of harmful pathogens [49], the most common of which, according to the World Health Organisation (WHO), are foodborne nontyphoidal strains of *Salmonella*, pathogenic strains of *Escherichia coli* (*E. coli*) and *Campylobacter* species. To gain perspective on the scale of this issue, the WHO estimates that 420,000 deaths occur each year due to consuming contaminated food and subsequent infection [50].

Although not as extensively characterised and discussed as the intestinal microbiota, the respiratory tract has frequent exposure to microbes due to the continuous intake of air for gas exchange [35]. In the upper respiratory tract, which includes the oral cavity, nasal cavity, nasopharynx, and oropharynx, there are large numbers of microbes present, commonly including *Staphylococcus*, *Corynebacterium*, and *Streptococcus*. The lower respiratory tract, including the trachea and lungs, frequently has *Prevotella*, *Streptococcus*, and *Veillonella* isolated [35]. The lung microbiome is especially fluid due to the constant intake and subsequent clearance of pathogens [35]. Despite aiding host defence against pathogens, overstimulation of the immune response within the respiratory tract can lead to sustained inflammation and disease. Moreover, it has been suggested that reduced microbial diversity in the lungs can correlate with increased susceptibility to secondary infection, for instance, by *Pseudomonas aeruginosa* (*Pseudomonas*) [35]. Whether this reduced microbial diversity drives colonisation and infection, or whether reduced microbial diversity is a consequence of increased abundance of the colonising strain, requires ongoing clarification. It should be noted that viruses, parasites, and fungi may also infect and inhabit the gut and lungs; however, this review will focus on the bacterial populations.

There is an increasingly recognised crosstalk between the gut microbiota and the lungs. The so-called 'gut-lung axis' is proposed to involve bi-directional translocation of bacteria or transport of LPS or short-chain fatty acids (bacterial metabolites produced via fermentation) between the intestine and airways enabled by the lymphatic system [51, 52]. Alterations in gut microbial populations occur during lung diseases such as asthma and chronic obstructive pulmonary disease (COPD) [51], exacerbations of which often lead to prescriptions of antibiotics that further disrupt the intestinal microbiome [53]. It is less clear whether initial changes in the intestinal microbiome are a cause or effect of respiratory conditions.

In the following sections, this review will delve deeper into the relationship between apoptosis, bacterial infection, and tissue physiology.

Sterile Signal-Induced Apoptosis

Apoptosis is a term that was first used in 1972 to describe controlled cell death as opposed to uncontrolled necrosis [54]. During apoptosis, a cascade of signals and events ultimately leads to the condensation and fragmentation of chromatin, membrane blebbing and formation of apoptotic bodies [55], and eventual efferocytosis (engulfment) by phagocytes [56]. Apoptosis features the use of specific cysteine proteases, termed 'caspases', that are further divided into the initiator caspases [2, 8–10], that then activate the executioner caspases [3, 6, 7], and cleave substrates leading to the characteristic features of apoptosis [57, 58]. Apoptosis is triggered via two distinct pathways, the 'intrinsic' and 'extrinsic' pathways.

The intrinsic pathway occurs in response to diverse events such as DNA damage, oxidative stress, pharmacological agents and other cellular stressors which lead to mitochondrial outer membrane permeabilisation (MOMP). Stress signals activate pro-apoptotic BH3-only Bcl-2 family proteins. BH3-only proteins are initiators of apoptosis and simultaneously activate pro-apoptotic proteins while inhibiting anti-apoptotic ones. Bcl-2 proteins such as PUMA and BIM directly and indirectly activate Bcl-2-associated X members such as BAX and BAK [59]. Bcl-X members then mediate MOMP, which leads to the release of pro-apoptotic cytochrome c, Omi, and SMAC (DIABLO) into the cytosol and the association of the apoptosome [59]. Accordingly, apoptosome-associated caspase 9 is activated and in turn activates executioner caspases 3 and 7, which drives cell death. SMAC and Omi induce apoptosis via the inhibition of otherwise anti-apoptotic effectors [60].

The extrinsic pathway involves caspase 8 and is triggered in response to external stimuli which signal through death receptors. Often, extrinsic pathway activation leads to intrinsic pathway activation, with caspase 8 capable of activating Bcl-2 family members.

In caspase 8-dependent apoptosis, stress signals are sensed via cell surface receptors in the tumour necrosis factor (TNF) receptor superfamily. Such 'death' receptors include Fas (aka CD95), TRAIL receptor 1 and 2 (aka DR4/DR5), and TNFR1. Upon receptor binding, assembly of the DISC complex—or in the context of TNFR/TNF, complex II—is initiated and ultimately leads to activation of caspase 8 or 10, which in turn activate caspases 3 and 7 [5, 61]. Both complexes comprise procaspase 8 and FADD, with the addition of TRADD or receptor-interacting protein kinase 1 in complex II [61]. FADD can additionally facilitate caspase 8-mediated apoptosis when TLRs are activated (often by bacterial lipoproteins) and nuclear kappa factor B (NF-kB) is inhibited, using MyD88 as an adaptor [62].

Bacterial-Induced Apoptosis

Bacterial intestinal infections are a huge global health burden. It is often advantageous for bacteria to utilise apoptosis to promote their virulence in favour of other forms of cell death, as apoptosis is typically regarded as an anti-inflammatory event [63]. The nontyphoidal strains of *Salmonella,* including *Salmonella enterica* subspecies *enterica* serovar Typhimurium (*Salmonella*), are foodborne pathogens that invade the intestinal epithelium and cause acute GI symptoms in humans [64]. *Salmonella* can differentially influence apoptosis depending upon the tissue setting and the type of host cell. *Salmonella* uses two type 3 secretion systems (T3SSs) encoded by *Salmonella* pathogenicity islands 1 and 2 (SPI1/2) to inject effector proteins into host cells and mediate host response [65]. For example, in THP-1-derived macrophages, deletion of *Salmonella* SPI-1 effectors SipB, SipD or SopD inhibits *Salmonella*-induced caspase 3 activation [66]. Moreover, inhibiting NF-kB before infection with *Salmonella* mutants lacking SipB/SipD restored caspase 3 activation, indicating these effectors may be able to induce macrophage apoptosis via inhibition of NF-kB. Inhibition of TLR4 in THP-1 cells similarly blocked *Salmonella*-induced apoptosis, whereas caspase 3 activation was not blocked by knockdown of TLR5 in THP-1 cells but was in HEK-293 T cells, indicating some cell-specific differences in the receptors required for *Salmonella*-induced apoptosis and a potential role for *Salmonella* LPS or flagellin in the activation of apoptosis [66].

In IECs, *Salmonella* can conversely inhibit apoptosis in the early stages of infection [64]. This is dependent upon SPI-1 effector SopB (SigD), with apoptosis resumed and caspase 3 activation restored when HeLa cells were infected with a *sopB* deletion strain. SopB activates the phosphoinositide 3-kinase pathway, revealed by depleting the Akt isoforms present in HeLa cells using siRNA and using an Akt phosphorylation-specific antibody. However, apoptosis can be re-initiated 18–24 h after *Salmonella* infection via the SPI-2 effector protein AvrA [64, 67]. One possible reason behind this delay may be to give *Salmonella* the time to establish a replication-suitable niche before apoptosis and represents an elaborate mechanism of influence over host cells that remains to be fully tested. Examining the careful control of cell death by *Salmonella* is a valuable example of bacterial manipulation and exploitation of the host immune response to promote virulence.

Shigella flexneri (*Shigella*) also invades IECs and causes GI disease and represents another example of a pathogen that can induce or inhibit cell death in different cell types [68]. *Shigella* inhibits caspase 3 activation and staurosporine-induced apoptosis in HeLa cells [69]. This inhibition relies upon T3SS activity, as a Δ*mxiE* or Δ*ipaB* mutant with dysfunctional T3SSs was unable to prevent apoptosis following staurosporine treatment. Distinct from early infection by *Salmonella*, which involves inhibition of cytochrome c release, *Shigella* prevents late-stage IEC apoptosis despite cytochrome c release and caspase 9 activation [69]. IpaD, another *Shigella* T3SS component, has been characterised to indirectly induce apoptosis in B lymphocytes [70] and macrophages [71]. Induction of macrophage apoptosis in vitro appeared to involve mitochondrial disruption, as wildtype *Shigella* or isolated IpaD were able to induce a loss of mitochondrial membrane potential in J774 cells; however, the precise molecular mechanism is yet to be defined [71]. With regards to B lymphocytes, IpaD can cause apoptosis in CL-01 B cells via TLR2 independently of *Shigella* invasion/effector translocation, which has been demonstrated using a combination of in vitro, in vivo and ex vivo models, although which caspases are involved is unknown [70]. This raises the question of whether other bacterial strains may exploit TLR pathways during infection and whether this occurs only in macrophages or in diverse cell types.

Pathogenic strains of *E. coli*, including enteropathogenic *E. coli* and enterohemorrhagic *E. coli* leave a characteristic 'attaching and effacing lesion' in the intestinal epithelium during infection, causing inflammation and diarrhoea [72]. *E. coli* is similar to *Salmonella* in that infected host cells display signs of apoptosis early on during infection, but the apoptotic cascade is seemingly blocked at later stages [73]. *E. coli* similarly uses a T3SS to inject effector proteins into host cells. Two such effectors are NleB1 and NleF, which prevent caspase cleavage and apoptosis in Fas (CD95)-ligand-treated cells [74]. The balance and actions of each effector are yet to be fully defined, although it is thought that NleB1 targets FADD using N-acetylglucosamine transferase activity to inhibit apoptosis [74]. Other *E. coli* effectors, NleH 1 and 2, interact with Bax-inhibitor-1 protein (B-I1) to prevent early apoptosis during infection, as knockdown of B-I1 ameliorates NleH-dependent anti-apoptotic activity [73]. NleH family members are similar to the *Shigella* effector OspG, yet, in contrast to OspG, which inhibits NF-kB, they were initially reported to increase NF-kB and TNF activity in mouse colonic mucosa [75]. However, in further research, NleH1, but not NleH2, was found to inhibit the NF-kB pathway via IKKβ phosphorylation of RPS3 [76]. Accordingly, *Citrobacter rodentium* (a mouse pathogen closely related to *E. coli*) infection of mice was studied. *C. rodentium* colonisation of mice and their subsequent mortality was reduced when mice were infected by mutant strains of *C. rodentium* lacking *nleH* compared to wildtype bacteria, with virulence restored when complemented by NleH1 but not NleH2 [76]. Given their additional function associating with cytoprotective BI-1, NleH effectors thus appear to have multifunctional roles.

In the case of the gram-positive pathogen *Listeria monocytogenes (Listeria),* which causes systemic infection but is commonly acquired via the GI tract, apoptosis is induced in hepatocytes following infection; however, it is unclear if this is protective or harmful to the host or bacterium [77]. The first study to report that *Listeria* induced apoptosis of hepatocytes suggested a beneficial consequence for the host due to the recruitment of neutrophils [78]. In contrast, studies on TRAIL-deficient mice, which lack a critical component of extrinsic apoptosis initiation, showed greater resistance to infection, with a level of bacteria present in the liver 10–100 times lower than that of the controls [79]. Mice deficient for BH3-only proteins such as Bim, Bid, and Noxa also showed an enhanced ability to eliminate bacterial infection, although this study focused more on the consequences for the spleen [80]. Given that mice deficient in some pro-apoptotic components exhibit resistance to *Listeria* compared to control mice, it is implied that apoptosis benefits the pathogen in this setting [77]. A more recent study, while not addressing apoptosis directly, reported that long-term *Listeria* infection disrupts hepatocyte innate immune functions, and this may leave *Listeria* better facilitated to establish persistent infection [81]; however, it remains unclear if the amount of hepatocyte apoptosis influences this disrupted innate immune function. Unlike hepatic cells, lymphocyte apoptosis is induced via two distinct mechanisms. Firstly, the high concentration of the bacterial effector Listeriolysin O (LLO) induces permeability of the lymphocyte vesicular compartment, resulting in the translocation of granzyme granules into the cytosol, triggering lymphocyte apoptosis [82]. Secondly, the production of type I interferons during infection with *Listeria*, mediated by the activation of STING [83, 84], also contributes to making T lymphocytes more sensitive to apoptosis [85]. Apoptosis of T lymphocytes and their efferocytosis by macrophages induces an immunosuppressive state via IL-10 secretion for the benefit of *Listeria*, as IL-10-deficient mice show increased resistance to infection [86].

Similar uncertainties exist surrounding the consequences of apoptosis during bacterial infection in the lung, which is also a large global issue; in 2019, the WHO reported that lower respiratory tract infections accounted for the fourth leading cause of death globally [87]. *Pseudomonas* are gram-negative bacteria that trigger apoptosis in lung epithelial cells (LECs) using its T3SS, which increases Fas ligand availability and, when this is bound to its receptor, triggers the apoptotic cascade [88, 89]. This has been shown both in vitro and in vivo. It seems that LEC apoptosis is a crucial mediator of host defence against *Pseudomonas,* as mice lacking Fas died as a result of sepsis after pulmonary infection vs. control mice, which lived [89]. In concordance, human bronchial epithelial cells engineered to lack cystic fibrosis (CF) transmembrane conductance regulator channels exhibited delayed apoptosis upon *Pseudomonas* infection, quantified by M30 staining, compared to those with functional channels [90]. CF patients (who lack functional CF transmembrane conductance regulator channels) are frequently colonised with *Pseudomonas* and experience particularly severe symptoms following *Pseudomonas* infection that often causes exacerbations of lung disease. Together, these findings highlight the potential importance of LEC apoptosis in efficient *Pseudomonas* clearance, though this remains to be tested.

Streptococcus pneumoniae (*S. pneumoniae*) are gram-positive bacteria which cause community-acquired pneumonia. *S. pneumoniae* characteristically induce rapid apoptosis of lung alveolar macrophages, identified via TUNEL staining, with alveolar macrophage apoptosis evident 6 h post *S. pneumoniae* infection [91]. Pan caspase inhibition, using zVAD-FMK, leads to increased mortality and bacteraemia in vivo, suggesting a protective function of apoptosis [91]. Certainly, in more recent research, alveolar macrophage apoptosis has been confirmed to enhance clearance of *Streptococcus* in wildtype mice vs. a CD68.hMcl-1 transgenic line that overexpressed Mcl-1 [92]. *S. pneumoniae* can also induce apoptosis in alveolar epithelial cells; however, in this cell type, apoptosis appears to be detrimental to the host. Epithelial cell apoptosis was induced due to hydrogen peroxide production, which is dependent upon *Streptococcus* pyruvate oxidase, and leads to DNA disruption and apoptosis [93]. In vivo, hydrogen peroxide production increased *Streptococcus* virulence [93].

Klebsiella pneumoniae (*K. pneumoniae*) are gram-negative bacteria that similarly can cause community-acquired pneumonia but are a more frequent cause of hospital-acquired pneumonia [94]. *K. pneumoniae* has been reported to induce apoptosis in several cell types, including murine alveolar macrophages in vivo [94], and HLF-1 and BEAS-2B lung cells in vitro [95]. In HLF-1 and BEAS-2B cells, *K. pneumoniae* infection increased levels of IL-6, Amp-activated protein kinase (AMPK), and caspase 9 mRNA, among others; however, these increases were preventable by treatment with melatonin [95]. Moreover, silencing AMPK prevented melatonin from suppressing inflammation and apoptosis in infected lung cells, indicating an involvement of AMPK in the melatonin-induced rescue from apoptosis following *K. pneumoniae* infection. The authors also employed the use of culturing 3D lung spheroids to test their findings and confirm that melatonin reduced *K. pneumoniae*-induced inflammation [95]. Interestingly, *Pseudomonas*-induced lung injury and subsequent apoptosis of epithelial cells are also prevented by metformin-induced AMPK activation in mice in vivo [96]. AMPK can be activated in situations of metabolic stress when ATP is depleted and regulates major metabolic pathways and mitochondrial biogenesis [96]. Altogether, these data suggest that AMPK is a promising therapeutic target to promote recovery from bacterial lung infection, apparently having a connection to inhibiting infection-induced apoptosis.

Staphylococcus aureus (S. aureus) are also gram-positive bacteria and, alongside being responsible for the majority of surgically encountered skin and soft tissue infections, cause pneumonia and food poisoning [97]. *S. aureus* has the ability to induce intrinsic and extrinsic apoptosis. In Jurkat cells, a lymphocyte cell line, *S. aureus* is responsible for inducing apoptosis via the intrinsic pathway [98]. This was revealed in a series of experiments in which an anti-α-toxin antibody was first found to block caspase 3 activation. Given caspase 3 can be activated by caspases 8 or 9, the authors then analysed extracts of *S. aureus*-treated Jurkat cells for their caspase 8 and 9 activity, which were both increased [98]. To hone in on the molecular pathway, they then investigated whether CD95 was crucial for *S. aureus*-induced apoptosis using Jurkat cells with CD95, FADD, or caspase 8 deficiencies. Knockout Jurkat cells had a similar level of death compared to wild-type Jurkats, indicating that these death receptors are not required. To confirm that the intrinsic pathway was in fact mediating apoptosis, Jurkat cells expressing antiapoptotic Bcl-2 were incubated with *S. aureus* or α-toxin, and cell death was inhibited dose-dependently. The α-toxin was also shown to activate caspase 3 in human lymphocytes and monocytes, although it was not confirmed whether this was via the intrinsic pathway [98].

Staphylococcal enterotoxin B (SEB), which is important in *S. aureus*-related diseases, can also induce apoptosis via the extrinsic pathway [99]. SEB was found to induce apoptosis in THP-1 monocytes via upregulating TNF, which can increase the expression of cell surface SEB receptor HLA-DRa in a positive feedback loop. During SEB treatment of THP-1 cells, caspase 3 and 8 activities were increased, and inhibitable by Z-VAD-FMK; however, caspase 9 activity was unchanged by SEB. qPCR then revealed increased mRNA levels of TNF and a trend towards increased MyD88 following THP-1 treatment with SEB, with no significant changes in Fas or TRAIL mRNA. Combining an anti-SEB and anti-TNF antibody reduced THP-1 apoptosis to a greater extent than treatment with anti-SEB alone, and furthermore, knockdown of TNFR1 or HLA-DRa also reduced SEB-induced apoptosis and TNF levels. This led to the proposition that reduced TNFR1 may downregulate HLA-DRa expression due to the role of TNF in regulating HLA-DRa [99]. Although there are many more studies investigating the various mechanisms of apoptosis induction by *S. aureus*, the examples above provide insight into the diversity of *S. aureus* toxins and their ability to induce cell death.

Finally, *Mycobacterium tuberculosis (Mtb)* is an airborne intracellular pathogen that causes the pulmonary disease tuberculosis, a leading cause of death globally [100]. Using the ESX-1 secretion system, *Mtb* has been reported to inhibit macrophage apoptosis [101]. The virulence factor MptpB mediated survival, suppressing macrophage apoptosis and blocking the NF-kB and MAPK pathways downstream of TLRs, to suppress pro-inflammatory cytokine expression in vitro [102]. However, more recent work reported some continuing apoptosis during *Mtb* infection, where the proportion of apoptosis was lower in *Casp8 Mlkl* double knockouts and *TNF* knockout *Mtb*-infected BMDMs, suggesting that some macrophages do undergo extrinsic apoptosis [103]. Moreover, apoptosis-deficient mice (caspase 8 and MLKL deficient) had reduced clearance of *Mtb*, demonstrating a host-protective function [103]. A following study examining gene expression in THP-1 macrophage subsets during *Mtb* infection found that compared to unstimulated macrophages, apoptosis-related genes were significantly higher in LPS- or IL-4-activated *Mtb*-infected macrophages [104]. This may explain some of the differences in macrophage apoptosis during *Mtb* infection as being dependent on what subset of macrophage is sampled; however, further work in vivo is clearly needed to fully define this mechanism.

The examples listed above highlight the complexities of defining whether or not apoptosis is to the benefit of the host or the microbe within the context of intestinal pathogens (Fig. 1) or respiratory pathogens (Fig. 2). Conflicting host is protective or detrimental phenotypes have been observed depending on the pathogen, tissue, cell type, kinetics, and specific activated host apoptosis effector molecules. Careful consideration must therefore be given when trying to separate phenotypes that are directly apoptosis-dependent versus changes in downstream inflammatory mediators that are perhaps unrelated to the death process. The use of genetic models targeting apoptotic executioner caspases, particularly in a cell type-specific and inducible manner, are valuable tools in making these distinctions [105, 106]. Furthermore, the interrelatedness of the different forms of programmed cell death [107] requires thorough controls to ensure that model systems deficient for various components of apoptosis do not instead shift towards an alternative form of programmed cell death.

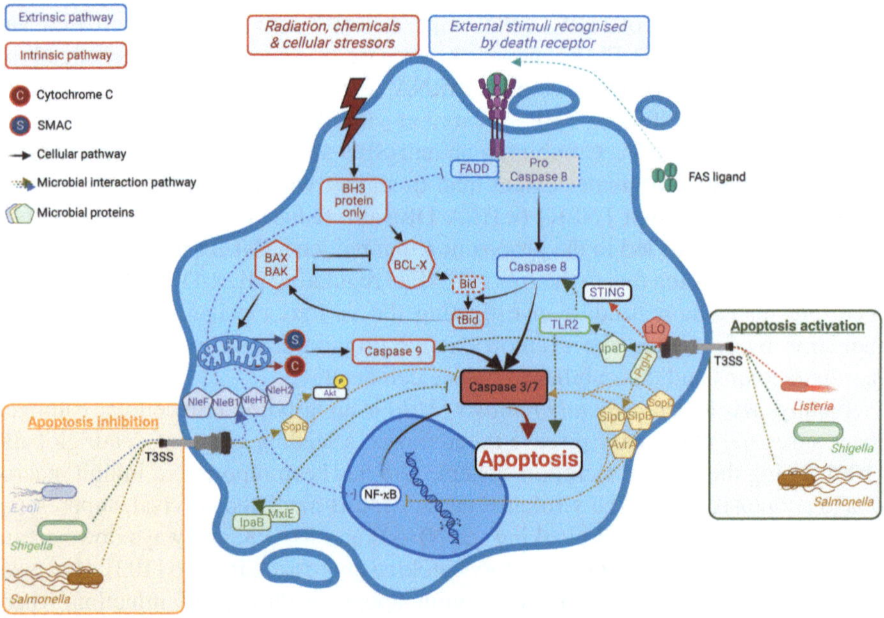

Fig. 1 The interplay of various GI bacteria with the host cell. Bacterial effector proteins and secretion systems from common GI bacteria that have been found to either inhibit or activate the various steps of intrinsic or extrinsic apoptosis

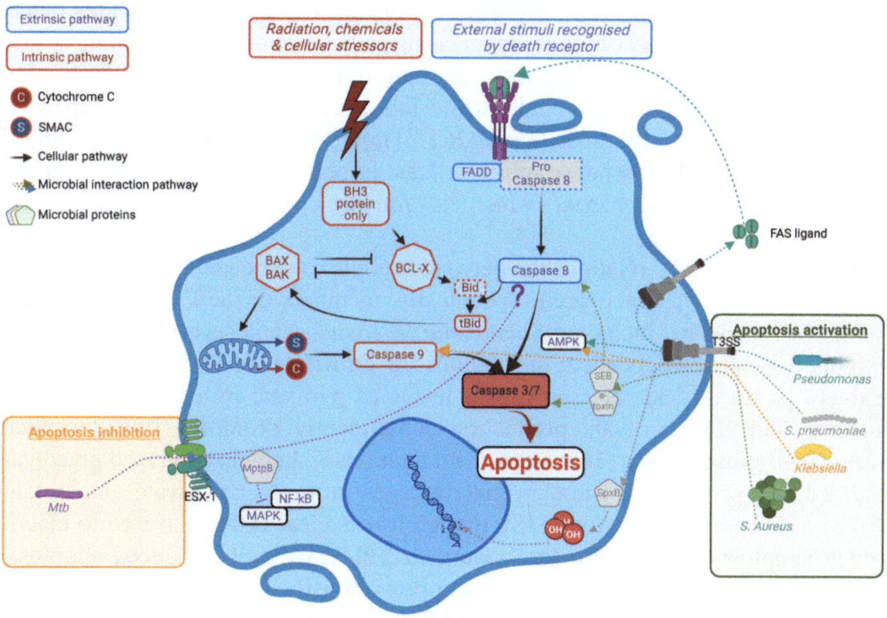

Fig. 2 The interplay of various respiratory bacteria with the host cell. Bacterial effector proteins and secretion systems from common respiratory bacteria that have been found to either inhibit or activate the various steps of intrinsic or extrinsic apoptosis

Apoptosis-Dependent Secreted Factors

Apoptotic cells secrete a number of factors, including proteins, metabolites, and extracellular vesicles [108]. Apoptotic cells are recognised, in part, via the release of 'find me' signals such as nucleotides and lysophosphatidylcholine to encourage phagocyte migration towards the corpse [108]. Aside from chemotactic properties, apoptotic metabolites have additional roles involving inflammation and cell clearance. Metabolite release is a controlled process, rather than a random dispersion of cellular contents, as membrane integrity is selectively controlled. The abundance of arginine, which is metabolised by macrophages to ornithine via the Arg1 enzyme, leads to the production of putrescine and optimises the efferocytosis of new apoptotic cells by these same macrophages. Indeed, putrescine increases the expression of Guanosine triphosphate-Exchange Factor Dbl, responsible for the activation of Rac1, which is involved in the regulation of actin dynamics [109], and therefore in the internalisation of apoptotic cells [110]. Although the majority of mRNAs are degraded during apoptosis, some are conserved or enriched, such as the mRNA encoding spermidine synthase [111], which facilitates the conversion of putrescine into spermidine, thereby reducing inflammation [112]. The plasma membrane channel Pannexin-1 (Panx1) mediates the selective release of certain small metabolites from apoptotic cells following activation by caspase cleavage [108], including spermidine and ATP, which aid phagocyte migration and cell clearance using the $P2Y_2$ receptor [113]. Moreover, apoptosis-dependent metabolites induce gene programmes in host tissues, suppressing inflammation and promoting wound healing in murine models of arthritis and lung transplant rejection [114]. Further, Panx1-dependent signalling regulates the induction of a macrophage-dependent reparative response in zebrafish (*Danio Rerio*) injury models and mouse LEC injury [115]. This reparative response relies, at least in part, on the induction of amphiregulin within macrophages, which is reliant upon Panx1-dependent ATP release [115].

Recently, a direct relationship between apoptosis-dependent metabolites and bacterial growth was uncovered. Specifically, members of the Enterobacteriaceae family of organisms, including *Salmonella*, *Klebsiella*, and several *E. coli* strains, exploited apoptosis-dependent metabolites for growth [116, 117]. Apoptotic IECs actively produced and released higher levels of pyruvate in a PKM2-dependent manner, and IEC pyruvate production subsequently promoted PflB-dependent *Salmonella* growth [116]. This aligns with *Salmonella* infection-induced glycolysis and reduced serine production in macrophages promoting pyruvate and lactate-driven *Salmonella* replication [118], though it is unclear if this is linked to macrophage apoptosis. In addition to *Salmonella* SPI-1 and SPI-2-induced apoptosis, pyruvate-fuelled *Salmonella* growth was observed using models of chemotherapy-induced apoptosis and A20-dependent TNF-driven apoptosis, suggesting conserved pyruvate production and release by IECs following a wide array of apoptotic triggers [116]. Cytotoxic chemotherapy treatment commonly induces intestinal apoptosis as well as an Enterobacteriaceae-enhanced dysbiosis [119, 120], thereby suggesting the possibility that apoptotic metabolites are being exploited by commensal bacteria to bloom during treatment and potentially impact tissue health. Furthermore, A20 is an anti-inflammatory protein implicated in IBD and inhibits TNF-induced apoptosis in several cell types, including IECs [121]. A20-deficient animals are hypersusceptible to TNF stimulation, with A20 deficient IECs undergoing rapid apoptosis in response to TNF signalling [122]. Given that TNF-induced apoptosis in A20-deficient IECs promoted Enterobacteriaceae growth in vitro and in vivo [116], this raises the possibility that increased apoptosis during IBD flare-ups could be partially responsible for clinically observed dysbiosis and/or intestinal symptoms. Interestingly, pyruvate is not produced or released from apoptotic thymocytes or macrophages [114], which suggests that the content of death-dependent communication events may be cell type and/or death trigger dependent. Future studies elucidating the metabolic reprogramming that different apoptotic cell types undergo in response to different death stimuli, particularly in vivo, would have a significant impact on the field.

At homeostasis, baseline levels of IEC apoptosis do not shape the microbiota [105]. These findings suggest that basal levels of apoptotic cell turnover produce insufficient levels of death-dependent metabolites or, alternatively, that the disposal of apoptotic cells (discussed below) is happening with such efficiency that these apoptotic cells remain segregated from the commensal microbes. Importantly, the same Panx1-dependent metabolites that help execute host cell–cell communication events also potently promote Enterobacteriaceae growth [114, 116]. This new revelation that both host and microbe seem to benefit from using the same apoptosis-dependent metabolites opens the door to the possibility that intestinal or respiratory bacteria, beyond the Enterobacteriaceae, are competing for soluble factors with host cells. If so, then this interspecies competition may be restricting the ability of apoptotic cells to communicate with their surrounding environments during infection and may be dampening critical tissue repair mechanisms. Thus, further research is required to better understand the mechanisms underlying these communication events and how the added biological stressor of an active bacterial infection influences both the dying corpse and the responding cell.

Apoptotic cells also release extracellular vesicles (EVs) that can influence host–host communication and are involved in many stages of efferocytosis. Notably, EVs form 'find-me' signals via the release of CX3CL1/fractalkine, which helps the chemotaxis of macrophages via their CX3CR1 receptor [113, 123]. EVs also promote their own internalisation by expressing several 'eat-me' signals such as phosphatidylserine (PtdSer) or ICAM-3 [124], which promote rapid internalisation of cell fragments. Additionally, EVs can play a role in regulating antigen presentation. This regulation can be positive, as observed in dendritic cells or macrophages producing EVs carrying the MHC-II molecule or bacterial antigens enabling T lymphocyte activation, as observed in *Mtb* infection [125–127]. But this regulation can also be negative, as observed with EVs generated by lymphoblasts, which decrease the presence of MHC II on the surface of dendritic cells [128]. The impact of EVs on efferocytosis therefore appears to vary depending on the apoptotic cell and the infectious state.

Finally, apoptotic cells produce and secrete proteins, most commonly cytokines and/or chemokines. TNF and IL-6 both signal via type 1 cytokine receptors, whereas the chemokine IL-8 uses G protein-coupled receptors [129]. TNF induces inflammation within the innate immune system and can have opposing roles; for example, induction of apoptosis of damaged IECs may spread the apoptotic cascade to healthy IECs if TNF reaches a high concentration [130]. TNF is secreted by macrophages in response to *Mtb* infection and is vital in host protection, aiding granuloma formation and inducing chemokine release by macrophages to attract further immune cells to the site of infection [131]. Anti-TNF antibody treatment accordingly disrupts granuloma formation and increases the risk of *Mtb* reactivation [132]. The role of IL-6 has been highlighted in mouse models, where IL-6 knockout mice infected with influenza A had increased bacterial burden and mortality upon coinfection with *S. pneumoniae* [133]. IL-6 deficient macrophages had reduced phagocytic ability and increased apoptosis in the lung [133]. On the other hand, IL-8 regulates neutrophil-mediated acute inflammation and is induced in IECs following infection with gram-negative *Vibrio cholerae* [134] and *E. coli* [135] and has been reported to be expressed temporally during *P. aeruginosa* infection in CF patients [136]. During early *P. aeruginosa* infection, IL-8 was expressed at low levels; however, IL-8 expression was much higher during chronic infection and led to sustained inflammation [136]. *P. aeruginosa* additionally induces IL-8 expression in LECs during infection of COPD patients to exacerbate the condition [137].

As highlighted above, there are both similarities and differences in the secreted factors that are produced during apoptosis and during infection. It will remain critically important for the field to better understand how the presence of a bacterial infection influences (i) the secreted signals that an apoptotic cell is producing; (ii) the ability of responding cells to recognise and utilise apoptosis-dependent secreted signals; and (iii) the efficiency with which the dying tissue is repaired.

Apoptotic Cell Clearance

Following apoptosis, phagocytes recognise 'eat me' signals on apoptotic cells, including PtdSer, which initiate engulfment [138] in a process termed efferocytosis [139]. Briefly, apoptotic cells are internalised by phagocytes (e.g., macrophages) into an efferosome and eventually degraded via efferosome-lysosome fusion and acidification of the efferosome [139, 140]. Removal of apoptotic material via efferocytosis is thought essential to prevent secondary necrosis, which involves the uncontrolled release of intracellular contents from apoptotic cells and consequently promoting inflammation [141, 142]. In contrast to phagocytosis of pathogens, efferocytosis is generally considered an immunologically silent event accompanied by the induction of tissue repair [56]. Traditionally, macrophages were thought to be the primary cells responsible for the engulfment of apoptotic IECs [143]; however, Paneth cells can additionally phagocytose microbes [10, 12] and can remove neighbouring apoptotic IECs via efferocytosis [144]. Similarly, healthy IECs can engulf their apoptotic neighbours [145]. One of the major hurdles in the field is the current inability to track the efferocytic process in real time in mammalian tissues. Recently, live cell tracking of efferocytosis in the *Drosophila* embryo revealed important cellular heterogeneity of apoptotic cell clearance during development [146, 147]. Application of such approaches to the mammalian intestinal tract would shed light on the respective contributions of each specialised cell type to the efferocytic process and, importantly, how this critical process is influenced by invading pathogens or the resident commensal microbe population.

As discussed, host cell death does not always increase the virulence of pathogens but instead can prevent intracellular replication and expose bacteria to host defences. Accordingly, pathogens may generate 'find-me' signals within infected cells in order to be efferocytosed by phagocytes following host cell death, allowing them to 'hide' inside phagocytes from the extracellular mileu [148]. *Listeria* is an example of such a pathogen, using efferocytosis to its advantage. By employing pore-forming toxins such as LLO to disrupt phagocytic plasma membranes and hijacking host actin machinery to exit the phagosome, *Listeria* releases bacteria-containing vesicles from phagocytes [149]. The inclusion of host PtdSer on these vesicles, recognised by macrophage Tim4 receptors, allows cell-to-cell *Listeria* spread in macrophages in vitro [149]. This process includes the incorporation of PtdSer alongside bacteria into membrane structures [149] and the uptake of actin-containing protrusions by neighbouring cells involves caveolin-mediated uptake of bacteria [150]. Tim4 has also recently been implicated in a link between antibiotics, the microbiome and efferocytosis. Tim4 was upregulated by microbiota-derived butyrate, which rescued defects in large peritoneal macrophage efferocytosis in germ-free, antibiotic-treated mice, which lack normal intestinal microbiota [151].

Staphylococcus aureus (*S. aureus*) frequently exists as a commensal bacterium in the nasopharynx, as a coloniser of the airways in chronic lung disease (such as CF and non-CF bronchiectasis), and also as a pathogen that can cause severe pneumonia [152, 153]. Mer receptor tyrosine kinase (MerTK), expressed primarily on macrophages, helps mediate efferocytic clearance of apoptotic cells and helps drive anti-inflammatory responses [142, 154]. MerTK is upregulated in the mouse lung during *S. aureus* infection and regulates SOCS1/3, ultimately repressing TLR2/6 signalling and MAPK/NF-kB-mediated inflammation that usually follows infection [155]. Very recently, multi-omic analyses (based on bulk RNA-seq, scRNA-seq data, and BioGPS.RNA) have shown that respiratory diseases such as COPD and asthma feature upregulated MerTK in human lungs, suggesting a similar MerTK-mediated repression of inflammation could occur [156].

MerTK can be proteolytically cleaved by ADAM17 to release soluble Mer, which inhibits efferocytosis in vitro [157]. Cleavage can be initially driven by LPS/TLR4 signalling in macrophages, suggesting that this may be implicated in bacterial infections [157]. Accordingly, LPS treatment of the lung increased soluble Mer while decreasing membrane-bound Mer in alveolar macrophages and lung tissue via enhanced Mer cleavage [158]. Pharmacologically restoring Mer expression using ADAM-17 inhibitor TAPI-0 rescued inflammation during LPS-induced acute lung injury in mice [159]. However, further research reported that soluble Mer expression was increased in patients with septic acute lung injury/acute respiratory distress syndrome, and that in 2 human cell lines (HL-60 and pulmonary aortic ECs), inhibiting Mer using UNC2250 inhibited LPS-driven increases in cell proliferation and endothelial dysfunction while increasing apoptosis [160]. UNC2250-induced Mer inhibition also reduced mortality in LPS-induced acute lung injury mouse models in this study [160]. This suggests that in this setting, Mer is detrimental to the host, not beneficial as previously found [159], a fact acknowledged by the authors and which may occur due to different methods of infection being used. Indeed, one study induced acute lung injury by inhalation of LPS [160] and the other used LPS intratracheally [159]. The role of Mer in bacterial lung infections, therefore, urgently requires clarification and may represent an important therapeutic target during bacterial-induced acute lung injury.

Within the intestine, ablation of MerTK in a mouse model of colorectal cancer increased pro-inflammatory cytokine expression and promoted the development of colitis-associated cancer, suggesting a beneficial role of MerTK in this setting, despite contributing to oncogenesis in other conditions [161]. In light of this, it could be questioned whether MerTK is a valuable target for reducing inflammation in IBD. However, recent research has reported that the apoptosis-induced MerTK/ERK/TGF-B1 pathway is overactive in fibrotic intestinal tissue of IBD (Crohn's disease) patients, and inhibition of MerTK reduced intestinal fibrosis in mouse models of colitis [162]. Therefore, it seems that the role of MerTK in the gut as well as the lung has opposing positive and negative influences and represents an ongoing area of study.

IL-17-producing T helper cells (Th17) are a subset of CD4+ T cells that secrete IL-17 to aid neutrophil recruitment to sites of infection [163]. Th17 cells have been revealed to be specifically induced in vivo by efferocytosis of infected apoptotic neutrophils or B cells, which resulted in the release of the cytokines (TGF-B and IL-6) necessary to expand the Th17 population [164]. Inhibition of apoptosis in the intestinal epithelium of mice infected with *C. rodentium* accordingly inhibited the Th17 immune response [164]. However, infection-induced apoptosis may lead to autoimmunity, which has been shown to occur with Th17 cells and can promote intestinal auto-inflammation and autoimmunity in certain scenarios [165]. Specifically, CD4+ T cells differentiate into Th17 cells in mice when activated by MHC-II self-antigen presentation following microbial (*C. rodentium*)-induced apoptosis, leading to colitis [165].

Salmonella possesses the ability to initiate phagocytic uptake by neutrophils using complement components [148]. Using the T3SS to puncture *Salmonella*-containing vacuoles, *Salmonella* induces complement-dependent 'find-me' signal release that triggers phagocytosis, allowing *Salmonella* to reside in pore-induced intracellular traps and avoid reactive oxygen species (ROS)-induced damage [148]. As parts of the complement system have been shown to bind apoptotic cells [166, 167], it is entirely possible that this mechanism could be used during apoptosis to mediate pathogen uptake. Similarly, non-professional phagocyte efferocytosis occurs in the neonatal intestine in mice [168]. Infected apoptotic enterocytes in the gut are removed by neighbouring IECs, exploiting 'eat me' adaptors and receptors including PtdSer, C1q, and CD36 with their corresponding genes upregulated in infected neonatal mouse intestines compared to adult animals [168]. Intriguingly, such non-professional efferocytosis has not been confirmed during infection with *E. coli*, suggesting pathogen-specific mechanisms of host defence, with additional influence by age as *Salmonella* is cleared via exfoliation in adult mice [168]. Non-professional efferocytosis has been reported in several tissues, including by airway epithelial cells in response to apoptosis in the lung [169]; therefore, it is very plausible that more instances of non-professional efferocytosis in response to gut and lung pathogens are yet to be discovered.

In the brain, opsonins released by microglia in response to LPS increased phagocytosis of *E. coli* in mouse and rat glial cultures [170]. The opsonins in question are calreticulin and galectin-3, which are also expressed by neutrophils and macrophages and can function as 'eat-me' signals during apoptosis [170]. Furthermore, MerTK is a receptor for galectin-3. Blocking MerTK partially inhibited LPS-mediated phagocytosis (binding of gram-negative bacterial surface LPS to TLR4) of *E. coli* [170], indicating the importance of this opsonin in phagocytosis. Given that calreticulin and galectin-3 are also found in humans and are upregulated in some disease states [170], it remains to be determined whether a similar phagocytic mechanism could be triggered in human tissues. Moreover, LPS is a core feature of gram-negative bacteria, indicating that perhaps a similar mechanism could occur with other organisms.

Mtb, as described above, infects macrophages, leading to neighbouring healthy macrophages efferocytosing infected cells in vivo and in vitro [171]. This is of benefit to the host and facilitates lysosomal death of *Mtb*, which is not achieved through apoptosis alone. Besides engulfment by other (uninfected) macrophages, dendritic cells also recognise and engulf infected macrophages by recognition of annexin 1 [172]. This aids defence against *Mtb,* as annexin 1-deficient mice have higher bacterial burdens than their wildtype counterparts [172]. Annexin 1 appears to aid antigen transfer from apoptotic vesicles to dendritic cells for presentation to T cells, leading to increased CD8+ T cells [172]. Additionally, *Mtb* suppressed annexin 1 expression in dendritic cells derived from human blood samples, indicating that this role may be conserved across species.

Macrophages are potent killers of *Pseudomonas*, presumably via lysosomal degradation, and when macrophages are exposed to *Pseudomonas* as a pre-stimulus, an increase in macrophage efferocytic ability is observed [173]. Strikingly, this increase is lower when macrophages are exposed to *Pseudomonas* plus apoptotic cells as a pre-stimulus. Given that apoptotic cells would be expected to increase efferocytosis, it was suggested that there may be some antagonism between bacteria and apoptotic cells, leading to the greatest efferocytic increase occurring when macrophages were exposed to bacteria alone. Although this requires further study, it was also reported that *Pseudomonas* infection induces pro-inflammatory gene expression in mouse bone marrow-derived macrophages (BMDMs), particularly a rise in IL-6, which is accordingly involved in the increased efferocytic ability of infected macrophages and has previously been linked to the induction of efferocytosis [173]. The involvement of IL-6 in the efferocytic differences seen in macrophages is yet to be determined. In another study, *Pseudomonas* was found to interact with the epithelial barrier at sites of apoptotic body release, where it attached to the apoptotic bodies and was efferocytosed and killed by surrounding epithelial cells [174].

BAI1 is a GPCR also implicated in efferocytosis and can recognise and bind PtdSer upstream of efferocytic protein ELMO [175]. Moreover, knockdown of BAI1 reduces efferocytosis [175]. In the gut, BAI1-mediated clearance of apoptotic cells reduced inflammation and colitis severity in transgenic mice overexpressing BAI1 either globally or in colonic epithelial cells [145]. BAI1 also recognises LPS, and downstream of this is critical for ROS-dependent bactericidal function against multiple gram-negative pathogens [176]. It has been reported that fibroblasts expressing BAI1 phagocytosed *E. coli* more effectively than fibroblasts lacking BAI1 due to knockout [176]. Using BMDMs from wildtype or BAI1-deficient mice, internalisation of *E. coli* was also reduced in BAI1-deficient mice. The study reported recognition of gram-negative bacteria only (as opposed to gram-positive) at the macrophage membrane by BAI1, discovered by using BMDMs expressing a BAI1 construct with an N-terminal HA tag [176]. BAI1-deficient BMDMs exhibited reduced Rac activity and ROS production in response to bacteria and, concordantly, reduced ability to kill *E. coli* compared to wildtype BMDMs. In a model of peritoneal infection, BAI-1 deficient mice were more susceptible to death upon bacterial infection due to reduced clearance [176], cumulatively suggesting that BAI1 aids bacterial clearance.

During *Salmonella* infection, BAI1 binds LPS to mediate phagocytosis [177]. Inhibition of BAI1 in murine J774 macrophages led to reduced internalisation of *Salmonella*, whereas overexpression of BAI1 enhanced engulfment [175]. It is currently unclear how one particular receptor, such as BAI1, can differentiate between two very distinct biological ligands (LPS versus PtdSer) to execute either ROS-dependent bactericidal function or efferocytosis of apoptotic corpses. Presumably, additional accessory molecules are required to make these distinctions; however, the identity and the mechanisms of these molecules remain unexplored. This particular example raises the interesting question of how the efferocytic and/or phagocytic process is 'distracted' or altered when confronted with opposing signals. Furthermore, although it is clear that BAI1 has a potential role in efferocytosis, it is less clear which phagocytes or cells it is expressed by [178], leaving another exciting avenue for future research.

Finally, various gut pathogens have effector proteins that can bind ELMO1 in the host to avoid macrophage engulfment and shelter in vacuoles [179]. Such pathogens are *Salmonella, Shigella,* and *E. coli*, and all of these have *WxxxE* motif-containing effectors that are able to bind a specific PH domain site on ELMO1, leading to activation of Rho GTPases and accordingly disruption of host actin dynamics [179]. Disrupting this effector-ELMO1 interaction reduced bacterial burden in J774 macrophages, and this represents an interesting crossover where a diverse range of bacteria utilise a conserved mechanism to avoid engulfment [179]. Whether a similar rescue occurs at the tissue level is yet to be confirmed, as is the involvement of apoptosis in these events.

Overall, despite not being an exhaustive list of pathogen-efferocytosis interactions during apoptosis, the examples described represent the broad range of scenarios in which microbes or hosts may influence efferocytic events for their advantage. Future studies assessing the ability of host cells to simultaneously cope with apoptotic corpses and infectious bacteria will be primarily driven by novel imaging tools in vivo and may identify critical choke points that tip the scales between host defence and pathogen adaptation.

Phagocytosis-Induced Cell Death

Although macrophages undergoing apoptosis are generally accepted to be detrimental to the host, neutrophils have a short life span and high turnover rate due to high levels of constitutive apoptosis. Neutrophil apoptosis has been well explored as an anti-inflammatory mechanism [29, 180, 181]. The inflammatory protein myeloid cell leukaemia 1 (Mcl-1) has been studied in neutrophils as a key apoptosis regulator and target that can be manipulated to reduce inflammation while leaving host bacterial defence intact [29]. Mcl-1 inhibition by a cyclin-dependent kinase (CDK) inhibitor or wogonin-induced apoptosis of human and mouse neutrophils without affecting macrophage apoptosis and without detriment to the phagocytosis of apoptotic neutrophils. The CDK inhibitor successfully increased neutrophil apoptosis even during the addition of LPS or other bacterial-derived products. In an

LPS-induced model of neutrophilic lung inflammation, Mcl-1 inhibition reduced inflammatory cells—predominantly neutrophils—and enhanced resolution of inflammation. Additionally, in an in vivo model of lung infection with *E. coli*, Mcl-1 down-regulation showed improved resolution of inflammation and enhanced bacterial clearance [29].

However, neutrophil apoptosis as a mechanism driving successful or failed bacterial clearance has been explored much less [182]. Neutrophils can also undergo apoptosis upon their engulfment of pathogens, called phagocytosis-induced cell death (PICD). In a study of the phagocytosis of different pathogens such as *Staphylococcus* and *Listeria* by human neutrophils, these pathogens increased neutrophil apoptosis after phagocytosis [183]. DNA microarray measurements showed an increase in the expression of apoptosis-related genes such as BAC and TNF but also an increase in genes encoding surface molecules involved in the recruitment of immune effector cells, while genes linked to the expression of surface molecules involved in chemotaxis were decreased [183].

A further study examining phagocytosis in human and mouse neutrophils in vitro reported that during antibody-independent phagocytosis of *E. coli* or *S. aureus*, class III PI3K was involved in ROS generation, which contributed to bacterial killing, although it was dispensable for phagocytosis itself [184]. More work in neutrophils uncovered that Resolvin E1 (RvE1), a lipid mediator, promoted PICD and that RvE1 aided resolution of *E. coli*-induced acute lung injury in mice and had anti-inflammatory actions [185]. This RvE1 rescue was inhibited by caspase inhibition with zVAD-FMK, indicating that the anti-inflammatory effects are related to increased neutrophil apoptosis [185]. In this case, therefore, the induction of PICD is a desirable outcome for the host.

In healthy volunteer human neutrophils, *Pseudomonas* induces apoptosis in vitro using its self-produced phenazine pyocyanin in a time- and concentration-dependent manner, with apoptosis reduced in a phenazine-deficient *Pseudomonas* strain [186]. This was confirmed in clinical isolates of *Pseudomonas*, in which pyocyanin production drove neutrophil apoptosis in culture. Anti-apoptotic LPS-treated and CF patient-derived neutrophils were unable to limit pyocyanin-induced apoptosis in co-culture experiments, whereas cAMP analogues and antioxidants successfully reduced apoptosis [186]. Due to the pro-inflammatory effects of pyocyanin and the fact that high concentrations of pyocyanin can induce necrotic neutrophil death in the lung, PICD may be regarded as a mechanism of pathogen evasion of host defence in this instance. In more recent research, pyocyanin was reported to induce apoptosis in cultured and murine neutrophils ex vivo via interacting with the mitochondrial respiratory chain [187]. This resulted in ROS release, activation of mitochondrial acid sphingomyelinase (ASM), formation of ceramide, and cytochrome c release. Interestingly, this pathway was abrogated in neutrophils deficient in ASM, and these neutrophils were resistant to *Pseudomonas*-induced death [187].

Pseudomonas is also reported to induce PICD in vivo and in vitro in natural killer cells when treated with a PI3K inhibitor [188]. The increased caspase 9-dependent apoptosis of natural killer cells exacerbated the metastasis of lung cancer cells in mice [188], indicating that PICD was contributing to this phenotype in these patients. In contrast, in J774 macrophages, *Pseudomonas* was phagocytosed and induced macrophage apoptosis [189]. In this case, PICD aided inhibition of intracellular growth of *Pseudomonas* and was key in host defence, although this wasn't tested in vivo as with the natural killer cell findings, perhaps explaining the opposing influences.

In human neutrophils, *Mtb* is phagocytosed and can induce neutrophil apoptosis [190]. Upon ingestion of the pathogen, neutrophils produce ROS to kill *Mtb*, and ROS promotes neutrophil apoptosis [190]. Moreover, efferocytosis of apoptotic neutrophils by human macrophages increases proinflammatory cytokine production by macrophages at a level higher than that seen when macrophages were exposed to *Mtb* alone. Despite phagocytosis-induced apoptosis in this setting contributing to a proinflammatory response from macrophages, the removal of *Mtb* via neutrophil apoptosis is still a crucial line of host defence to prevent infection spreading, and proinflammatory TNF is thought to be necessary to control *Mtb* infection.

E. coli has been reported to inhibit PICD in RAW 267.4 and BMDMs upon infection, achieved via activation of NF-kB [191]. When NF-kB was inhibited in macrophages, increased apoptosis following *E. coli* engulfment was observed. However, the phagocytic uptake of *E. coli,* shown by flow cytometry, was unchanged. Alongside this, inhibition of NF-kB led to reduced antigen-specific T cell activation, suggesting an impaired adaptive immune response, and the ability of macrophages to express costimulatory molecules was imapired [191]. This suggests that although inhibition of PICD by *E. coli* reduces bacterial killing, the activation of NF-kB appears important in mounting an effective immune response, making it difficult to determine an overall positive or negative result of PICD.

In summary, PICD adds another layer to the apoptosis tug of war between host defence and pathogen virulence, showing that beyond pathogens inducing apoptosis in different cell types with different outcomes, the mechanism of inducing apoptosis itself—namely, whether this occurs following phagocytosis—is also a key consideration.

Applicability to Other Forms of Cell Death

Apoptosis is perhaps the best characterised pathway of programmed cell death; however, it is certainly not the only form [2]. Whether it is inflammatory cell death pathways such as pyroptosis and necroptosis or iron-mediated ferroptosis [2], each programmed cell death pathway can apply to the three stages described above: (i) some specific trigger induces pathway-dependent molecular execution of death; (ii) the dying corpse secretes factors that communicate with the surrounding tissue; and (iii) the host will clear the dying corpse to minimise inflammation and damage to the healthy tissue. In the case of bacterial infection, *Salmonella*, for instance, induces pyroptosis in infected IECs [192, 193] and pyroptosis drives the release of dying IECs into the lumen of the gut [194], which puts the corpse in close proximity with the bacterial community. Furthermore, the fact that during caspase inhibition, *Salmonella*-induced necroptosis may occur highlights the complexity of rerouting cell death pathways away from apoptosis [195, 196]. With novel imaging capabilities to detect some of these lytic forms of programmed cell death in real-time [147] and similar functions being observed for the mediators secreted during lytic death as previously shown during apoptosis [197], future research studying these steps of the cell death process may reveal new layers to host-microbe interactions during infection. Furthermore, insights into how infection influences these processes during apoptosis could be more broadly applicable to other forms of programmed cell death and represent a blossoming field of research.

Summary

This review aims to provoke thought surrounding the sheer volume of research surrounding the field of apoptosis during host-microbe interactions and, more importantly, highlight the questions yet to be answered. Perhaps one of the most frequently mentioned points throughout this review is the vast array of methods that different bacterial species or strains may use to induce or inhibit apoptosis in different cell types and whether or not this is to the advantage of the host or microbe. These seemingly contradictory findings underscore the importance that there is no one-size-fits-all approach likely to be useful when exploring this field, whereas research applied to one cell, tissue, or bacteria may have relevance to another, it is also entirely likely that each finding exists only in a specific niche. Given the challenges we highlight above, we propose that a re-focus on the downstream consequences of apoptosis during infection, including secreted factors and corpse uptake, rather than the molecular machinery that executes the death process, may be of value to the field. As evidence continues to mount that inhibiting specific apoptotic executioners re-routes the cell into an alternative form of programmed cell death, focusing on the next steps of the apoptotic process may help resolve which components of apoptosis benefit the host, which components benefit the microbe, and which components benefit both.

Acknowledgements We apologise to any authors whose work we omitted due to space constraints. AD is supported by the Ker Memorial PhD Studentship. CDL is supported by an MRC Clinician Scientist award (MR/X019314/1) and an MRC Programme Grant (MR/W019264/1). CJA is supported by a Wellcome Trust Career Development Award (225923/Z/22/Z). For the purpose of open access, the authors have applied a Creative Commons Attribution (CC BY) licence to any Author Accepted Manuscript version arising from this submission.

References

1. Yuan, J., & Ofengeim, D. (2023). A guide to cell death pathways. *Nature Reviews Molecular Cell Biology*. https://doi.org/10.1038/s41580-023-00689-6
2. Galluzzi, L., et al. (2018). Molecular mechanisms of cell death: Recommendations of the nomenclature committee on cell death 2018. *Cell Death and Differentiation, 25*, 486–541. https://doi.org/10.1038/s41418-017-0012-4

3. Costigan, A., Hollville, E., & Martin, S. J. (2023). Discriminating between apoptosis, necrosis, necroptosis, and ferroptosis by microscopy and flow cytometry. *Current Protocols, 3*. https://doi.org/10.1002/cpz1.951
4. Park, W., et al. (2023). Diversity and complexity of cell death: A historical review. *Experimental & Molecular Medicine, 55*, 1573–1594. https://doi.org/10.1038/s12276-023-01078-x
5. Tang, D., Kang, R., Berghe, T. V., Vandenabeele, P., & Kroemer, G. (2019). The molecular machinery of regulated cell death. *Cell Research, 29*, 347–364. https://doi.org/10.1038/s41422-019-0164-5
6. Bertheloot, D., Latz, E., & Franklin, B. S. (2021). Necroptosis, pyroptosis and apoptosis: An intricate game of cell death. *Cellular & Molecular Immunology, 18*, 1106–1121. https://doi.org/10.1038/s41423-020-00630-3
7. Smith, K. G., Kamdar, A. A., & Stark, J. M. (2019). *Kendig's* disorders of the respiratory tract in children (8th ed., eds Robert William Wilmott et al., pp. 120–133.e122). Elsevier.
8. Panwar, H., et al. (2021). Gastrointestinal stress as innate defence against microbial attack. *Journal of Applied Microbiology, 130*, 1035–1061. https://doi.org/10.1111/jam.14836
9. Negroni, A., Cucchiara, S., & Stronati, L. (2015). Apoptosis, necrosis, and necroptosis in the gut and intestinal homeostasis. *Mediators of Inflammation, 2015*, 250762. https://doi.org/10.1155/2015/250762
10. Kong, S., Zhang, Y. H., & Zhang, W. (2018). Regulation of intestinal epithelial cells properties and functions by amino acids. *BioMed Research International, 2018*, 2819154. https://doi.org/10.1155/2018/2819154
11. Gustafsson, J. K., & Johansson, M. E. V. (2022). The role of goblet cells and mucus in intestinal homeostasis. *Nature Reviews Gastroenterology & Hepatology, 19*, 785–803. https://doi.org/10.1038/s41575-022-00675-x
12. Wallaeys, C., Garcia-Gonzalez, N., & Libert, C. (2023). Paneth cells as the cornerstones of intestinal and organismal health: A primer. *EMBO Molecular Medicine, 15*, e16427. https://doi.org/10.15252/emmm.202216427
13. Bergstrom, K. S. B., et al. (2010). Muc2 protects against lethal infectious colitis by disassociating pathogenic and commensal bacteria from the colonic mucosa. *PLoS Pathogens, 6*, e1000902. https://doi.org/10.1371/journal.ppat.1000902
14. Gribble, F. M., & Reimann, F. (2019). Function and mechanisms of enteroendocrine cells and gut hormones in metabolism. *Nature Reviews Endocrinology, 15*, 226–237. https://doi.org/10.1038/s41574-019-0168-8
15. Arora, T., Vanslette, A. M., Hjorth, S. A., & Bäckhed, F. (2021). Microbial regulation of enteroendocrine cells. *Med, 2*, 553–570. https://doi.org/10.1016/j.medj.2021.03.018.
16. Snoeck, V., Goddeeris, B., & Cox, E. (2005). The role of enterocytes in the intestinal barrier function and antigen uptake. *Microbes and Infection, 7*, 997–1004. https://doi.org/10.1016/j.micinf.2005.04.003
17. Hegarty, L. M., Jones, G.-R., & Bain, C. C. (2023). Macrophages in intestinal homeostasis and inflammatory bowel disease. *Nature Reviews Gastroenterology & Hepatology*. https://doi.org/10.1038/s41575-023-00769-0
18. Domanska, D., et al. (2022). Single-cell transcriptomic analysis of human colonic macrophages reveals niche-specific subsets. *Journal of Experimental Medicine, 219*. https://doi.org/10.1084/jem.20211846
19. Schulz, O., & Pabst, O. (2013). Antigen sampling in the small intestine. *Trends in Immunology, 34*, 155–161. https://doi.org/10.1016/j.it.2012.09.006
20. Stagg, A. J. (2018). Intestinal dendritic cells in health and gut inflammation. *Frontiers in Immunology, 9*. https://doi.org/10.3389/fimmu.2018.02883
21. Maynard, C. L., & Weaver, C. T. (2009). Intestinal effector T cells in health and disease. *Immunity, 31*, 389–400. https://doi.org/10.1016/j.immuni.2009.08.012
22. Belkaid, Y. (2007). Regulatory T cells and infection: A dangerous necessity. *Nature Reviews Immunology, 7*, 875–888. https://doi.org/10.1038/nri2189

23. Sonnenberg, G. F., & Hepworth, M. R. (2019). Functional interactions between innate lymphoid cells and adaptive immunity. *Nature Reviews Immunology, 19*, 599–613. https://doi.org/10.1038/s41577-019-0194-8
24. Ruysseveldt, E., Martens, K., & Steelant, B. (2021). Airway basal cells, protectors of epithelial walls in health and respiratory diseases. *Frontiers in Allergy, 2*. https://doi.org/10.3389/falgy.2021.787128
25. Ye, Q., & Bankova, L. G. (2022). Brush cells fine-tune neurogenic inflammation in the airways. *The Journal of Clinical Investigation, 132*. https://doi.org/10.1172/JCI161439
26. Perniss, A. et al. (2020). Chemosensory cell-derived acetylcholine drives tracheal mucociliary clearance in response to virulence-associated formyl peptides. *Immunity, 52*, 683–699. e611. https://doi.org/10.1016/j.immuni.2020.03.005.
27. Hiemstra, P. S., Tetley, T. D., & Janes, S. M. (2019). Airway and alveolar epithelial cells in culture. *European Respiratory Journal, 54*, 1900742. https://doi.org/10.1183/13993003.00742-2019
28. Croasdell Lucchini, A., Gachanja, N. N., Rossi, A. G., Dorward, D. A., & Lucas, C. D. (2021). Epithelial cells and inflammation in pulmonary wound repair. *Cells, 10*. https://doi.org/10.3390/cells10020339
29. Lucas, C. D., et al. (2014). Downregulation of Mcl-1 has anti-inflammatory pro-resolution effects and enhances bacterial clearance from the lung. *Mucosal Immunology, 7*, 857–868. https://doi.org/10.1038/mi.2013.102
30. Felton, J. M., Lucas, C. D., Rossi, A. G., & Dransfield, I. (2014). Eosinophils in the lung – Modulating apoptosis and efferocytosis in airway inflammation. *Frontiers in Immunology, 5*, 302. https://doi.org/10.3389/fimmu.2014.00302
31. Bain, C. C., Lucas, C. D., & Rossi, A. G. (2022). Pulmonary macrophages and SARS-Cov2 infection. *International Review of Cell and Molecular Biology, 367*, 1–28. https://doi.org/10.1016/bs.ircmb.2022.01.001
32. Marwick, J. A., et al. (2013). Oxygen levels determine the ability of glucocorticoids to influence neutrophil survival in inflammatory environments. *Journal of Leukocyte Biology, 94*, 1285–1292. https://doi.org/10.1189/jlb.0912462
33. Wang, Y. T., et al. (2023). Metabolic adaptation supports enhanced macrophage efferocytosis in limited-oxygen environments. *Cell Metabolism, 35*, 316–331.e316. https://doi.org/10.1016/j.cmet.2022.12.005
34. Martín, R., et al. (2013). Role of commensal and probiotic bacteria in human health: A focus on inflammatory bowel disease. *Microbial Cell Factories, 12*, 71. https://doi.org/10.1186/1475-2859-12-71
35. Natalini, J. G., Singh, S., & Segal, L. N. (2023). The dynamic lung microbiome in health and disease. *Nature Reviews Microbiology, 21*, 222–235. https://doi.org/10.1038/s41579-022-00821-x
36. Rinninella, E., et al. (2019). What is the healthy gut microbiota composition? A changing ecosystem across age, environment, diet, and diseases. *Microorganisms, 7*. https://doi.org/10.3390/microorganisms7010014
37. Honda, K., & Littman, D. R. (2016). The microbiota in adaptive immune homeostasis and disease. *Nature, 535*(7610), 75–84. https://doi.org/10.1038/nature18848
38. Montassier, E., et al. (2015). Chemotherapy-driven dysbiosis in the intestinal microbiome. *Alimentary Pharmacology and Therapeutics*. https://doi.org/10.1111/apt.13302
39. Shin, N.-R., Whon, T. W., & Bae, J.-W. (2015). Proteobacteria: Microbial signature of dysbiosis in gut microbiota. *Trends in Biotechnology, 33*(9), 496–503. https://doi.org/10.1016/j.tibtech.2015.06.011
40. Gevers, D., et al. (2014). The treatment-naive microbiome in new-onset crohn's disease. *Cell Host & Microbe, 15*, 382–392. https://doi.org/10.1016/j.chom.2014.02.005
41. Lupp, C., et al. (2007). Host-mediated inflammation disrupts the intestinal microbiota and promotes the overgrowth of enterobacteriaceae. *Cell Host & Microbe, 2*, 119–129. https://doi.org/10.1016/j.chom.2007.06.010

42. Walker, A. W., & Hoyles, L. (2023). Human microbiome myths and misconceptions. *Nature Microbiology, 8*, 1392–1396. https://doi.org/10.1038/s41564-023-01426-7
43. Ni, J., Wu, G. D., Albenberg, L., & Tomov, V. T. (2017). Gut microbiota and IBD: Causation or correlation? *Nature Reviews Gastroenterology & Hepatology, 14*, 573–584. https://doi.org/10.1038/nrgastro.2017.88
44. Sonis, S .T. (2004). The pathobiology of mucositis. *Nature Reviews Cancer, 4*(4), 277–284. https://doi.org/10.1038/nrc1318
45. van Vliet, M. J., Harmsen, H. J. M., de Bont, E. S. J. M., & Tissing, W. J. E. (2010). The role of intestinal microbiota in the development and severity of chemotherapy-induced mucositis. *PLoS Pathogens, 6*, 1–7. https://doi.org/10.1371/JOURNAL.PPAT.1000879
46. Wei, L., Wen, X. S., & Xian, C. J. (2021). Chemotherapy-induced intestinal microbiota dysbiosis impairs mucosal homeostasis by modulating toll-like receptor signaling pathways. *International Journal of Molecular Sciences, 22*. https://doi.org/10.3390/IJMS22179474
47. Pammi, M., et al. (2017). Intestinal dysbiosis in preterm infants preceding necrotizing enterocolitis: A systematic review and meta-analysis. *Microbiome, 5*. https://doi.org/10.1186/s40168-017-0248-8
48. Zhu, W., et al. (2018). Precision editing of the gut microbiota ameliorates colitis. *Nature, 553*, 208–211. https://doi.org/10.1038/nature25172
49. Bäumler, A. J., & Sperandio, V. (2016). Interactions between the microbiota and pathogenic bacteria in the gut. *Nature, 535*, 85–93. https://doi.org/10.1038/nature18849
50. World Health Organization [WHO]. (2015). *WHO estimates of the global burden of foodborne diseases*. https://apps.who.int/iris/bitstream/handle/10665/199350/9789241565165_eng.pdf?sequence=1. Access 21 Jul 2020.
51. Zhang, D., et al. (2020). The cross-talk between gut microbiota and lungs in common lung diseases. *Frontiers in Microbiology, 11*. https://doi.org/10.3389/fmicb.2020.00301
52. Budden, K. F., et al. (2017). Emerging pathogenic links between microbiota and the gut–lung axis. *Nature Reviews Microbiology, 15*, 55–63. https://doi.org/10.1038/nrmicro.2016.142
53. Patangia, D. V., Anthony Ryan, C., Dempsey, E., Paul Ross, R., & Stanton, C. (2022). Impact of antibiotics on the human microbiome and consequences for host health. *Microbiology, 11*, e1260. https://doi.org/10.1002/mbo3.1260
54. Kerr, J. F., Wyllie, A. H., & Currie, A. R. (1972). Apoptosis: A basic biological phenomenon with wide-ranging implications in tissue kinetics. *British Journal of Cancer, 26*, 239–257. https://doi.org/10.1038/bjc.1972.33
55. Ketelut-Carneiro, N., & Fitzgerald, K. A. (2022). Apoptosis, pyroptosis, and necroptosis — Oh my! The many ways a cell can die. *Journal of Molecular Biology, 434*, 167378. https://doi.org/10.1016/j.jmb.2021.167378
56. Yin, C., & Heit, B. (2021). Cellular responses to the efferocytosis of apoptotic cells. *Frontiers in Immunology, 12*, 631714. https://doi.org/10.3389/fimmu.2021.631714
57. McIlwain, D. R., Berger, T., & Mak, T. W. (2013). Caspase functions in cell death and disease. *Cold Spring Harbor Perspectives in Biology, 5*, a008656. https://doi.org/10.1101/cshperspect.a008656
58. Parrish, A. B., Freel, C. D., & Kornbluth, S. (2013). Cellular mechanisms controlling caspase activation and function. *Cold Spring Harbor Perspectives in Biology, 5*. https://doi.org/10.1101/cshperspect.a008672
59. Tait, S. W., & Green, D. R. (2013). Mitochondrial regulation of cell death. *Cold Spring Harbor Perspectives in Biology, 5*. https://doi.org/10.1101/cshperspect.a008706
60. Wang, C., & Youle, R. J. (2009). The role of mitochondria in apoptosis*. *Annual Review of Genetics, 43*, 95–118. https://doi.org/10.1146/annurev-genet-102108-134850
61. Dickens, L. S., Powley, I. R., Hughes, M. A., & MacFarlane, M. (2012). The 'complexities' of life and death: Death receptor signalling platforms. *Experimental Cell Research, 318*, 1269–1277. https://doi.org/10.1016/j.yexcr.2012.04.005
62. Aliprantis, A. O., Yang, R. B., Weiss, D. S., Godowski, P., & Zychlinsky, A. (2000). The apoptotic signaling pathway activated by toll-like receptor-2. *The EMBO Journal, 19*, 3325–3336. https://doi.org/10.1093/emboj/19.13.3325

63. Grassmé, H., Jendrossek, V., & Gulbins, E. (2001). Molecular mechanisms of bacteria induced apoptosis. *Apoptosis, 6*, 441–445. https://doi.org/10.1023/A:1012485506972
64. Knodler, L. A., Finlay, B. B., & Steele-Mortimer, O. (2005). The salmonella effector protein sopb protects epithelial cells from apoptosis by sustained activation of Akt. *Journal of Biological Chemistry, 280*, 9058–9064. https://doi.org/10.1074/jbc.M412588200
65. Lawrence, A.-L. E., et al. (2021). Salmonella enterica serovar typhimurium SPI-1 and SPI-2 shape the global transcriptional landscape in a human intestinal organoid model system. *MBio, 12*. https://doi.org/10.1128/mbio.00399-00321
66. Lin, H.-H., et al. (2021). Activation of apoptosis by salmonella pathogenicity Island-1 effectors through both intrinsic and extrinsic pathways in salmonella-infected macrophages. *Journal of Microbiology, Immunology and Infection, 54*, 616–626. https://doi.org/10.1016/j.jmii.2020.02.008
67. Collier-Hyams, L. S., et al. (2002). Cutting edge: salmonella AvrA effector inhibits the key proinflammatory, anti-apoptotic NF-κB pathway1. *The Journal of Immunology, 169*, 2846–2850. https://doi.org/10.4049/jimmunol.169.6.2846
68. Faherty, C. S., & Maurelli, A. T. (2008). Staying alive: Bacterial inhibition of apoptosis during infection. *Trends in Microbiology, 16*, 173–180. https://doi.org/10.1016/j.tim.2008.02.001
69. Clark, C. S., & Maurelli, A. T. (2007). Shigella flexneri inhibits staurosporine-induced apoptosis in epithelial cells. *Infection and Immunity, 75*, 2531–2539. https://doi.org/10.1128/iai.01866-06
70. Nothelfer, K., et al. (2014). B lymphocytes undergo TLR2-dependent apoptosis upon Shigella infection. *The Journal of Experimental Medicine, 211*, 1215–1229. https://doi.org/10.1084/jem.20130914
71. Arizmendi, O., Picking, W. D., & Picking, W. L. (2016). Macrophage apoptosis triggered by IpaD from shigella flexneri. *Infection and Immunity, 84*, 1857–1865. https://doi.org/10.1128/iai.01483-15
72. Kaper, J. B., Nataro, J. P., & Mobley, H. L. T. (2004). Pathogenic Escherichia coli. *Nature Reviews Microbiology, 2*, 123–140. https://doi.org/10.1038/nrmicro818
73. Hemrajani, C., et al. (2010). NleH effectors interact with Bax inhibitor-1 to block apoptosis during enteropathogenic Escherichia coli infection. *Proceedings of the National Academy of Sciences of the United States of America, 107*, 3129–3134. https://doi.org/10.1073/pnas.0911609106
74. Li, S., et al. (2013). Pathogen blocks host death receptor signalling by arginine GlcNAcylation of death domains. *Nature, 501*, 242–246. https://doi.org/10.1038/nature12436
75. Hemrajani, C., et al. (2008). Role of NleH, a type III secreted effector from attaching and effacing pathogens, in colonization of the bovine, ovine, and murine gut. *Infection and Immunity, 76*, 4804–4813. https://doi.org/10.1128/iai.00742-08
76. Pham, T. H., et al. (2012). Functional differences and interactions between the Escherichia coli type III secretion system effectors NleH1 and NleH2. *Infection and Immunity, 80*, 2133–2140. https://doi.org/10.1128/iai.06358-11
77. McDougal, C. E., & Sauer, J. D. (2018). Listeria monocytogenes: The impact of cell death on infection and immunity. *Pathogens, 7*. https://doi.org/10.3390/pathogens7010008
78. Rogers, H. W., Callery, M. P., Deck, B., & Unanue, E. R. (1996). Listeria monocytogenes induces apoptosis of infected hepatocytes. *Journal of Immunology, 156*, 679–684.
79. Calame, D. G., Mueller-Ortiz, S. L., Morales, J. E., & Wetsel, R. A. (2014). The C5a anaphylatoxin receptor (C5aR1) protects against listeria monocytogenes infection by inhibiting type 1 IFN expression. *The Journal of Immunology, 193*, 5099–5107. https://doi.org/10.4049/jimmunol.1401750
80. Margaroli, C., et al. (2016). Role of proapoptotic BH3-only proteins in listeria monocytogenes infection. *European Journal of Immunology, 46*, 1427–1437. https://doi.org/10.1002/eji.201545857
81. Descoeudres, N., et al. (2021). An immunomodulatory transcriptional signature associated with persistent listeria infection in hepatocytes. *Frontiers in Cellular and Infection Microbiology, 11*, 761945. https://doi.org/10.3389/fcimb.2021.761945

82. Carrero, J. A., Vivanco-Cid, H., & Unanue, E. R. (2008). Granzymes drive a rapid Listeriolysin O-induced T cell Apoptosis1. *The Journal of Immunology, 181*, 1365–1374. https://doi.org/10.4049/jimmunol.181.2.1365
83. Sauer, J.-D., et al. (2011). The N-ethyl-N-nitrosourea-induced goldenticket mouse mutant reveals an essential function of sting in the in vivo interferon response to listeria monocytogenes and cyclic dinucleotides. *Infection and Immunity, 79*, 688–694. https://doi.org/10.1128/iai.00999-10
84. Woodward, J. J., Iavarone, A. T., & Portnoy, D. A. (2010). c-di-AMP secreted by intracellular Listeria monocytogenes activates a host type I interferon response. *Science, 328*, 1703–1705. https://doi.org/10.1126/science.1189801
85. Carrero, J. A., Calderon, B., & Unanue, E. R. (2004). Type I interferon sensitizes lymphocytes to apoptosis and reduces resistance to Listeria infection. *The Journal of Experimental Medicine, 200*, 535–540. https://doi.org/10.1084/jem.20040769
86. Dai, W. J., Köhler, G., & Brombacher, F. (1997). Both innate and acquired immunity to Listeria monocytogenes infection are increased in IL-10-deficient mice. *Journal of Immunology, 158*, 2259–2267.
87. Global Health Estimates 2020: Deaths by cause, age, sex, by country and by region, 2000–2019. Geneva: World Health Organization 2020.
88. Deshpande, R., & Zou, C. (2020). Pseudomonas aeruginosa induced cell death in acute lung injury and acute respiratory distress syndrome. *International Journal of Molecular Sciences, 21*. https://doi.org/10.3390/ijms21155356
89. Grassmé, H., et al. (2000). CD95/CD95 ligand interactions on epithelial cells in host defense to Pseudomonas aeruginosa. *Science, 290*, 527–530. https://doi.org/10.1126/science.290.5491.527
90. Cannon, C. L., Kowalski, M. P., Stopak, K. S., & Pier, G. B. (2003). Pseudomonas aeruginosa-induced apoptosis is defective in respiratory epithelial cells expressing mutant cystic fibrosis transmembrane conductance regulator. *American Journal of Respiratory Cell and Molecular Biology, 29*, 188–197. https://doi.org/10.1165/rcmb.4898
91. Dockrell, D. H., et al. (2003). Alveolar macrophage apoptosis contributes to pneumococcal clearance in a resolving model of pulmonary infection 1. *The Journal of Immunology, 171*, 5380–5388. https://doi.org/10.4049/jimmunol.171.10.5380
92. Preston, J. A., et al. (2019). Alveolar macrophage apoptosis–associated bacterial killing helps prevent murine pneumonia. *American Journal of Respiratory and Critical Care Medicine, 200*, 84–97. https://doi.org/10.1164/rccm.201804-0646OC
93. Rai, P., et al. (2015). Streptococcus pneumoniae secretes hydrogen peroxide leading to DNA damage and apoptosis in lung cells. *Proceedings of the National Academy of Sciences, 112*, E3421–E3430. https://doi.org/10.1073/pnas.1424144112
94. Cano, V., et al. (2015). lebsiella pneumoniae survives within macrophages by avoiding delivery to lysosomes. *Cellular Microbiology, 17*, 1537–1560. https://doi.org/10.1111/cmi.12466
95. Jiang, W., Liu, J., Zhao, X., & Yang, W. (2022). Melatonin ameliorates lung cell inflammation and apoptosis caused by Klebsiella pneumoniae via AMP-activated protein kinase. *Inflammopharmacology, 30*, 2345–2357. https://doi.org/10.1007/s10787-022-01073-0
96. Becker, E., et al. (2023). AMPK activation improves recovery from pneumonia-induced lung injury via reduction of er-stress and apoptosis in alveolar epithelial cells. *Respiratory Research, 24*, 185. https://doi.org/10.1186/s12931-023-02483-6
97. Qadan, M., & Cheadle, W. G. (2009). Common microbial pathogens in surgical practice. *Surgical Clinics of North America, 89*, 295–310. https://doi.org/10.1016/j.suc.2008.09.002
98. Bantel, H., et al. (2001). A-Toxin is a mediator of Staphylococcus aureus–induced cell death and activates caspases via the intrinsic death pathway independently of death receptor signaling. *Journal of Cell Biology, 155*, 637–648. https://doi.org/10.1083/jcb.200105081
99. Zhang, X., et al. (2016). Positive FEEDBACK CYCle of TNFα promotes staphylococcal enterotoxin B-induced THP-1 cell apoptosis. *Frontiers in Cellular and Infection Microbiology, 6*. https://doi.org/10.3389/fcimb.2016.00109

100. Talbot, E. A., & Raffa, B. J. (2015). *Molecular medical microbiology (Second Edition)* (eds Tang, Y.- W. et al., pp. 1637–1653). Academic Press.
101. FitzGerald, E. S., Luz, N. F., & Jamieson, A. M. (2020). Competitive cell death interactions in pulmonary infection: Host modulation versus pathogen manipulation. *Frontiers in Immunology, 11*. https://doi.org/10.3389/fimmu.2020.00814
102. Fan, L., et al. (2018). MptpB promotes mycobacteria survival by inhibiting the expression of inflammatory mediators and cell apoptosis in macrophages. *Frontiers in Cellular and Infection Microbiology, 8*. https://doi.org/10.3389/fcimb.2018.00171
103. Stutz, M. D., et al. (2021). Macrophage and neutrophil death programs differentially confer resistance to tuberculosis. *Immunity, 54*, 1758–1771.e1757. https://doi.org/10.1016/j.immuni.2021.06.009
104. Yang, Z., et al. (2024). Identification and validation of genes related to macrophage polarization and cell death modes under mycobacterium tuberculosis infection. *Journal of Inflammation Research, 17*, 1397–1411. https://doi.org/10.2147/jir.S448372
105. Ghazavi, F., et al. (2022). Executioner caspases 3 and 7 are dispensable for intestinal epithelium turnover and homeostasis at steady state. *Proceedings of the National Academy of Sciences, 119*, e2024508119–e2024508119. https://doi.org/10.1073/PNAS.2024508119
106. Nozaki, K., et al. (2022). Caspase-7 activates ASM to repair gasdermin and perforin pores. *Nature*, 1–8. https://doi.org/10.1038/s41586-022-04825-8
107. Christgen, S., et al. (2020). Identification of the PANoptosome: A molecular platform triggering Pyroptosis, apoptosis, and necroptosis (PANoptosis). *Frontiers in Cellular and Infection Microbiology, 10*, 237–237. https://doi.org/10.3389/fcimb.2020.00237
108. Medina, C. B., & Ravichandran, K. S. (2016). Do not let death do us part: 'find-me' signals in communication between dying cells and the phagocytes. *Cell Death and Differentiation, 23*, 979–989. https://doi.org/10.1038/cdd.2016.13
109. D'Ambrosi, N., Rossi, S., Gerbino, V., & Cozzolino, M. (2014). Rac1 at the crossroad of actin dynamics and neuroinflammation in amyotrophic lateral sclerosis. *Frontiers in Cellular Neuroscience, 8*. https://doi.org/10.3389/fncel.2014.00279
110. Yurdagul, A., Jr. et al. (2020). Macrophage metabolism of apoptotic cell-derived arginine promotes continual efferocytosis and resolution of injury. *Cell Metabolism, 31*, 518–533.e510. https://doi.org/10.1016/j.cmet.2020.01.001
111. Lam, A., et al. (2017). Role of apoptosis and autophagy in tuberculosis. *American Journal of Physiology. Lung Cellular and Molecular Physiology, 313*, L218–l229. https://doi.org/10.1152/ajplung.00162.2017
112. Madeo, F., Eisenberg, T., Pietrocola, F., & Kroemer, G. (2018). Spermidine in health and disease. *Science, 359*, eaan2788. https://doi.org/10.1126/science.aan2788
113. Elliott, M. R., et al. (2009). Nucleotides released by apoptotic cells act as a find-me signal to promote phagocytic clearance. *Nature, 461*, 282–286. https://doi.org/10.1038/nature08296
114. Medina, C. B., et al. (2020). Metabolites released from apoptotic cells act as tissue messengers. *Nature, 580*, 130–135. https://doi.org/10.1038/s41586-020-2121-3
115. Lucas, C. D., et al. (2022). Pannexin 1 drives efficient epithelial repair after tissue injury. *Science Immunology, 7*, eabm4032. https://doi.org/10.1126/sciimmunol.abm4032
116. Anderson, C. J., et al. (2021). Microbes exploit death-induced nutrient release by gut epithelial cells. *Nature, 596*(7871), 262–267. https://doi.org/10.1038/s41586-021-03785-9
117. Anderson, C. J., et al. (2024). Metabolite-based inter-kingdom communication controls intestinal tissue recovery following chemotherapeutic injury. *Cell Host & Microbe*. https://doi.org/10.1016/j.chom.2024.07.026
118. Jiang, L., et al. (2021). Salmonella Typhimurium reprograms macrophage metabolism via T3SS effector SopE2 to promote intracellular replication and virulence. *Nature Communications, 12*, 879. https://doi.org/10.1038/s41467-021-21186-4
119. Elting, L. S., et al. (2003). The burdens of cancer therapy. Clinical and economic outcomes of chemotherapy-induced mucositis. *Cancer, 98*, 1531–1539. https://doi.org/10.1002/cncr.11671

120. O'Reilly, M., Mellotte, G., Ryan, B., & O'Connor, A. (2020). Gastrointestinal side effects of cancer treatments. *Therapeutic Advances in Chronic Disease, 11*, 2040622320970354. https://doi.org/10.1177/2040622320970354
121. Martens, A., & van Loo, G. (2020). A20 at the crossroads of cell death, inflammation, and autoimmunity. *Cold Spring Harbor Perspectives in Biology, 12*. https://doi.org/10.1101/cshperspect.a036418
122. Vereecke, L., et al. (2010). Enterocyte-specific A20 deficiency sensitizes to tumor necrosis factor-induced toxicity and experimental colitis. *Journal of Experimental Medicine*. https://doi.org/10.1084/jem.20092474
123. Truman, L. A., et al. (2008). CX3CL1/fractalkine is released from apoptotic lymphocytes to stimulate macrophage chemotaxis. *Blood, 112*, 5026–5036. https://doi.org/10.1182/blood-2008-06-162404
124. Torr, E. E., et al. (2012). Apoptotic cell-derived ICAM-3 promotes both macrophage chemoattraction to and tethering of apoptotic cells. *Cell Death and Differentiation, 19*, 671–679. https://doi.org/10.1038/cdd.2011.167
125. Muhsin-Sharafaldine, M.-R., et al. (2016). Procoagulant and immunogenic properties of melanoma exosomes, microvesicles and apoptotic vesicles. *Oncotarget, 7.*
126. Schaible, U. E., et al. (2003). Apoptosis facilitates antigen presentation to T lymphocytes through MHC-I and CD1 in tuberculosis. *Nature Medicine, 9*, 1039–1046. https://doi.org/10.1038/nm906
127. Winau, F., et al. (2006). Apoptotic vesicles crossprime CD8 T cells and protect against tuberculosis. *Immunity, 24*, 105–117. https://doi.org/10.1016/j.immuni.2005.12.001
128. Fehr, E.-M., et al. (2013). Apoptotic-cell-derived membrane vesicles induce an alternative maturation of human dendritic cells which is disturbed in SLE. *Journal of Autoimmunity, 40*, 86–95. https://doi.org/10.1016/j.jaut.2012.08.003
129. Turner, M. D., Nedjai, B., Hurst, T., & Pennington, D. J. (2014). Cytokines and chemokines: At the crossroads of cell signalling and inflammatory disease. *Biochimica et Biophysica Acta (BBA) – Molecular Cell Research, 1843*, 2563–2582. https://doi.org/10.1016/j.bbamcr.2014.05.014
130. Patankar, J. V., Bubeck, M., Acera, M. G., & Becker, C. (2023). Breaking bad: Necroptosis in the pathogenesis of gastrointestinal diseases. *Frontiers in Immunology, 14*, 1203903. https://doi.org/10.3389/fimmu.2023.1203903
131. Sasindran, S. J., & Torrelles, J. B. (2011). Mycobacterium tuberculosis infection and inflammation: What is beneficial for the host and for the bacterium? *Frontiers in Microbiology, 2*. https://doi.org/10.3389/fmicb.2011.00002
132. Mezouar, S., Diarra, I., Roudier, J., Desnues, B., & Mege, J.-L. (2019). Tumor necrosis factor-alpha antagonist interferes with the formation of granulomatous multinucleated giant cells: New insights into mycobacterium tuberculosis infection. *Frontiers in Immunology, 10*. https://doi.org/10.3389/fimmu.2019.01947
133. Gou, X., et al. (2019). IL-6 during influenza-Streptococcus pneumoniae co-infected pneumonia-a protector. *Frontiers in Immunology, 10*, 3102. https://doi.org/10.3389/fimmu.2019.03102
134. Sarkar, M., Bhowmick, S., Casola, A., & Chaudhuri, K. (2012). Interleukin-8 gene regulation in epithelial cells by vibrio cholerae: Role of multiple promoter elements, adherence and motility of bacteria and host MAPKs. *The FEBS Journal, 279*, 1464–1473. https://doi.org/10.1111/j.1742-4658.2012.08539.x
135. Subramanian, S., et al. (2008). Characterization of epithelial IL-8 response to inflammatory bowel disease mucosal E. Coli and its inhibition by mesalamine. *Inflammatory Bowel Diseases, 14*, 162–175. https://doi.org/10.1002/ibd.20296
136. Phuong, M. S., Hernandez, R. E., Wolter, D. J., Hoffman, L. R., & Sad, S. (2021). Impairment in inflammasome signaling by the chronic Pseudomonas aeruginosa isolates from cystic fibrosis patients results in an increase in inflammatory response. *Cell Death & Disease, 12*, 241. https://doi.org/10.1038/s41419-021-03526-w

137. Nakamoto, K., et al. (2019). Pseudomonas aeruginosa-derived flagellin stimulates IL-6 and IL-8 production in human bronchial epithelial cells: A potential mechanism for progression and exacerbation of COPD. *Experimental Lung Research, 45*, 255–266. https://doi.org/10.1080/01902148.2019.1665147
138. Arandjelovic, S., & Ravichandran, K. S. (2015). Phagocytosis of apoptotic cells in homeostasis. *Nature Immunology, 16*, 907–917. https://doi.org/10.1038/ni.3253
139. Morioka, S., Maueröder, C., & Ravichandran, K. S. (2019). Living on the edge: Efferocytosis at the Interface of homeostasis and pathology. *Immunity, 50*, 1149–1162. https://doi.org/10.1016/j.immuni.2019.04.018
140. McCubbrey, A. L., & Curtis, J. L. (2013). Efferocytosis and lung disease. *Chest, 143*, 1750–1757. https://doi.org/10.1378/chest.12-2413
141. Martin, C. J., Peters, K. N., & Behar, S. M. (2014). Macrophages clean up: Efferocytosis and microbial control. *Current Opinion in Microbiology, 17*, 17–23. https://doi.org/10.1016/j.mib.2013.10.007
142. Felton, J. M., et al. (2018). Mer-mediated eosinophil efferocytosis regulates resolution of allergic airway inflammation. *The Journal of Allergy and Clinical Immunology, 142*, 1884–1893.e1886. https://doi.org/10.1016/j.jaci.2018.01.029.
143. Cummings, R. J., et al. (2016). Different tissue phagocytes sample apoptotic cells to direct distinct homeostasis programs. *Nature, 539*, 565–569. https://doi.org/10.1038/nature20138
144. Shankman, L. S., et al. (2021). Efferocytosis by Paneth cells within the intestine. *Current Biology, 31*, 2469–2476.e2465. https://doi.org/10.1016/j.cub.2021.03.055
145. Lee, C. S., et al. (2016). Boosting apoptotic cell clearance by colonic epithelial cells attenuates inflammation in vivo. *Immunity, 44*, 807–820. https://doi.org/10.1016/j.immuni.2016.02.005
146. Raymond, M. H., et al. (2022). Live cell tracking of macrophage efferocytosis during drosophila embryo development in vivo. *Science, 375*, 1182–1187. https://doi.org/10.1126/science.abl4430
147. Davidson, A. J., Heron, R., Das, J., Overholtzer, M., & Wood, W. (2024). Ferroptosis-like cell death promotes and prolongs inflammation in Drosophila. *Nature Cell Biology*. https://doi.org/10.1038/s41556-024-01450-7
148. Hiyoshi, H., et al. (2022). Virulence factors perforate the pathogen-containing vacuole to signal efferocytosis. *Cell Host & Microbe, 30*, 163–170.e166. https://doi.org/10.1016/j.chom.2021.12.001
149. Czuczman, M. A., et al. (2014). Listeria monocytogenes exploits efferocytosis to promote cell-to-cell spread. *Nature, 509*, 230–234. https://doi.org/10.1038/nature13168
150. Dhanda, A. S., et al. (2020). Listeria monocytogenes exploits host Caveolin for cell-to-cell spreading. *MBio, 11*. https://doi.org/10.1128/mbio.02857-02819, https://doi.org/10.1128/mbio.02857-19
151. Saavedra, P. H. V., et al. (2024). Broad-spectrum antibiotics disrupt homeostatic efferocytosis. *Nature Metabolism*. https://doi.org/10.1038/s42255-024-01107-7
152. Liu, C., et al. (2019). Mast cells participate in regulation of lung-gut axis during Staphylococcus aureus pneumonia. *Cell Proliferation, 52*, e12565. https://doi.org/10.1111/cpr.12565
153. Pidwill, G. R., Gibson, J. F., Cole, J., Renshaw, S. A., & Foster, S. J. (2020). The role of macrophages in Staphylococcus aureus infection. *Frontiers in Immunology, 11*, 620339. https://doi.org/10.3389/fimmu.2020.620339
154. Behrens, E. M., et al. (2003). The mer receptor tyrosine kinase: Expression and function suggest a role in innate immunity. *European Journal of Immunology, 33*, 2160–2167. https://doi.org/10.1002/eji.200324076
155. Zahoor, A., et al. (2020). MerTK negatively regulates Staphylococcus aureus induced inflammatory response via SOCS1/SOCS3 and mal. *Immunobiology, 225*, 151960. https://doi.org/10.1016/j.imbio.2020.151960
156. Liu, S., et al. (2024). Big data analytics for MerTK genomics reveals its double-edged sword functions in human diseases. *Redox Biology*, 103061. https://doi.org/10.1016/j.redox.2024.103061

157. Thorp, E., et al. (2011). Shedding of the Mer tyrosine kinase receptor is mediated by ADAM17 protein through a pathway involving reactive oxygen species, protein kinase Cδ, and p38 mitogen-activated protein kinase (MAPK). *The Journal of Biological Chemistry, 286*, 33335–33344. https://doi.org/10.1074/jbc.M111.263020
158. Lee, Y.-J., et al. (2012). Inhibiting Mer receptor tyrosine kinase suppresses STAT1, SOCS1/3, and NF-κB activation and enhances inflammatory responses in lipopolysaccharide-induced acute lung injury. *Journal of Leukocyte Biology, 91*, 921–932. https://doi.org/10.1189/jlb.0611289
159. Choi, J.-Y., et al. (2013). Upregulation of Mer receptor tyrosine kinase Signaling attenuated lipopolysaccharide-induced lung inflammation. *Journal of Pharmacology and Experimental Therapeutics, 344*, 447–458. https://doi.org/10.1124/jpet.112.199778
160. Fukatsu, M., et al. (2022). The suppressive effects of Mer inhibition on inflammatory responses in the pathogenesis of LPS-induced ALI/ARDS. *Science Signaling, 15*, eabd2533. https://doi.org/10.1126/scisignal.abd2533
161. Bosurgi, L., et al. (2013). Paradoxical role of the proto-oncogene Axl and Mer receptor tyrosine kinases in colon cancer. *Proceedings of the National Academy of Sciences, 110*, 13091–13096. https://doi.org/10.1073/pnas.1302507110
162. Liu, J., et al. (2024). Osteopontin regulation of MerTK+ macrophages promotes Crohn's disease intestinal fibrosis. *iScience*, 110226. https://doi.org/10.1016/j.isci.2024.110226
163. Karaji, N., & Sattentau, Q. J. (2017). Efferocytosis of pathogen-infected cells. *Frontiers in Immunology, 8*. https://doi.org/10.3389/fimmu.2017.01863
164. Torchinsky, M. B., Garaude, J., Martin, A. P., & Blander, J. M. (2009). Innate immune recognition of infected apoptotic cells directs TH17 cell differentiation. *Nature, 458*, 78–82. https://doi.org/10.1038/nature07781
165. Campisi, L., et al. (2016). Apoptosis in response to microbial infection induces autoreactive TH17 cells. *Nature Immunology, 17*, 1084–1092. https://doi.org/10.1038/ni.3512
166. Heesterbeek, D. A. C., Angelier, M. L., Harrison, R. A., & Rooijakkers, S. H. M. (2018). Complement and bacterial infections: From molecular mechanisms to therapeutic applications. *Journal of Innate Immunity, 10*, 455–464. https://doi.org/10.1159/000491439
167. Trouw, L. A., Blom, A. M., & Gasque, P. (2008). Role of complement and complement regulators in the removal of apoptotic cells. *Molecular Immunology, 45*, 1199–1207. https://doi.org/10.1016/j.molimm.2007.09.008
168. Zhang, K., et al. (2024). Non-professional efferocytosis of salmonella-infected intestinal epithelial cells in the neonatal host. *Journal of Experimental Medicine, 221*. https://doi.org/10.1084/jem.20231237
169. Juncadella, I. J., et al. (2013). Apoptotic cell clearance by bronchial epithelial cells critically influences airway inflammation. *Nature, 493*, 547–551. https://doi.org/10.1038/nature11714
170. Cockram, T. O. J., Puigdellívol, M., & Brown, G. C. (2019). Calreticulin and Galectin-3 opsonise bacteria for phagocytosis by microglia. *Frontiers in Immunology, 10*, 2647. https://doi.org/10.3389/fimmu.2019.02647
171. Martin, C. J., et al. (2012). Efferocytosis is an innate antibacterial mechanism. *Cell Host & Microbe, 12*, 289–300. https://doi.org/10.1016/j.chom.2012.06.010
172. Tzelepis, F., et al. (2015). Annexin1 regulates DC efferocytosis and cross-presentation during mycobacterium tuberculosis infection. *The Journal of Clinical Investigation, 125*, 752–768. https://doi.org/10.1172/JCI77014
173. Jäger, A. V., et al. (2021). The inflammatory response induced by Pseudomonas aeruginosa in macrophages enhances apoptotic cell removal. *Scientific Reports, 11*, 2393. https://doi.org/10.1038/s41598-021-81557-1
174. Capasso, D., et al. (2016). Elimination of Pseudomonas aeruginosa through Efferocytosis upon binding to apoptotic cells. *PLoS Pathogens, 12*, e1006068. https://doi.org/10.1371/journal.ppat.1006068
175. Park, D., et al. (2007). BAI1 is an engulfment receptor for apoptotic cells upstream of the ELMO/Dock180/Rac module. *Nature, 450*, 430–434. https://doi.org/10.1038/nature06329

176. Billings, E. A., et al. (2016). The adhesion GPCR Bai1 mediates macrophage Ros production and microbicidal activity against gram-negative bacteria. *Science Signaling, 9*, ra14-ra14. https://doi.org/10.1126/scisignal.aac6250
177. Das, S., et al. (2011). Brain angiogenesis inhibitor 1 (BAI1) is a pattern recognition receptor that mediates macrophage binding and engulfment of gram-negative bacteria. *Proceedings of the National Academy of Sciences of the United States of America, 108*, 2136–2141. https://doi.org/10.1073/pnas.1014775108
178. Hsiao, C.-C., van der Poel, M., van Ham, T. J., & Hamann, J. (2019). Macrophages do not express the phagocytic receptor BAI1/ADGRB1. *Frontiers in Immunology, 10*. https://doi.org/10.3389/fimmu.2019.00962
179. Anandachar, M. S., et al. (2023). Diverse gut pathogens exploit the host engulfment pathway via a conserved mechanism. *Journal of Biological Chemistry, 299*, 105390. https://doi.org/10.1016/j.jbc.2023.105390
180. Lucas, C. D., et al. (2013). Flavones induce neutrophil apoptosis by down-regulation of Mcl-1 via a proteasomal-dependent pathway. *The FASEB Journal, 27*, 1084–1094. https://doi.org/10.1096/fj.12-218990
181. Dorward, D. A., et al. (2017). The cyclin-dependent kinase inhibitor AT7519 accelerates neutrophil apoptosis in sepsis-related acute respiratory distress syndrome. *Thorax, 72*, 182–185. https://doi.org/10.1136/thoraxjnl-2016-209229
182. DeLeo, F. R. (2004). Modulation of phagocyte apoptosis by bacterial pathogens. *Apoptosis, 9*, 399–413. https://doi.org/10.1023/B:APPT.0000031448.64969.fa
183. Kobayashi, S. D., et al. (2003). Bacterial pathogens modulate an apoptosis differentiation program in human neutrophils. *Proceedings of the National Academy of Sciences of the United States of America, 100*, 10948–10953. https://doi.org/10.1073/pnas.1833375100
184. Anderson, K. E., et al. (2008). CD18-dependent activation of the neutrophil NADPH oxidase during phagocytosis of Escherichia coli or Staphylococcus aureus is regulated by class III but not class I or II PI3Ks. *Blood, 112*, 5202–5211. https://doi.org/10.1182/blood-2008-04-149450
185. El Kebir, D., Gjorstrup, P., & Filep, J. G. (2012). Resolvin E1 promotes phagocytosis-induced neutrophil apoptosis and accelerates resolution of pulmonary inflammation. *Proceedings of the National Academy of Sciences, 109*, 14983–14988. https://doi.org/10.1073/pnas.1206641109
186. Usher, L. R., et al. (2002). Induction of neutrophil apoptosis by the Pseudomonas aeruginosa exotoxin Pyocyanin: A potential mechanism of persistent Infection1. *The Journal of Immunology, 168*, 1861–1868. https://doi.org/10.4049/jimmunol.168.4.1861
187. Managò, A., et al. (2015). Pseudomonas aeruginosa Pyocyanin induces neutrophil death via mitochondrial reactive oxygen species and mitochondrial acid sphingomyelinase. *Antioxidants & Redox Signaling, 22*, 1097–1110. https://doi.org/10.1089/ars.2014.5979
188. Chung, J. W., et al. (2009). Pseudomonas aeruginosa eliminates natural killer cells via phagocytosis-induced apoptosis. *PLoS Pathogens, 5*, e1000561. https://doi.org/10.1371/journal.ppat.1000561
189. Zhang, J., Jiang, R., Wang, W., Takayama, H., & Tanaka, Y. (2013). Apoptosis are induced in J774 macrophages upon phagocytosis and killing of Pseudomonas aeruginosa. *Cellular Immunology, 286*, 11–15. https://doi.org/10.1016/j.cellimm.2013.10.006
190. Perskvist, N., Long, M., Stendahl, O., & Zheng, L. (2002). Mycobacterium tuberculosis promotes apoptosis in human neutrophils by activating Caspase-3 and altering expression of Bax/Bcl-xL via an oxygen-dependent Pathway1. *The Journal of Immunology, 168*, 6358–6365. https://doi.org/10.4049/jimmunol.168.12.6358
191. Groesdonk, H. V., Schlottmann, S., Richter, F., Georgieff, M., & Senftleben, U. (2006). Escherichia coli prevents phagocytosis-induced death of macrophages via classical NF-kappaB signaling, a link to T-cell activation. *Infection and Immunity, 74*, 5989–6000. https://doi.org/10.1128/iai.00138-06
192. Knodler, L. A., et al. (2014). Noncanonical Inflammasome activation of Caspase-4/Caspase-11 mediates epithelial defenses against enteric bacterial pathogens. *Cell Host & Microbe, 16*, 249–256. https://doi.org/10.1016/j.chom.2014.07.002

193. Sellin, M. E., et al. (2014). Epithelium-intrinsic NAIP/NLRC4 Inflammasome drives infected enterocyte expulsion to restrict salmonella replication in the intestinal mucosa. *Cell Host & Microbe, 16*, 237–248. https://doi.org/10.1016/j.chom.2014.07.001
194. Fattinger, S. A., et al. (2023). Gasdermin D is the only Gasdermin that provides protection against acute Salmonella gut infection in mice. *Proceedings of the National Academy of Sciences, 120*. https://doi.org/10.1073/pnas.2315503120
195. Robinson, N., et al. (2012). Type I interferon induces necroptosis in macrophages during infection with salmonella Enterica Serovar Typhimurium. *Nature Immunology, 13*, 954–962. https://doi.org/10.1038/ni.2397
196. Satkovich, J., et al. (2020). RIPK3-dependent recruitment of low-inflammatory myeloid cells does not protect from systemic salmonella infection. *MBio, 11*, 1–7. https://doi.org/10.1128/MBIO.02588-20
197. Mehrotra, P., et al. (2024). Oxylipins and metabolites from pyroptotic cells act as promoters of tissue repair. *Nature, 631*, 207–215. https://doi.org/10.1038/s41586-024-07585-9

For Better or Worse: The Impact of Necrotic Cell Death Modalities

Peter Vandenabeele, Marcus Conrad, and Adam Wahida

Abstract

Cellular stress, infection, and inflammation lead to various forms of cell death. Depending on the stimulus, cell type, and cellular conditions, different modes of regulated cell death might be engaged. These include apoptosis, necroptosis, and pyroptosis, which are driven by genetically programmed mechanisms, and ferroptosis, a type of metabolic cell death. The outcome of these distinct cell death modalities is the activation of specific pore-forming mechanisms: caspase-3-mediated cleavage of gasdermin E in secondary necrosis following apoptosis (also classified as pyroptosis), RIPK3-mediated phosphorylation of MLKL in necroptosis, and caspase-1/11/4/5-mediated cleavage of GSDMD during pyroptosis. In the case of ferroptosis, a metabolic cell death modality driven by imbalances in iron, lipid, and redox metabolism, the plasma membrane also becomes

P. Vandenabeele (✉)
VIB-UGent Center for Inflammation Research, Ghent, Belgium

Department of Biomedical Molecular Biology, Ghent University, Ghent, Belgium
e-mail: Peter.Vandenabeele@irc.vib-ugent.be

M. Conrad
Institute of Metabolism and Cell Death, Helmholtz Munich, Neuherberg, Germany

Translational Redox Biology, Technical University of Munich (TUM), TUM Natural School of Sciences, Garching, Germany
e-mail: marcus.conrad@helmholtz-munich.de

A. Wahida
VIB-UGent Center for Inflammation Research, Ghent, Belgium

Department of Biomedical Molecular Biology, Ghent University, Ghent, Belgium

Institute of Metabolism and Cell Death, Helmholtz Munich, Neuherberg, Germany
e-mail: adam.wahida@irc.vib-ugent.be

C. D. Gregory et al. (eds.), *Apoptosis and Other forms of Cell Death in the Regulation of Inflammation in Health and Disease*, Advances in Experimental Medicine and Biology 1481, https://doi.org/10.1007/978-3-031-92785-0_8

permeabilized due to oxidative modifications of acyl chains in phospholipids. On top of the pore-forming mechanisms, NINJ1 detects cellular swelling ("oncosis") and triggers a massive plasma membrane rupture as a final stage of the cellular cataclysm, releasing large molecules and intracellular contents. Understanding the mechanisms of regulated necrotic cell death through signaling pathways or by disrupting metabolic networks offers tangible targeting strategies to enhance or reduce cell death processes and associated subroutines in various diseases, including cancer, ischemia/reperfusion conditions, inflammation, and degenerative diseases. Besides the molecular biology, we will concentrate this chapter on the effects of necroptosis, pyroptosis, and ferroptosis in cancer and inflammatory pathologies in the brain, intestine, and skin.

Keywords

Regulated necrotic cell death · Metabolic cell death · Necroptosis · Pyroptosis · Ferroptosis · MLKL · GSDM · NINJ1 · Necrotic cell death in cancer · Necrotic cell death in pathological immunity

Abbreviations

ACSL3	Acyl-CoA synthetase long-chain family member 3
ACSL4	Acyl-CoA synthetase long-chain family member 4
AIM2	Absent in melanoma 2
ALOX	Arachidonate 5-lipoxygenase
AML	Acute myeloid leukemia
APAP1	Apoptosis protease-activating factor 1
ASC	Apoptosis-associated speck-like protein containing a CARD
ATP	Adenosintriphosphat
BAK	Bcl-2 homologous antagonist
BAX	Bcl-2-associated X protein
BCL-2	B-cell lymphoma 2
CARD	Caspase activation and recruitment domains
CASP	Caspase
cDC1s	Conventional type 1 dendritic cells
CHX	Cycloheximide
CYB5R1	Cytochrome B5 reductase 1
DAMPs	Damage-associated molecular patterns
DHFR	Dihydrofolate reductase
ESCRT-III	Endosomal sorting complexes required for transport
FADD	FAS-associated death domain protein
FSP1	Ferroptosis suppressor protein 1
GPX4	Glutathione peroxidase 4
GSDM	Gasdermin
GSDMD	Gasdermin D

GSDME	Gasdermin E
GSH	Glutathione
HOIL	Haem-oxidized IRP2 ubiquitin ligase-1
HOIP	HOIL-1-interacting protein
IAPs	Inhibitors of apoptosis protein
IKK	Inhibitor of KappaB kinase
IL-18	Interleukin-18
IL-1R1	Interleukin receptor 1
IL-1β	Interleukin-1β
LOX	Lipoxygenase
LPCAT3	Lysophosphatidylcholine acyltransferase 3
LPS	Lipopolysaccharide
LUBAC	Linear ubiquitin chain assembly complex
MAC	Membrane attack complex
MAPK	Mitogen-activated protein
MBOAT1	Membrane-bound O-acyltransferase domain-containing proteins 1
MBOAT2	Membrane-bound O-acyltransferase domain-containing proteins 2
MIMP	Mitochondrial inner membrane permeabilization
MK2	Mitogen kinase 2
MLKL	Mixed lineage kinase domain-like pseudokinase
MOMP	Mitochondrial outer membrane permeabilization
MUFA	Monounsaturated fatty acids
NAD	Nicotinamide adenine dinucleotide
NAP1	Nucleosome assembly protein 1
NEMO	NF-κB essential modulator
NF-κB	Nuclear factor kappa-light-chain-enhancer of activated B cells
NINJ1	Ninjurin-1
NLRC4	Nod-like receptor CARD domain-containing protein 4
NLRP	Nucleotide-binding oligomerization domain, leucine-rich repeat, and pyrin domain-containing protein
NOX	NADPH oxidase
NRF2	Nuclear factor erythroid 2-related factor 2
P2X7	Purinergic receptor P2X 7
PAF	Platelet-activating factor
PAMPs	Pathogen-associated molecular patterns
PGE2	Prostaglandin E2
PIPs	Phosphatidylinositol phosphates
PLOOH	Phospholipid hydroperoxides
PMR	Plasma membrane rupture
POR	Cytochrome P450 oxidoreductase
PS	Phosphatidylserine
psK	Pseudokinase
PTM	Posttranslational modification
PUFA	Polyunsaturated fatty acids
PYCARD	PYD and CARD domain containing

PYRIN	Death domain motif
RHIM	RIP homology interacting motif
RIG-I	Retinoic acid-inducible gene I
RIPK1	Receptor-interacting serine/threonine-protein kinase 1
RIPK3	Receptor-interacting serine/threonine-protein kinase 3
RIPKs	RIP kinases
SHARPIN	SHANK-associated RH domain interactor
STING	Stromal interaction molecule 1
TAB	TGF-beta-activated kinase 1 binding protein
TAK1	Transforming growth factor beta-activated kinase 1
TANK11	TRAF family member-associated NFKB activator 11
TBK1	TANK binding kinase 1
TLR	Toll-like receptor
TNF	Tumor necrosis factor
TNFR1	Tumor necrosis factor receptor 1
TRADD	Tumor necrosis factor receptor type 1-associated death domain protein
TRAF2	TNF receptor associated factor 2
TRIF	Toll/interleukin-1 receptor domain-containing adapter-inducing interferon-β
TWIK2	Potassium channel
ZBP1	Z-DNA-binding protein
4HBD	4 helical binding domain

Introduction

We must go almost 175 years back for the first recognition that the behavior of cells, including cell death, could be the basis of many pathologies. In a lecture given in 1858 [1], the famous German pathologist and biologist Rudolf Virchow (1821–1902) proposed that cell death could be a basis for cellular pathology and that two types of cell death existed: necrobiosis versus necrosis. The former type is "*always here to deal with a gradual decay and death, a dissolution … Necrobiosis is death brought on by (altered) life, a spontaneous wearing out of living parts, the destruction and annihilation consequent on life, natural as opposed to violent death (mortification) … But the idea of necrosis really does not offer any analogy to these processes …*" [2]. For an extensive review of the multifaceted nineteenth-century research on naturally occurring cell death and related phenomena, we refer to Clarke and Clarke [3], although Virchow is not mentioned. This dichotomy between necrobiosis (which, retrospectively, we would now call physiological cell death or programmed cell death in the context of development, homeostasis, and pathologies) and necrosis (by cellular injury or accidental cell death) formed the basis for a division between apoptosis as a programmed cell death mode (in the words of Kerr, Wyllie and Currie in 1972 and 1980) "*that can be initiated and inhibited by a variety of environmental stimuli, both physiological and pathological*" [4] and necrosis, characterized by "*swelling proceeding to rupture of plasma and organelle membranes and dissolution of organized structure and the profound differences on inflammation*" [5]. *It (necrosis) results from injury by agents, such as toxins and ischemia, affects cells in groups rather than singly, and evokes exudative inflammation when it develops* in vivo. *The other morphological pattern (apoptosis) is characterized by condensation of the cell with maintenance of organelle integrity and the formation of surface protuberances that separate as membrane-bounded globules; in tissues, these are phagocytosed and digested by resident cells, there being no associated inflammation* [5]. These profound differences in morphology and intercellular outcome have fostered many research projects on the impact of non-lytic and immunologically silent cell death by apoptosis and lytic and pro-inflammatory cell death by necrosis.

The forward genetics using ethylmethanesulfonate-based random mutagenesis of germ progenitor cells in *C. elegans* revealed the genes that constitute the core apoptotic pathway: the proapoptotic *egl-1*, *ced-3*, and *ced-4* and the anti-apoptotic *ced-9* [6, 7]. The identification of the mammalian counterparts of these *C. elegans* genes confirmed the evolutionary conservation of the apoptotic program in mammals (in line with Nietzsche's "*Ye have made your way from the worm to man, and much within you is still worm*" in Thus Spoke Zarathustra, Chapter Five—Death of the Western God): the pro-apoptotic BH3-only proteins Bik, Bid, Harakiri, and Bad (EGL1-like) [8], BAX (no ortholog in *C. elegans*) [9], the IL-1β converting enzyme or ICE, the prototype of the caspases (CED-3-like) [10], APAF-1 (CED-4-like) [11], and the anti-apoptotic BCL-2 proteins (CED-9-like) [12]. The interaction between the three classes of BCL-2 family members—specifically, the activation of proapoptotic BH3-only proteins due to cellular stress—blocks anti-apoptotic BCL-2 family members and activates proapoptotic molecules like BAX, BAK, and BOK. This process leads to mitochondrial outer membrane permeabilization (MOMP) and the release of mitochondrial factors, including cytochrome c. The latter contributes to the formation and activation of APAF1, forming the apoptosome complex to which caspase-9 becomes recruited and activated, further proteolytically activating downstream caspase-3 and caspase-7. Already from the first use of the term apoptosis as "*a basic biological phenomenon with wide-ranging implications in tissue kinetics*, this mechanism of controlled cell deletion has been connected with the recognition and uptake by phagocytosis of *membrane-bound, ultrastructurally well-preserved fragments*" (apoptotic bodies) [4]. The forward genetic screening in *C. elegans* identified both the programmed cell death genes and the genetic machinery of phagocytosis of dying cells [13]. This landmark discovery emphasizes the intricate relationship between apoptotic dying cells and phagocytosis [14, 15], a process coined as efferocytosis [16] from the Greek "to bury" (see also Chapters "Efferocytosis in Homeostasis" and "Of Tingible Bodies and Starry Skies: Control of Normal and Malignant Tissues Through Apoptotic Cell-Dependent Communication (ACDC)"). From the beginning, it was noticed that the engagement of efferocyte receptors by ligands or ligand-bridging molecule complexes by macrophages inhibits the pro-inflammatory cytokine production through autocrine/paracrine mechanisms involving TGF-β, PGE2, and PAF [17], confirming the histological description of apoptotic cells and the association with the absence of inflammation described 18 years before [5].

Non-lytic Apoptotic Cell Death and Lytic Necrotic Cell Death: Mechanisms of Plasma Membrane Permeabilization (PMP)

Historically, three premises drove cell death research and its connection with inflammation: (1) the dichotomy between physiological necrobiosis (or apoptosis) and injury-driven necrosis, proposing the latter as an "accidental process" [1]; (2) the observation of apoptotic and necrotic cell morphologies in vivo and their differential association with phagocytosis and inflammation [5]; and (3) the discovery of the genes that control programmed cell death by apoptosis in *C. elegans* and the identification of the human orthologues (see above and Chap. Molecular Cell Biology of Apoptosis in Health and Disease). This has culminated in the idea that cell death can be non-lytic and largely immunologically silent (apoptosis) or lytic and pro-inflammatory (necrosis). Although lysis during necrosis was long thought to be a result of a passive process leading to loss of plasma membrane integrity, in the last two decades, it became more and more apparent that also necrotic cell death can be a highly regulated process by the discovery of the role of the kinase activities of RIPK1 and RIPK3 in necroptosis [18–21], and the discovery of pyroptosis as a lytic cell death modality in bacteria-infected macrophages driven by caspase-1 and -11 activation [22, 23]. However, how the plasma membrane is permeabilized in the case of necrosis was long considered to result from a loss of energy-dependent processes at the plasma membrane due to a decrease of intracellular ATP [24, 25] or overactivation of PARP leading to excessive consumption of NAD^+ [26]. Indeed, maintaining the selective membrane integrity of the plasma membrane and organelle membranes is one of the hallmarks of the origin and maintenance of life [27], requiring a large part of the cellular energy.

Cells have several plasma membrane repair mechanisms that recognize damaged and disorganized or leaky phospholipid bilayers. These include counteraction of lipid peroxidation by glutathione peroxidase 4 (GPX4), vesicle recruitment leading to exocytosis of damaged membrane, intercellular vesicle fusion beneath the damaged membrane during patching, lysosomal degradation following endocytosis of damaged membranes, actin-mediated contraction of the damaged area, annexin-mediated plugging out of breached membrane areas, annexin-mediated constriction of the leaky membrane area, and ESCRT-mediated excision of permeabilized plasma membrane areas [27]. Since all passively disrupted cells exhibit a ruptured plasma membrane, which is also the key feature of the necrotic phenotype, necrotic cell death has long been defined as a form of uncontrolled passive cell death as an indirect process. Energy depletion during necrotic cell death interfering with these membrane repair systems or the energy-dependent functioning of ion channels was thus a likely working hypothesis to occur during regulated necrosis [25, 28]. However, a direct mechanism involving protein pores is exemplified in many pathogenic bacteria that produce β-barrel pore-forming toxins (PFTs) for plasma membrane permeabilization from the outside [29].

In mammalian cells, β-barrel-shaped perforins [30] play a role in cancer immunosurveillance by cytotoxic T lymphocytes and NK cells (PFN1, lumen size of the pore 11 nm) [31], in macrophages for intracellular bacterial attacks (PFN2) (lumen size of the pore 7.5–9 nm) [32, 33], and during the endocytic escape of antigens during cross-presentation by dendritic cells (PFN2) [34]. Finally, the complement proteolytic cascade leads to the formation of the C5b-C9 membrane attack complex (MAC), which also forms a β-barrel pore (lumen size of 3 nm) [35–37]. It is remarkable that despite the vast existence of pore-forming proteins in microbes and immune cells, the discovery of directly acting pore-forming proteins still surprised the cell death field: RIPK3-dependent phosphorylation of MLKL in necroptosis [38, 39], caspase-3-mediated proteolytic activation of GSDME in secondary necrosis [40, 41], and inflammatory caspase-1/11/4/5 GSDMD in inflammasome-driven pyroptosis [42, 43]. In the case of ferroptosis, no PFPs have been identified so far. Still, ferroptosis is elicited by blocking a plasma membrane repair mechanism either indirectly by blocking the cystine/glutamate antiporter system x_c^-, resulting in reduced intracellular levels of GSH required for redox reactions, or directly by blocking GPX4 activity for reducing oxidized phospholipids [44–47]. Local foci of low-level lipid peroxidation affect membrane permeability by lipid pore formation [48–50].

Surprisingly, on top of the previously mentioned pore-forming mechanisms, there is an additional process executed by NINJ1, causing plasma membrane rupture (PMR) [51]. NINJ1 reacts to cellular swelling, called oncosis, which is historically one of the hallmarks of necrotic cell death since apoptosis is associated with cell shrinkage [52]. Instead of pore formation, an oligomerizing membrane protein, NINJ1, critically regulates plasma membrane rupture (PMR) as the final cataclysmic event in lytic cell death, including secondary necrosis, necroptosis, pyroptosis, and ferroptosis [51, 53]. Oncosis is sensed and causes a conformational change of dimeric NINJ1, resulting in an amphipathic surface that multimerizes and solubilizes the plasma membrane, causing rupture [54, 55]. PMR results in the release of large intracellular molecules that augment the inflammatory response [51]. NINJ1 by itself is sufficient for mechanical force-induced plasma membrane rupture (PMR) without the need to stimulate any cell death programs [56]. Thus, rather than being a cell death mechanism by itself, NINJ1 could operate as a subroutine that senses the swelling of cells due to imbalances in the functioning of ion channels and ion exchangers associated with necrotic cell death processes. Upon its activation, it results in PMR. In that sense, the NINJ1-mediated PMR has functionally grafted on necrotic cell death modalities as a "death-after-death" mechanism (see later).

What Will Be the Focus of This Chapter?

Cell death is not autonomous and operates in a social context of turnover during steady state, infection, injury, and regeneration. The molecular mechanisms underlying plasma membrane permeabilization and rupture may also have therapeutic potential in controlling barrier loss during infection and inflammation and preventing cell death-associated inflammation. Moreover, the immunogenicity of dying and dead cells is extremely important in cancer therapy and autoimmune diseases. The release of intracellular contents resulting from plasma membrane permeability change and rupture plays a key role in the induction of immune responses. The distinct permeabilizing mechanisms may enable differential environmental communication by releasing diverse molecules. On the other hand, parallel or sequential PFPs permit redundancy, plasticity, and backup of cell death-associated processes.

The existence of PFPs that are highly regulated at the level of activation of an inactive precursor by phosphorylation (MLKL), proteolysis (GSDMs), and lipid peroxidation, and the existence of the cataclysmic event by NINJ1 as a common endpoint suggest that the activation and recruitment to the plasma membrane of MLKL and GSDMs may not always function as the ultimate endpoint. In that respect, a more profound study of the interaction between these pore-forming mechanisms and plasma membrane repair mechanisms is required to better understand the dynamics of necrotic cell death modalities. ESCRT-III-mediated membrane repair mechanisms can fine-tune or even revert MLKL and GSDM pore formation [57, 58], allowing differential release of the type and amount of intracellular proteins and subsequent intercellular communication during physiological and pathophysiological processes. Besides direct targeting of pore-forming mechanisms, the interaction between pore formation and membrane repair mechanisms opens therapeutic opportunities for modulation by small molecule compounds, enhancing or inhibiting the effect of PFPs.

It should be noted that cell death studies have often focused on the analysis of one cell death modality by excluding (or denying) others. However, the sharing of different regulatory molecules such as caspase-8, RIPK1 and RIPK3 kinases, and inflammasome strongly interconnects pore-forming mechanisms of different cell death modalities (apoptosis, necroptosis, and pyroptosis). This provides redundancy and plasticity of pore-forming mechanisms affecting the fate of a single cell but also leading to a modus operandi in which different cell death modalities may coexist in a population of challenged cells. This cell death interconnectivity also integrates distinct pathways of cellular stress, inflammation, and infection and has profound physiological and pathophysiological implications [59].

The different PFPs may show restricted expression patterns in various cell types. Therefore, in vivo, the implication of other cell types may contribute to waves and dynamics of distinct cell death modes. Thus, PANoptosis, as a cell-autonomous integrated outcome during infection and inflammation towards pyroptosis, apoptosis, and necroptosis [60], may reflect the behavior of a population of cells. The waves and dynamics of different cell death modalities in a population of various cell types may be crucial for adaptive responses during inflammation and infection. Current data suggest that different types of necrotic cell death are differentially immunogenic. This suggests that plasma membrane permeabilization and plasma membrane rupture by themselves are not sufficient but that other subroutines occur that are associated with and dictate the immunogenicity of cell death. Indeed, cell death stimuli also engage specific subroutines such as activation of NF-κB and MAPK, leading to the production of cytokines and chemokines and the release of mtDNA by a process called MIMP (mitochondrial inner membrane permeabilization) and leading to cGAS-STING-induced production of cytokines, chemokines, and IFNs [61]. These cytokines, chemokines, and IFNs associated with cell death induction in a population of cells, alongside the DAMPs released by lytic cell death mechanisms, determine the immunogenicity of the cell death process. This cooperativity between cytokines, chemokines, IFNs, and DAMPs may play a crucial role in immunogenic cell death and inflammatory, autoimmune, and degenerative pathologies [62–64].

Necroptosis

Necroptosis is a caspase-independent form of regulated necrosis characterized by cell swelling (oncosis), plasma membrane permeabilization, and release of the intracellular content containing DAMPs, metabolites, and subcellular materials (ribosomes inter alia). Necroptosis is directly engaged during infection and inflammation by death domain-containing TNF receptor family members (TNFR1, TRAILR1/DR4, TRAILR2/DR5, DR3, and Fas), TLR3 and ZBP1, TNF family (TNF, TRAIL, FasL, and DR3), the TRIF-domain-containing TLR3 and TLR4, and the RIP Homology Interacting Motif (RHIM) domain-containing nucleic acid sensor ZBP1 [65]. The necroptotic pathway is dispensable for embryonic development and homeostasis under non-challenging conditions. Indeed, mice with kinase-dead RIPK1 [66, 67], kinase-deficient RIPK3 [68], or deficient MLKL mice [69] do not show any spontaneous phenotype. Because necroptosis is dispensable for development and homeostasis and is engaged during infectious, inflammatory, and degenerative diseases, targeting RIPK1 and RIPK3 kinases raised substantial interest among pharma and biotech industries [70, 71]. The core necroptotic pathway consists of three proteins: serine/threonine proteases RIPK1 and RIPK3 and the terminal executioner mixed lineage kinase domain-like (MLKL) pseudokinase. Central to the initiation and propagation of necroptosis are four proteins that contain an RHIM motif that allows homotypic interaction and higher-order aggregation [72, 73]: RIPK1, RIPK3, TRIF (an adaptor involved in TLR signaling), and ZBP1, a cytosolic sensor for z-RNA [74]. The converging axis of the different necroptosis-eliciting ligands is RIPK3, and its downstream target is mixed lineage kinase domain-like (MLKL) pseudokinase, which, after phosphorylation, is recruited to the plasma membrane [39].

TNFR1-Mediated Survival or Cell Death

Given the critical role of TNF in inflammatory diseases, we focus on the signaling of this cytokine. Interestingly, TNF binding to TNFR1 can have multiple outcomes. Although the name and the discovery of TNF as a "tumor necrosis factor" suggest the opposite, cell death induction is not the default activity of TNF. Depending on the cell type and the cellular context, TNF binding to TNFR1 leads either to cell survival (and transcriptional activation) or cell death (by apoptosis or necroptosis) [75]. RIPK1 is a crucial determinant in this cellular outcome. As a scaffold, RIPK1 is subject to many posttranslational modifications (e.g., phosphorylation, ubiquitylation) that determine its function as a platform for survival signaling by activating the NF-kB and MAPK pathways—its kinase activity is dispensable for this process [68]. As a kinase, RIPK1-dependent autophosphorylation drives cell death by apoptosis or necroptosis [76].

How Does Regulation of RIPK1 Exert These Opposing Functions?

Whether TNF binding to TNFR1 goes toward cell survival versus cell death depends on forming two distinct consecutive protein complexes [77] (Fig. 1). First, a membrane-bound complex I is formed that consists of TNFR1, TRADD, RIPK1, TRAF2, the ubiquitin ligases cellular inhibitor of apoptosis protein 1/2 (cIAP1/2), and the linear ubiquitin chain assembly complex (LUBAC; which is composed of HOIL1, HOIP, and SHARPIN). These E3 ligases create a network of ubiquitin chains on Lys residues of RIPK1 that recruit kinases for the canonical NF-κB signaling (on the M1-linked linear ubiquitin chains generated by LUBAC) and MAPK signaling pathway (on the K63-linked ubiquitin chains generated by IAP1/2). RIPK1 kinase activity is negatively regulated by phosphorylation by IKKα/IKKβ (S25) [78], TBK1 (T189) [79, 80], and MK2 (S320 and S335 in hRIPK1 and S321 and S336 in mRIPK1) [81–83] consolidating the platform function of RIPK1 and functioning as IKK- and TBK1-driven survival checkpoints.

The next stage is the formation of a cytosolic complex II [77], though the exact mechanism of the transition from receptor-bound complex I to cytosolic complex II remains unclear. The complex I-dependent survival response to TNF can be switched to apoptosis by cotreatment with a protein synthesis inhibitor, cycloheximide (CHX), or by cotreatment with an IAP inhibitor. Apparently, the former type of apoptosis is independent of RIPK1 kinase activity mediated by complex IIa, while the latter is RIPK1 kinase activity-dependent, driven by the formation of complex IIb [84]. This suggests that CHX addition interferes with the translation of survival factors induced by NF-κB and MAPK pathways, which forms a second checkpoint for RIPK1-independent apoptosis. Complex IIb begins with the binding of FADD to TRADD and RIPK1 (components from complex I), auto-activating phosphorylation of RIPK1 at positions S161 and S166 [18, 85], and recruitment and activation of caspase-8. This results in the induction of RIPK1-kinase activity-dependent apoptosis in complex IIb [86, 87]. Interaction of the RIPK1 death domains results in its dimerization of the RIPK1 kinase domains and autophosphorylation [88], the license to kill by apoptosis or necroptosis. The primary substrate of RIPK1 appears to be RIPK1 itself [18].

Fig. 1 (continued) (MLKL*), **(12)** converting MLKL from a closed to an open structure, resulting in oligomerization (tetramer in human, trimer in mouse), membrane recruitment, insertion, and oligomerization. **(13)** TLR3 and TLR4 ligand binding lead to the recruitment of the RHIM domain-containing adaptor TRIF, which binds and activates RIPK3*, further initiating the formation of a necrosome complex leading to MLKL activation. **(14)** The nucleic acid zRNA sensing ZBP1 zRNA initiates directly RIPK3 recruitment and activation. **(15)** MLKL*-mediated cellular swelling (oncosis) triggers a PMR-inducing mechanism independent of NINJ1

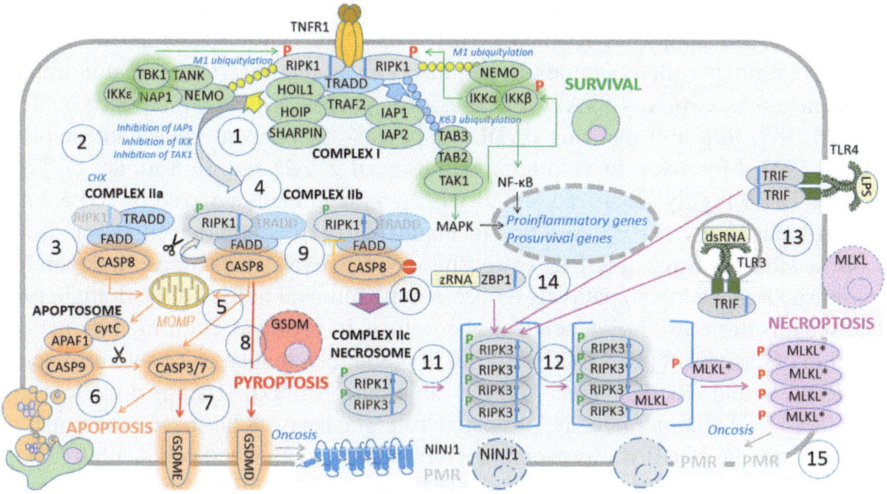

Fig. 1 Multiple cell death modalities. TNFR1-mediated survival (green) and cell death pathways (apoptosis, orange; pyroptosis, red; necroptosis, pink). **(1)** The binding of TNF to its cognate receptor induces the formation of a plasma membrane-bound complex I (TNFR1-TRADD-RIPK1), in which RIPK1 functions as a scaffold that is kept in its survival mode by ubiquitylation (yellow and blue chains) and trans-phosphorylation (red "P"). K63-ubiquitylation by IAP1/2 creates a platform on which the TAB1/TAB2/TAK1 complex is recruited, leading to the activation of MAPK pathways contributing to survival and inflammatory gene expression. M1-linear ubiquitylation by HOIL1/HOIP/SHARPIN (LUBAC complex) creates a platform for the recruitment of the NEMO/IKKα/IKKβ complex and the TBK1 complex (NEMO/TANK1/NAP1/TBK1/IKKε) also contributing to a.o. NF-κB-dependent survival and inflammatory gene expression. The RIPK1-dependent scaffold function is checked by several transphosphorylation events (red "P") mediated by a.o. IKK and TBK1. **(2)** By interfering in the survival mechanisms of complex I, it can be turned into a cytosolic complex II with different flavors: complex IIa, complex IIb, and the necrosome (sometimes called complex IIc). **(3)** When translation is inhibited [by addition of translation inhibitor cycloheximide (CHX) or defect in NF-κB signaling], complex IIa is formed, leading to FADD/Casp8-mediated apoptosis. **(4)** Pharmacological inhibition of IAPs, IKK, or TAK1 of RIPK1 in complex I leads to the formation of a cytotoxic complex II that requires the RIPK1 kinase activity to kill either by apoptosis or by necroptosis. In complex IIb RIPK1 undergoes activation resulting in autophosphorylation (green "P") and RIPK1* kinase-dependent FADD/Casp8-dependent apoptosis. **(5)** Casp8 either causes mitochondrial outer membrane permeabilization (MOMP), leading to the generation of the apoptosome complex (Cyto C/APAF1/Casp9), leading to Casp3/7 proteolytic activation, or Casp8 directly activates Casp3/7. **(6)** The latter propagates apoptotic features such as phosphatidylserine (PS) exposure and membrane blebbing and prepares the apoptotic cell for efferocytosis. **(7)** Some chemotherapeutic drugs cause Casp3/7, whether or not preceded by apoptotic features, to cleave GSDME (depending on its expression levels), leading eventually to pyroptosis and plasma membrane rupture by NINJR. **(8)** In conditions of Yersinia infection, TNF-mediated caspase-8 activation can also lead to pyroptosis involving GSDMD [262]. **(9)** Casp8 in complex-II also functions as a checkpoint for cell death that controls its substrate RIPK1* levels by proteolysis. **(10)** Inhibition of caspase-8 in complex II leads to the accumulation of activated RIPK1*. The RIP homotypic interaction motif (RHIM) domain in RIPK1* (blue stripe in RIPK1, RIPK3, ZBP1, and TRIF) causes recruitment and autophosphorylation of RIPK3*. **(11)** RIPK3 forms large β-amyloid-like necrosome complexes on which MLKL, already interacting with non-activated RIPK3 (MLKL-RIPK1 dimer in the human system) or in a kiss-and-run mode (monomeric MLKL in the mouse system), is phosphorylated in the C-terminal pseudokinase domain

In a third checkpoint, the caspase-8 activity is decisive in the outcome between RIPK1-induced apoptosis (complex IIb) or RIPK1-induced necroptosis (necrosome, sometimes called complex IIc). It has been an old observation that inhibition of caspase-8 by CrmA or zVAD-fmk sensitizes necrotic cell death induced by TNF or FasL [89, 90], and that this FADD-caspase-8-independent cell death involved RIPK1 [91]. Moreover, in vivo administration of zVAD-fmk in a model of TNF-induced SIRS highly sensitized lethality [92], a phenomenon driven by RIPK3 and RIPK1 kinase activity [93]. As a first mechanism to explain this sensitization, it was proposed that cleavage of RIPK3 (by caspase-8) inactivates the caspase-independent apoptosis (read necrosis) pathway by proteolytic removal of the kinase domain [94]. The sensitization by TNF-dependent necroptosis in vivo was confirmed by genetic ablation of caspase-8 or FADD, resulting in perinatal lethality that could be rescued by RIPK1 and RIPK3 and TNFR1 ablation [95–97]. Using non-cleavable *Ripk1*$^{D325A/D325A}$ mutant mice, the protective role of caspase-8 has been pinpointed on its ability to cleave RIPK1, preventing TNF-mediated enhanced cell death by apoptosis and necroptosis during embryonic development [98, 99]. In line with this paradoxical protective role of caspase-8 against excessive cell death, mutations that prevent RIPK1 cleavage by caspase-8 have been associated with autoinflammatory diseases characterized by enhanced RIPK1-mediated apoptosis and necroptosis [100, 101]. Besides these three checkpoints, other cellular processes, such as ATG9-mediated unconventional autophagy independent of LC3 [102] and poly-ADP ribosylation-mediated proteasomal degradation [103], limit complex II accumulation and consequent cell death.

Ripk1-deficient mice die perinatally but can be rescued by combined inhibition of TRADD-FADD-caspase-8-mediated apoptosis and RIPK3-MLKL-dependent necroptosis [104–108]. This demonstrates the absolute necessity for the presence of the RIPK1 scaffold for development. On the other hand, transgenic mice expressing kinase-inactive RIPK1 are viable, do not develop spontaneous pathology, and are protected from inflammation and tissue injury in different experimental disease models [66–68, 109]. Altogether, the kinase-independent scaffold RIPK1 inhibits cell death and safeguards survival and tissue homeostasis, while its kinase activity promotes apoptosis and necroptosis, causing inflammatory pathologies. In humans, loss of function mutations of RIPK1 do develop normally but suffer from lymphopenia, infections, arthritis, combined immunodeficiency, and inflammatory bowel diseases [110–112], while the hemizygous mutations of the caspase-8-dependent cleavage site in RIPK1 cause autoinflammation and immunodeficiency [100, 101]. The more stringent requirement for RIPK1 in mice versus humans may be due to species-specific differences in the expression patterns of RIPK1 or more sensitized cell death pathways in the absence of RIPK1. We should be aware of this in interpreting and translating findings of experimental cell death and inflammation models in mice to human pathologies.

TLR-Mediated Necroptosis

Besides the death domain receptors, apoptosis and necroptosis signaling pathways can also be initiated by TLR3 [113] and TLR4 [114] (Fig. 1). TLR3 activation is induced by binding dsRNA in the luminal part of the endosomes, while the signaling complexes are oriented to the cytosol. TLR4 recognizes extracellular bacterial lipopolysaccharide (LPS), which is internalized in the endosomal compartment, and the recruited signaling complexes are also oriented to the cytosol. Akin to the above-described TNFR1 signaling, the outcome of TLR3 and TLR4 is either pro-survival and pro-inflammatory gene expression, apoptosis, or necroptosis. This illustrates a common emerging theme of death ligand and PAMP recognition shared by the outcome of ZBP1 signaling (see below), viz., the receptor-mediated formation of multiple complexes that generate very different cellular fates. Both TLR3 and TLR4 recruit at the cytosolic-oriented part of the receptor TRIF, a Toll/IL-1 receptor (TIR) and RHIM domain-containing adaptor-inducing interferon β, engaging RIPK1 and RIPK3 by homotypic RHIM domain interactions RIPK1 and RIPK3 [113, 115]. RIPK1 can further recruit TRADD-TRAF2-cIAP1/2 and FADD-caspase-8 [116]. The latter cleaves and inactivates NEDD4-binding protein 1 (N4BP1), a suppressor of cytokine production, licensing TRIF-independent TLRs to produce higher levels of inflammatory cytokines [117]. Deficiency of IAPs or LUBAC results in robust induction of TRIF-mediated apoptosis [118–120]. Under conditions of caspase-8 inhibition, TLR3 and TLR4 induce RIPK3-mediated necroptosis, which can be inhibited by blocking RIPK1 kinase activity, similar to TNFR1 signaling [67, 114].

ZBP1-Mediated Necroptosis

The IFN-inducible ZBP1 senses the presence of Z-shaped nucleic acids by two N-terminal Zα domains and recruits RIPK3 directly by its two intermediate RHIM domains [121]. Interestingly, in the case of ZBP1, RIPK1 is dispensable for the activation of RIPK3 and necroptosis induction [104], but apparently, it competes with the recruitment of RIPK3 [122]. ZBP1 causes skin inflammation by inducing RIPK3-mediated necroptosis and RIPK1-caspase-8-mediated apoptosis in keratinocytes [123]. The FADD-caspase-8 axis prevents spontaneous ZBP1 expression and necroptosis induction [124] and also keeps in check RIPK3-caspase-8-mediated apoptosis [125]. The checkpoint mechanism is probably similar to TNFR1 signaling in complex II [98, 99], caspase-8-mediated cleavage of RIPK1, since cells from uncleavable *Ripk1*$^{D325A/D325A}$ mutant mice show enhanced ZBP1-mediated cell death [126]. Along the same line, mutation of the RIPK1 RHIM causes perinatal lethality that is rescued by genetic deficiency of RIPK3, MLKL, or ZBP1, revealing a critical role of the RIPK1 RHIM in suppressing ZBP1-mediated necroptosis [122, 127, 128]. Moreover, transgenic mice expressing the orthologue R588E mutant lacking a functional death domain (DD) required for RIPK1 recruitment [129] show enhanced inflammation-sensitized signaling of ZBP1 or TRIF towards RIPK3. This demonstrates the importance of RIPK1-DD to restrain both ZBP1 and TRIF during homeostasis and inflammation [130]. Besides necroptosis and apoptosis, ZBP1 was initially identified as an inducer of type I IFN production mediated by association and activation of TBK1 serine/threonine kinase. Subsequent phosphorylation of its target, IRF3, results in the transcription of type I IFN [131, 132] and acts as an activator of NF-κB by RHIM-mediated association of RIPK1 and RIPK3 [133, 134]. Altogether, the dampening role of caspase-8-mediated cleavage of RIPK1 on cell death mediated by TNF, ZBP1, and TRIF (TLR3 and TLR4) is clinically manifested by the existence of patients with the CRIA (cleavage-resistant RIPK1-induced autoinflammation) syndrome [100, 101].

Activation of MLKL

The necroptosis-inducing pathways involving TNFR1-TRADD-RIPK1-RIPK3, TLR3/TLR4-TRIF-RIPK3, and ZBP1-RIPK3 eventually result in RHIM-mediated oligomerization and activation of RIPK3, forming an amyloid signaling complex, termed the necrosome [73]. Within this necrosome, RIPK3 undergoes autophosphorylation and activation [20, 21, 135], leading to the phosphorylation of MLKL, the executioner of necroptosis [39, 69, 136]. Under basal conditions, MLKL is present in the cytosol in an inert form that contains two domains: an N-terminal four-helix bundle domain (4HBD, H1-H4) and a C-terminal pseudo-kinase domain (psK) connected by a flexible brace region (H5-H6) (Fig. 2a). The 4HBD is responsible for the cytotoxic activity, while the psK domain and the brace domain control the cytotoxic 4HBD domain [136]. Following phosphorylation by RIPK3 of the activation loop of the psK domain (S345, S347, and T349 in mouse MLKL, and T357 and S358 in human MLKL), a transition to the killer form is triggered [39, 136]. This suggests that perturbation of the psK domain may cause a switch between an inactive and an active conformer. Human MLKL and RIPK3 form a stable heterodimer in cells before necroptosis induction, with inactive RIPK3 keeping hMLKL in an "open inactive state" [39, 137]. Phosphorylated and phosphomimetic mutants of hMLKL adopt the "closed active state" conformation in crystal structures [138, 139], favoring the release from RIPK3 and the formation of

Fig. 2 (continued) pores consist of α-helices (α-PFP) or β-strands (β-PFP). GSDMs are an example of a barrel stave β-PFP. In the toroidal pores, the amphipathic PFPs interact with the plasma membrane, inducing a pore lined by a positive curvature of the membrane (+) in which PFPs alternate with phospholipids, creating a protein-lipid pore. In the carpet model, lipid pores are induced by the interaction of amphipathic protein stretches, resulting in positive curvature in the membrane and half-micelles and micelles lined by the protein stretches. (**d**) Model of MLKL activation and recruitment (modified from [143]). Basal MLKL::RIPK3 heterodimer (human) and monomer MLKL (mouse) are recruited to the RIPK3* necrosome complex, resulting in the phosphorylation of MLKL. Single phospho-MLKL forms a dimer and tetramer (human) or a trimer (mouse), resulting in a positioning of the 4HBD that allows electrostatic interaction with negatively charged PIPs at the inner leaf of the plasma membrane (orange circles in phospholipid layer). As no membrane-inserted MLKL structure is available, the exact mechanism of MLKL pore formation is still elusive. Amphipathic 4HBD stretches elicit a toroidal mechanism protein-lipid formation (Modified from [143])

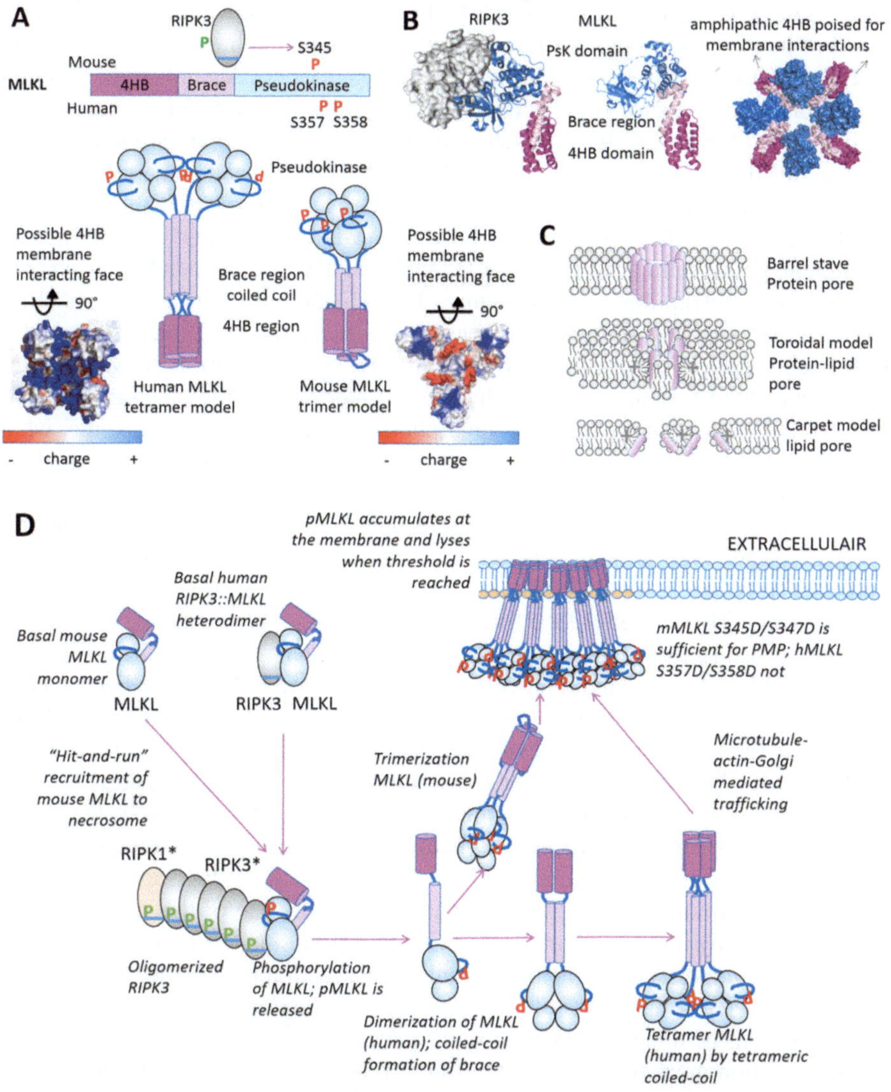

Fig. 2 Plasma membrane permeabilization by MLKL. (**a**) The MLKL molecule contains an N-terminal four-helix bundle domain (4HBD) (pink), an intermediate brace region (Lila), and a C-terminal pseudokinase (psK) domain (blue). RIPK3 phosphorylates the psK domain at S357/S358 in hMLKL and S345 in mMLKL. Upon phosphorylation by RIPK3*, the brace region of MLKL forms coiled-coil-driven tetramers (human) and presumed trimers (mouse) by which the 4HD is poised for electrostatic interaction with the plasma membrane of positively charged residues from the 4HBD and the negatively charged phosphatidylinositol phospholipids (PIPs) at the interleaf of the plasma membrane (modified from [143]). (**b**) Structure of RIPK3::hMLKL heterodimer (PDP entry 4 m69). Structure of hMLKL [4HBD (pink), brace (lila), psK (blue)]: monomer (PDB entry 4BTF) and modeled tetramer (based on PDB entry 4BTF). (**c**) Mechanisms of plasma membrane permeabilization by pore-forming proteins. In the barrel-stave model, the lumen of the pores is lined by pore-forming proteins (PFP), in which the transmembrane segments of the

pseudokinase back-to-back dimers [139, 140]. There are fundamental differences between human and mouse MLKL regarding the activation mechanism [138, 141, 142]. In mouse MLKL, back-to-back dimerization does not depend on the adoption of the "closed active state" conformation. This implies that mMLKL is more readily active, as evidenced by the fact that phosphomimetic mutations are sufficient to activate mMLKL, which is not the case for hMLKL [138] and some other species [143]. Also, a mutation in the activation loop helix that interferes with K219:Q343 salt bridge in mMLKL results in the form of mMLKL that causes readily plasma membrane permeabilization (mMLKLQ343A) in a RIPK3-independent way [136]. In contrast, disruption of the same K230:Q356 salt bridge in hMLKL (hMLKLQ356A) does not result in constitutive activation [142], suggesting additional regulation levels for hMLKL. The difference in activation between mouse and human MLKL is also reflected by hMLKL forming stable heterodimers with inactive RIPK3 and being released upon phosphorylation in the necrosome, while preformed mMLKL dimers are activated by active RIPK3 in the necrosome by a hit-and-run mechanism [138]. R30E MLKL is the only constitutive activating mutation identified in hMLKL, relying on disengagement of the N-terminal membrane-permeabilizing four-helix bundle domain from the central autoinhibitory brace helix [144].

Following RIPK3-mediated activation, MLKL forms oligomers whose stoichiometry has been a subject of debate, ranging from trimers, tetramers, hexamers, and octamers [143]. An AlphaFold model for phosphorylated hMLKL predicts an elongated coiled-coil-driven tetramerization by heptad repeats present in the brace region, the functional importance of which has been validated by mutational analysis [138, 139, 143]. For mMLKL, a brace domain-driven oligomerization has also been proposed, but rather involving trimer formation [145–147] (Fig. 2a). Accordingly, ColabFold generates a coiled-coil-driven trimer with a more compact structure than the hMLKL tetramer (12 nm for the mMLKL trimer versus 20 nm for the hMLKL tetramer). However, the functional importance of this trimer has not been validated by mutational analysis [143]. All MLKL models in different species argue for the importance of the brace domain in coiled-coil oligomerization, while the cytotoxic 4HB domain is not implicated in this process.

The reasons behind differences in stoichiometry (trimer vs. tetramer) and underlying activation mechanisms—whether more readily activated by phosphorylation (mMLKL) or requiring additional regulation (hMLKL)—among vertebrate MLKL orthologues remain unclear and require further study [141]. These differences may reflect evolutionary pressures, given the role of necroptosis in antiviral responses, and have even led to the absence of MLKL in certain species, such as carnivores and marsupials. Likely, this absence reflects counterselection of the lethal consequence of massive necroptosis as a battlefront of extensive counteradaptation associated with viral infection [74, 148–150]. However, despite the differences in the stoichiometry of oligomerization and precise activation mechanism of MLKL, the common theme is that RIPK3 phosphorylates MLKL as the downstream executioner of necroptosis driven by three major RHIM-driven pathways (death domain receptors/

RIPK1, TLR3/TLR4/TRIF, and ZBP1), leading to the targeting and permeabilization of the plasma membrane [145, 151–154].

In the first instance, activated and oligomerized MLKL is recruited to the plasma membrane and exerts its cytotoxic activity. Recruitment to the plasma membrane requires a positively charged patch in the 4HBD interacting with negatively charged phosphatidylinositol phosphate areas in the plasma membrane [152, 154, 155]. Furthermore, inositol-polyphosphate multikinase (IPMK) and inositol-tetrakisphosphate 1-kinase (ITPK1) are essential for necroptosis by generating cytosolic inositol pentakisphosphate (IP5) and inositol hexakisphosphate (IP6) [156]. The IP5/IP6 interaction with the positively charged 4HBD displaces H6 in the brace domain, unplugs the 4HBD, and initiates the brace domain-mediated oligomerization [155]. Accordingly, a mutant in H6 ($mMLKL^{D139V}$) is sufficient to generate constitutively active mMLKL independent of RIPK3 [157]. By single-cell imaging approaches, the order of MLKL events has been documented (Fig. 2b). First, phosphorylated MLKL assembles into higher-order cytosolic complexes, presumably in the microsomes. Next, MLKL co-traffics with tight junction proteins to the cell periphery via Golgi-microtubule-actin-dependent mechanisms and co-accumulates at the plasma membrane with tight-junction proteins as heterogeneous micron-sized hotspots [158].

In the absence of structural information on MLKL in the membrane-bound states, several models have been proposed on how MLKL would indirectly contribute to or directly mediate plasma membrane permeabilization. These mechanisms include the indirect activation of transmembrane Ca^{2+} channels (TRPM7, transient receptor potential melastatin related 7) [146], the direct functioning as a cation channel with H1, H2, H3, H5, and H6 as transmembrane segments and H4 located in the cytoplasm becoming permeable to Mg^{2+} rather than Ca^{2+} in the presence of Na^{+} and K^{+} [159], or a plug-release system in which H6 from the brace upon phosphorylation of the psK domain enables the unfolding of the 4HBD in the membrane [155]. This 4HBD unfolding in the membrane would expose the amphipathic features of the α-helices. This may lead to three non-exclusive mechanisms, including (1) a carpet-like permeabilization due to lipid-interacting amyloid-like structures, (2) a membrane thinning and tension leading to toroidal pore-like mechanisms, or (3) an induction of a positive phospholipid curvature and permeabilization by local foci of high levels of PIPs [160] (Fig. 2c). Interestingly, MLKL-mediated cytotoxicity apparently can still be counterbalanced by interaction with the ESCRT-III secretory system, generating vesicles that can remove membrane-targeted MLKL. The subsequent release of phosphorylated MLKL via extracellular vesicles prolongs cell survival [57, 161] and may result in prolonged cytokine production [57]. Whether MLKL accumulation at the plasma membrane also requires an additional plasma membrane rupture mechanism (PMR) mechanism still remains to be determined. Interestingly, in contrast to NINJ1-mediated PMR as an additional endpoint of secondary necrosis, pyroptosis, and ferroptosis [51, 53, 162], NINJ1 seems not to be implicated in PMR following necroptosis [53].

Pyroptosis

Pyroptosis is a regulated necrotic cell death modality characterized by two molecular features: the formation of a cytosolic sensor complex coined the inflammasome [163], and the permeabilization of the plasma membrane leading to necrotic cell death (Fig. 3a). The name pyroptosis is based on the association between this cell death modality and the production by infected cells of inflammatory cytokines (IL-1β and IL-18) by caspase-1 [164]. Ten years later in 2015, the precise mechanism of plasma membrane permeabilization was discovered as a caspase-1- and caspase-4/5/11-driven proteolytic activation of GSDMD [42, 43]. With the finding of GSDME being activated by the apoptotic executioner caspase-3, the definition of pyroptosis was broadened as a regulated necrotic cell death modality executed by gasdermins [40, 41]. By doing so, this definition also includes the process of secondary necrosis following apoptosis in the absence of phagocytosis [165, 166]. Interestingly, caspase-8 has also been reported to activate GSDMD conditions such as *Yersinia* infection [167–169]. Though the caspase-1-, caspase-11/4/5-, caspase-8-, and caspase-3-mediated pyroptosis have a common endpoint, viz., the proteolytic activation of gasdermins, the pathways and the associated cellular responses causing pyroptosis may be very different and may generate differential intercellular responses upon lysis of the cells. This implies that pyroptosis as a necrotic cell death modality probably functions as a flag that covers many cargoes.

Inflammasomes

The inflammasome is a cytosolic sensor complex that recruits and activates procaspase-1 by proximity-induced dimerization, leading to the proteolytic maturation of pro-IL-1β [163]. The canonical inflammasome is typically associated with the proteolytic processing of pro-IL-1β by caspase-1, predominantly occurring in macrophages. There are many flavors of canonical inflammasomes. During infection and cellular stress, multiple types of nucleotide-binding oligomerization-domain-like receptors (NLRs) in the cytosol directly or indirectly sense the presence of PAMPs

Fig. 3 (continued) Caspase-11/4/5 proteolytically activates GSDMD. Potassium efflux through TWIK2 and ATP influx through P2X7 contribute to pyroptosis induction. Pro-IL1β mRNA is transcriptionally induced by TLRs, TNFR1, IL-1R1, and IFNAR, a process called "priming." GSDMD pores mediate the release of IL-1β and IL-18. Pyroptosis is also associated with cellular swelling (oncosis) that is sensed by the NINJ1, resulting in swelling (**b**) GSDMD and E contain an N-terminal membrane-permeabilizing domain (N-GSDM) (orange) and a C-terminal repressor domain (C-GSDM) (blue) connected by a protease-sensitive loop. The N-terminal domain is kept inactive by the C-terminal domain. Release of the N-GSDM domain from the CTD via proteolytic cleavage essentially primes the NTD to undergo restructuring, leading to a 4-stranded antiparallel β-sheet that can self-associate upon interaction with the plasma membrane. These form lines and arcs that can fuse to membrane slits or rings. The latter form barrel-stave pores with defined stoichiometry (33 GSDMD subunits and an internal diameter of 21.5 nm). The globular domains of GSDMD form a cytosolic rim

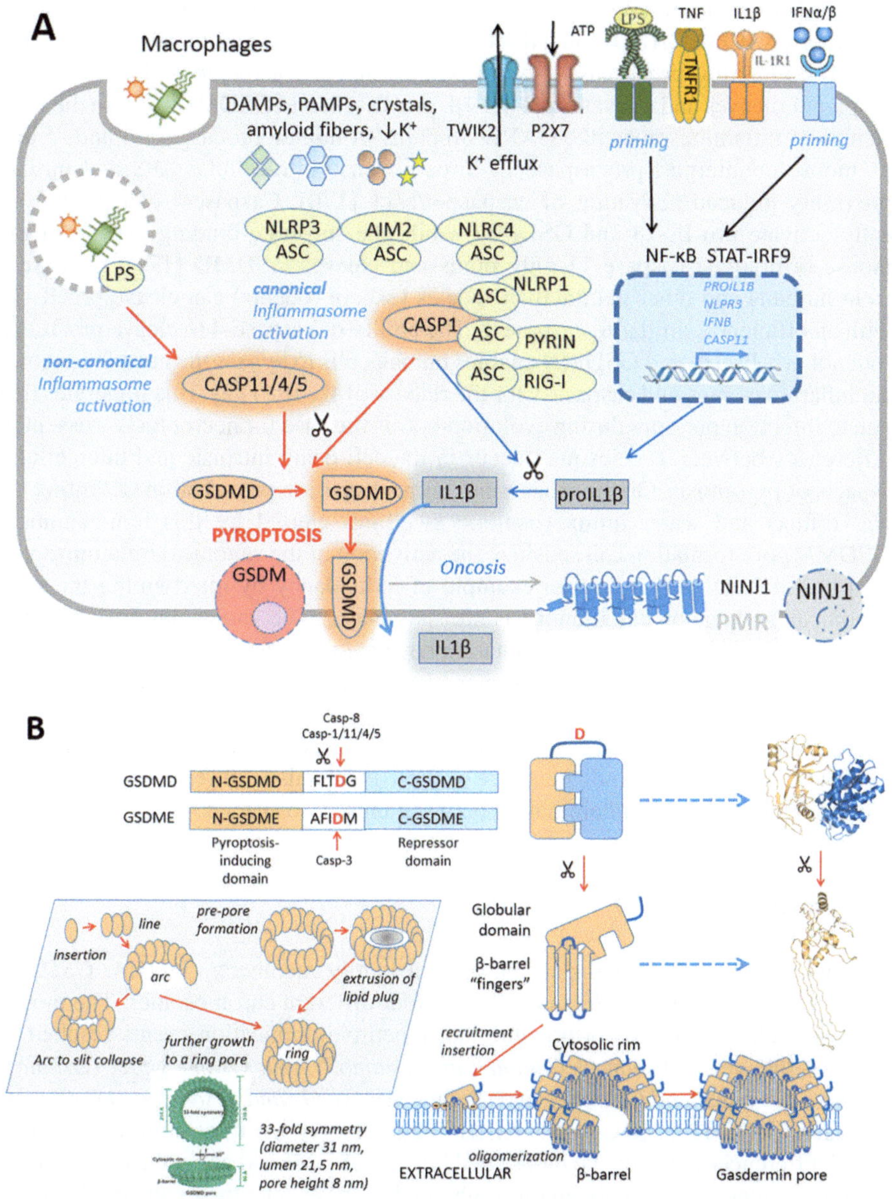

Fig. 3 Pyroptosis and GSDMD-mediated pore formation. (**a**) Inflammasome-induced pyroptosis combines cytokine induction (a.o. pro-IL-1β) with cell death induction and GSDMD pore activation. Inflammatory caspase-1/11 (mouse) and caspase-1/4/5 eventually proteolytically activate GSDMD. There are two main classes of inflammasomes: the canonical and the non-canonical. The canonical inflammasome sensors NLRP3, AIM2, NLRC4, NLRP1, PYRIN, and RIG-I recruit and activate procaspase-1 through the adaptor ASC. Caspase-1 proteolytically cleaves and activates GSDMD and pro-IL1β, leading to pyroptosis. In the noncanonical pathway, cytosolic gram-negative bacteria or LPS molecules directly activate caspase-11 (or caspase-4/5 in humans).

and DAMPs, start to oligomerize, and bind bipartite CARD-domain-containing adaptors (ASC/PYCARD and NLRC4). These, in turn, recruit and activate the CARD-containing procaspase-1 by proximity-induced activation. Next, activated caspase-1 cleaves and activates pro-IL-1β, pro-IL-18, and GSDMD [68]. In the non-canonical inflammasomes, the CARD domains of human procaspase-4 and -5 and its mouse counterpart procaspase-11 directly sense intracellular LPS and initiate proximity-induced activation of caspase-4/5/11 [170]. Caspase-4 can proteolytically activate pro-IL-18 and GSDMD, while the substrate-binding exosite of the mouse orthologue caspase-11 only binds and cleaves GSDMD [171]. Caspase-4 from humans and other mammalian species (except rodents) can cleave pro-IL-18 with an efficiency similar to caspase-1. The ability of caspase-4 to cleave pro-IL-18 (but not pro-IL-1β) and GSDMD enables human cells to bypass the need for canonical inflammasomes and caspase-1 for the release of IL-18 [172]. This illustrates that due to infection pressure during evolution—as is the case for necroptosis—essential differences between species may occur in the cell death intrinsic and intercellular aspects of pyroptosis. On the other hand, intracellular ion perturbation (K^+ influx vs. Na^+ efflux) and water influx (osmotic swelling) caused by this non-canonical GSDMD pore formation can sensitize the activation of the canonical inflammasome [43] in a parallel way, again an example of redundancy by intertwining the non-canonical and canonical inflammasome pathways. The molecular insights into inflammasome composition and signaling have led to mechanistic insights for therapeutic targeting [173, 174]. However, as illustrated above for the differences between the mouse (caspase-11) and human (caspase-4) orthologues in pro-IL-18 processing and the differences in the effect of RIPK1 deficiency between mouse and human [100, 101], we should be aware of these mechanistic differences in the regulation of cell death and inflammation between these two species.

Activation of GSDMD

The human genome contains six gasdermin family members: *GSDMA, GSDMB, GSDMC, GSDMD, GSDME,* and *PJVK* on four different chromosomes. The mouse genome lacks a GSDMB orthologue, but repetitive duplication events resulted in three *Gsdma* genes (*Gsdma1, Gsdma2, and Gsdma3*), four *Gsdmc* genes (*Gsdmc1, Gsdmc2, Gsdmc3, and Gsdmc4*), and single genes for *Gsdmd, Gsdme, and Pjvk*, raising again the need to be cautious on the interpretation and translation of the functional differences between the mouse and human GSDMs and about the evolutionary selective forces that have driven this high yield of these gene losses and duplications [175]. The biology of the GSDM proteins remained enigmatic for a long time, as they show a very diverse expression pattern in the gastrointestinal tract and the skin; therefore, they are coined "gasdermins." Mutations have been connected with a wide variety of diseases, such as alopecia (m*Gsdma3*), asthma (h*GSDMB*), hearing loss (*hGSDME* or *DFNA5* and *PJVK* or *DFNB59*), and cancer (h*GSDME*) [176, 177]. The fundamental understanding of their role and mechanism of action came with the discovery that GSDMD, through proteolytic activation, was the executioner of canonical and non-canonical inflammasome-mediated pyroptosis [42, 43, 178], and that GSDME was activated by caspase-3 during secondary necrosis [40, 41]. In resting human and mouse cells, GSDMs are inactive pro-proteins in the cytoplasm consisting of an N-terminal ($GSDM^{NT}$) and a C-terminal domain ($GSDM^{CT}$), connected by a flexible linker. Aspartate-specific cleavage within the central linker liberates $GSDM^{NT}$ from the inhibitory $GSDM^{CT}$, resulting in the recruitment to the plasma membrane and the consecutive plasma membrane permeabilization by $GSDM^{NT}$. We further focus on GSDMD, given its importance in inflammasome-driven pyroptosis. The crystal structure between dimeric active human caspase-1 and full-length murine GSDMD revealed an extensive interaction surface between both. This is formed by the engagement of the tetrapeptide target sequence in the GSDMD linker ($LLSD^{276}$-G^{277}) by the catalytic site of caspase-1 and the interaction of a conserved hydrophobic patch of the $GSDMD^{CT}$ with an exosite outside the catalytic site of caspase-1. This exosite consists of an anti-parallel β-sheet formed by the active site loops between the large and small subunits of caspase-1 [179]. A similar hydrophobic exosite is used in the case of the non-canonical LPS-activated and oligomerized caspase-4/11 in the recognition of GSDMD [180]. Interestingly, in both cases, the substrate binding does not require the cleavage site recognition in the linker region, demonstrating that the exosite is sufficient to determine the substrate recognition by inflammatory caspase. Moreover, this exosite is not required for the binding of pro-IL-1β by caspase-1, emphasizing the specificity of its role in GSDMD recognition and cleavage [179, 181]. On top of the exosite recognition, the binding of dimeric caspase-4/11 to the hydrophobic patch of $GSDMD^{CT}$ creates a second autoprocessing event at D270 (besides D289 in caspase-4 or D285 in caspase-11) to remove an inhibitory interdomain linker, generating a small p10 subunit resulting in higher affinity binding and cleavage of GSDMD [180, 181]. These substrate-specific exosites open new perspectives for the development of drugs targeting particular substrates of the inflammatory caspases rather than all their substrates.

After proteolytic cleavage, the $GSDMD^{NT}$ fragment is recruited to the inner leaflet of the plasma membrane, involving several essential patches exposed on the $GSDMD^{NT}$ fragment that bind negatively charged phospholipids, including phosphoinositide phosphates, phosphatidylserine, phosphatidic acid, and cardiolipin [182–184] (Fig. 3b). In the next step, oligomerization, the lateral interface between adjacent $GSDMD^{NT}$ units involves the α1 helices that align head-to-tail and H-bonds between the transmembrane β-hairpins [185]. Depending on the methods used, two models have been proposed for how individual $GSDMD^{NT}$ monomers proceed to pore formation [181]. One model is based on cryo-EM structures where a fully assembled prepore with the final pore stoichiometry was detectable without membranes. In such a prepore, the interacting $GSDMD^{NT}$ units still have globular conformation similar to the protomer's auto-inhibited state [185]. In the presence of membranes, this pre-pore undergoes an intensive conformational change with the formation of a curled β-sheet structure deeply inserted in the inner leaflet of the membrane. This process forms a transmembrane β-barrel protein pore with a 33-fold symmetry and an inner pore size of 21 nm [185]. The surrounded lipid plug is expelled as a vesicle or bicelle, creating a hydrophilic lumen [181] (Fig. 3b). The "pre–pore–pore" transition model suggests that oligomerization would occur at the membrane before the extensive conformational change needed for membrane insertion [185]. However, time-resolved atomic force microscopy revealed a more dynamic model in which $GSDMD^{NT}$ small intermediates penetrate the membranes following β-sheet formation and further oligomerize to form multimeric line- and arc-shaped structures with open membrane edges. These lines and arcs can grow and fuse to ring-shaped structures representing the β-barrel protein pores [186, 187] (Fig. 3b). Recently, the role of specific cysteine residues in the targeting and oligomerization of GSDMD was revealed. Palmitoylation of C192 and the presence of the negatively charged phospholipids target GDMD to the membrane, while C39/C57/C192-enable low-number oligomerization (dimeric and trimeric building blocks) before membrane insertion and formation of higher-order oligomerization [188].

Ferroptosis

Identifying distinct regulated cell death pathways offers exciting potential for manipulating cellular life and death decisions to prevent diseases and aid tissue regeneration [189]. While apoptosis and later non-apoptotic cell death modalities, as discussed above, have been extensively studied for decades, recent work has uncovered regulatory mechanisms and signaling pathways for a novel form of regulated necrotic cell death, called ferroptosis [44, 190]. This pathway holds significant untapped potential for developing new treatments for degenerative diseases and cancer. Ferroptosis is an iron-dependent cell death process associated with uncontrolled lipid peroxidation, leading to membrane rupture and subsequent cell death (Fig. 4). It is increasingly recognized as one of the most disease-relevant cell death pathways, alongside apoptosis, necroptosis, and pyroptosis. Ferroptosis is unique in that it is typically induced only under pathological conditions, such as ischemia/reperfusion injury (IRI), organ damage, and certain neurodegenerative diseases. Conversely, inducing ferroptosis can be a powerful strategy to combat therapy-resistant and metastasizing cancers. Older literature also mentions "oxytosis" as a cell death modality in neuronal cell death that shares many characteristics with ferroptosis [191] both in the way it can be triggered (i.e., glutamate-mediated inhibition of the cystine/glutamate antiporter [192] leading to cellular glutathione [GSH] exhaustion) and protected (by lipophilic antioxidants such as vitamin E [193]). This resemblance suggests that *"oxytosis and ferroptosis should be regarded as two names for the same cell death pathway"* [191], or at least that diving into the oxytosis literature may enrich current ferroptosis research.

Ferroptosis: A Metabolic Cell Death

Mechanistically, the cyst(e)ine/GSH/GPX4 axis and the ferroptosis suppressor protein-1 (FSP1)/CoQ_{10}/vitamin K axis are central to preventing ferroptosis. GPX4 is

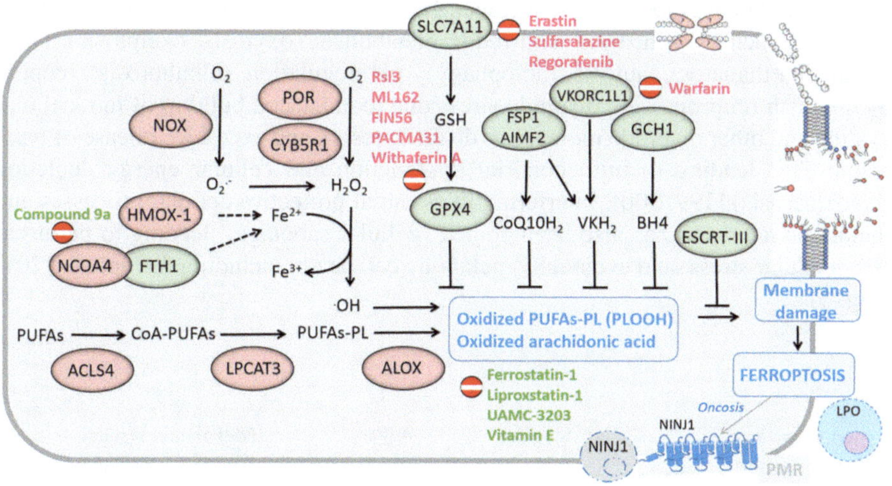

Fig. 4 Ferroptosis. The proteins and compounds that negatively regulate ferroptosis are indicated in green, and those that promote ferroptosis are indicated in red. The promotion of ferroptosis involves enhanced PUFA synthesis (ACSL4, acyl-CoA synthetase long-chain family member 4; LPCAT3, lysophosphatidylcholine acyltransferase 3; ALOX, lipoxygenase) [194, 262], enhanced oxygen radical production (POR, cytochrome P450 oxidoreductase; CYB5R1, cytochrome B5 reductase 1; NOX, NADPH oxidases) [262, 263], and enhanced availability of free iron (HMOX-1, heme oxygenase-1) [264], propagating Fenton reactions on PUFA phospholipids (PUFA-PL) and ALOX-dependent formation of oxidized PUFA-PL (PLOOH) and oxidized arachidonic acid. The availability of free iron in the lysosomes is propagated by NCOA4 (nuclear receptor coactivator 4)-mediated ferritinophagy [265]. The generated PLOOH are reduced by the GSH pathway involving the system x_c^- Cystine-Glutamate antiporter (also called SLC7A11, solute carrier family 7 member 11), GPX4 (glutathione peroxidase 4) [47, 202], FSP1/AIMF2 (ferroptosis suppressor protein-1/apoptosis-inducing factor mitochondria-associated 2)-mediated formation of $CoQ10H_2$ (Coenzyme $Q10H_2$ or ubiquinol) [195, 196], vitamin K epoxide reductase complex subunit 1 L1 (VKORC1L1) formation of reduced vitamin K (VKH_2) [197], and GCH1 (GTP cyclohydrolase)-mediated formation of BH4 (tetrahydrobiopterin) [204, 205], preventing the accumulation of excessive LPO (lipid peroxidation). ESCRT-III continuously repairs PLOOH-mediated membrane damage [266]. Lipid peroxidation promotes the formation of areas with permeabilized plasma membrane, resulting in small-diameter lipid openings. Possibly, although still speculative, there may also happen recruitment and activation of yet-to-be-identified amphipathic membrane proteins that would allow stabilization or enlargement of protein-lipid pores by a toroidal or carpet mechanism. Oncosis following ferroptosis activates NINJ1, resulting in PMR

crucial in limiting uncontrolled lipid peroxidation and preventing cell rupture at the level of phospholipid hydroperoxides [45, 194]. FSP1 [195, 196] and extra-mitochondrial CoQ10 or vitamin K [197] suppress the lipid peroxidation chain reaction by quenching phospholipid radicals, such as alkoxyl and peroxyl radicals, in esterified fatty acyl chains. Further research has highlighted the involvement of cellular iron handling, energy metabolism, and hypoxia signaling in ferroptotic cell death [198]. These findings suggest that ferroptosis is a metabolic form of cell death with multiple nodes susceptible to pharmacological intervention, offering new avenues for therapeutic development. Recently, we have experienced an explosion of necrotic cell death modalities that can emerge from interfering in crucial metabolic processes such as iron, lipid, and redox metabolism (oxytosis, oxeiptosis), DNA repair (parthanatos), autosis (autophagy), pH regulation (alkaliptosis), copper metabolism (cuproptosis), disulfide stress due to enhanced buildup of intracellular cystine and other disulfide molecules (disulfidptosis), and excessive release of lysosomal Zn^{2+} leading to mitochondrial dysfunction and cellular energy depletion (lysozincrosis) [199, 200]. Interfering in essential protective cellular processes and metabolic networks has also been coined "cellular sabotage," leading to perturbation, cellular stress, and eventually metabolic cell death, including ferroptosis [201] (Fig. 5).

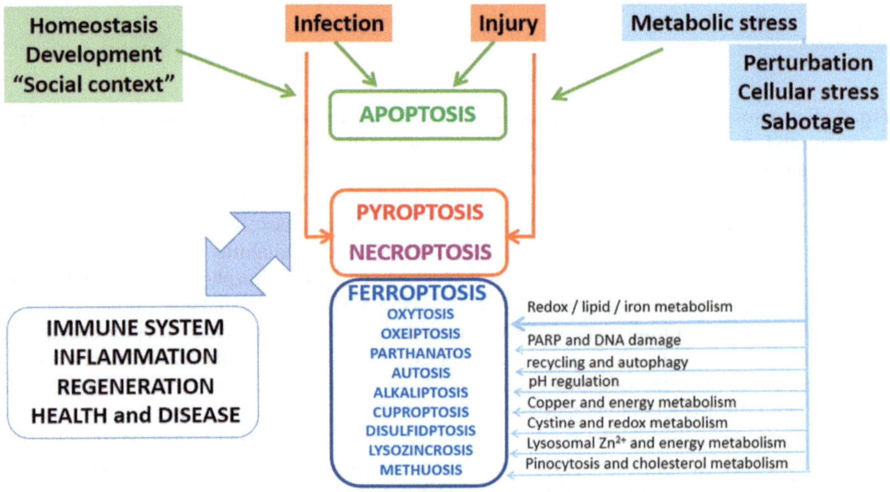

Fig. 5 Metabolic cell death modalities. Different regulated cell death modalities are associated with development, homeostasis, infection, and injury: apoptosis, pyroptosis, and necroptosis. Many additional cell death modalities are related to perturbation of crucial subcellular (e.g., DNA repair, autophagy, and lysosomal degradation) and metabolic processes (redox, lipid, pH, iron, copper, and Zn^{2+}). Ferroptosis, oxytosis, and oxeiptosis involve redox impairment and ROS generation. Due to DNA damage, parthanatos results from overactivation of poly ADP ribose polymerase 1 (PARP1). Autosis is a consequence of dysregulated autophagy. Cuproptosis results from Cu^{2+} disbalance and consecutive impaired mitochondrial energy metabolism. Disulfidptosis is due to cystine accumulation and impaired redox metabolism. Lysozincrosis involves the release of Zn^{2+} and consecutive impaired mitochondrial energy metabolism. Methuosis is associated with enhanced pinocytosis, which results in large vacuoles and cholesterol accumulation. Interfering in essential protective cellular processes and metabolic networks has also been coined "cellular sabotage," leading to perturbation, cellular stress, and eventually metabolic cell death [201]. Cell death occurring in a particular intercellular constellation modulates cellular responses of the immune system, inflammation, regeneration, and health and disease processes in general

Mechanisms of Ferroptosis Suppression

Cells employ various defense mechanisms to mitigate lipid peroxidation and prevent ferroptosis. The primary anti-ferroptotic pathway involves the cysteine/GSH)/GPX4 system [45, 46, 202]. GPX4, a selenoprotein, reduces harmful (phospho)lipid hydroperoxides ((P)LOOHs) to their redox chemically inert alcohols using GSH as a cofactor. Since GPX4 activity preferably requires GSH as an electron donor, GSH biosynthesis is crucial to ferroptosis inhibition. Cysteine, derived from cystine (the oxidized, dimeric form), is transported into cells via system x_c^- and subsequently used for GSH synthesis. Consequently, pharmacological or genetic disruptions in the cysteine/GSH/GPX4 axis induce ferroptosis. Inhibitors of GPX4 and system x_c^- are commonly used to trigger ferroptosis in experimental models [199].

FSP1, which is encoded by the *apoptosis-inducing factor mitochondria-associated 2* (*AIFM2*) gene, operates independently from GPX4 and GSH to prevent ferroptosis [190, 195, 196]. FSP1 uses NAD(P)H to reduce extramitochondrial ubiquinone (coenzyme Q_{10}, CoQ_{10}) or vitamin K into their reduced forms, ubiquinol and vitamin K hydroquinone, respectively [197, 203]. These molecules serve as radical-trapping antioxidants (RTAs), neutralizing lipid peroxyl radicals that drive lipid peroxidation and thus inhibit ferroptosis. Another pathway that suppresses ferroptosis involves GTP cyclohydrolase 1 (GCH1) and tetrahydrobiopterin (BH4) [204, 205]. GCH1 catalyzes the rate-limiting step in BH4 synthesis, which acts as an endogenous RTA and is recycled by dihydrofolate reductase (DHFR). Vitamin E, a potent natural lipophilic RTA, protects cells from lipid peroxidation and ferroptosis.

The cholesterol biosynthesis pathway, specifically the mevalonate pathway, also plays a role in ferroptosis suppression [206]. Intermediate metabolites, such as squalene and 7-dehydrocholesterol, counteract phospholipid autoxidation, inhibiting ferroptosis. Furthermore, mevalonate pathway metabolites are essential for CoQ_{10} synthesis and for the stabilization of selenocysteinyl-tRNA, which is critical for GPX4 expression, respectively.

The Role of Polyunsaturated Fatty Acids (PUFAs) and Iron in Ferroptosis

Peroxidized polyunsaturated fatty acids (PUFAs) esterified in phospholipid bilayers and their lipid radical derivatives are key initiators of ferroptosis [194]. Cells undergoing ferroptosis exhibit high levels of oxidized phospholipids, including phosphatidylcholines (PCs) and phosphatidylethanolamines (PEs) containing PUFA chains. Since PUFAs are inherently more susceptible to lipid peroxidation than monounsaturated (MUFAs) and saturated fatty acids (SFAs), the ratio of PUFAs to MUFAs in cellular membranes is a key determinant of ferroptosis sensitivity. Increased PUFA levels in the plasma membrane heighten ferroptosis susceptibility, whereas elevated MUFA levels confer resistance.

The incorporation of PUFAs into phospholipids is facilitated by acyl-CoA synthetase long-chain family member 4 (ACSL4) and lysophosphatidylcholine acyltransferase 3 (LPCAT3) [207]. ACSL1 enhances the incorporation of conjugated PUFAs, such as α-eleostearic acid, which is highly prone to peroxidation. In contrast, high expression of stearoyl-CoA desaturase (SCD1), acyl-CoA synthetase long-chain family member 3 (ACSL3), and membrane-bound O-acyltransferase domain-containing proteins 1 and 2 (MBOAT1 and MBOAT2) promotes MUFA synthesis, thereby increasing ferroptosis resistance [208].

Iron plays a crucial role in ferroptosis, though its precise function in ferroptosis execution, particularly in vivo, remains debated. A small fraction of redox-active iron within the labile iron pool is sufficient to catalyze the Fenton reaction, generating lipid alkoxyl and hydroxyl radicals that initiate the lipid peroxidation chain reaction. Consequently, increased labile iron levels enhance ferroptosis susceptibility, necessitating well-controlled regulation of iron homeostasis.

Interconnectivity of Cell Death Pathways

The different types of programmed cell death (PCD) are classified as non-lytic (apoptosis) and lytic cell death modalities (pyroptosis, necroptosis, ferroptosis) based on the engagement of plasma membrane permeabilizing mechanisms. The pathways leading to these different cell death modalities are often presented as linear, independently operating pathways that, in the absence of efferocytosis, eventually accumulate in the activation of pore-forming mechanisms. More specifically, these include caspase-dependent proteolytic activation of GSDMs in secondary necrosis and pyroptosis, RIPK3-mediated phosphorylation of MLKL in necroptosis, and iron-dependent lipid peroxidation and disintegration of the plasma membrane during ferroptosis. Interestingly, the plasma membrane permeabilization by necroptosis, secondary necrosis, pyroptosis, and ferroptosis is associated with oncosis, the ballooning of the dead cells, and eventually followed by plasma membrane rupture (PMR). The latter was thought to be the consequence of osmosis, a passive process involving water influx. Recently, the discovery of ninjurin-1 (NINJ1) as the driver of PMR radically changed this view [51]. PMR is an active process resulting from the still unresolved mechanism of NINJ1 activation, which serves as an additional cell death endpoint following pyroptosis and ferroptosis, as a kind of "death after death" mechanism. The mechanism of PMR following ballooning during necroptosis has not been identified yet since NINJ1 deficiency does not affect the occurrence and kinetics of PMR during necroptosis [51, 53, 162], suggesting the existence of an alternative mechanism. Altogether, this implies that lytic cell death by secondary necrosis, pyroptosis, necroptosis, and ferroptosis converge on a similar cataclysmic endpoint, suggesting that crucial features and functions might be associated with this PMR as a typical "death after death" mechanism in the absence of efferocytosis. Consequently, the intracellular content, including whether or not intracellular pathogens in case of infection, is released and becomes available as damage-associated molecular patterns (DAMPs) and pathogen-associated molecular patterns (PAMPs) that alert the immune system to danger [209]. However, not only is "death after death" a common mechanism of lytic cell death modalities, but also, at the levels of the signal transduction pathways, intense interconnectivity has been described between the cell death pathways at different levels. This is the subject of the next section.

Parallel Signaling, Interconnectivity, and Backup of Cell Death Pathways

With growing insights into the regulation of PCD signaling pathways, a model emerges of various PCD pathways that is highly interconnected at multiple levels (Fig. 6). This interconnectivity implies flexibility and modulation of the cell death outcomes by different stimuli or conditions of cellular stress and infection. The interconnected regulation of PCD pathways also enables plasticity and adaptivity to various stress and infection conditions by operating as a system with alternative backup cell death fates in case specific pathways are modulated or blocked. Already with the discovery that necrotic cell death induced by TNF could be sensitized in cells overexpressing CrmA or in the presence of zVAD-fmk [89], it was obvious that TNF-induced necrotic cell death was, strictly speaking, not really caspase-independent but rather negatively regulated by caspases. Later, it became clear that caspase-8 negatively affected RIPK1- and RIPK3-mediated necroptosis [93], demonstrating that necroptosis is not caspase-independent but rather suppressed by caspase-8. Eventually, this concept was also shown to operate in vivo, as witnessed by the complete rescue of embryonic lethality of *Casp8*-deficient embryos once the *Ripk3* or *Mlkl* necroptotic genes were genetically deleted [95, 96, 210, 211]. Thus, caspase-8 functions as an initiator caspase of extrinsic apoptosis and as a negative regulator of necroptosis. The mechanism involves caspase-8-mediated cleavage of RIPK1, controlling both excessive RIPK1 kinase-dependent apoptosis and necroptosis [98, 99].

Another concept revealed at the beginning of necrotic cell death research was the principle that inhibition of apoptotic cell death allowed the occurrence of alternative cell death modalities. Indeed, in a model of anti-Fas-induced apoptosis, the presence of caspase inhibitors initially blocked apoptotic cell death, as expected, but the cells died with delayed kinetics by anti-Fas-induced necroptosis [90]. This observation suggests that necroptosis could function as a backup cell death modality if apoptosis fails. These results also demonstrated for the first time that the same stimulus can lead to different pathways depending on the availability of proteolytically active caspase-8. The switch between apoptosis and necroptosis has been elaborated in detail for cell death induced by various TNF family members (TNF, FasL, TRAIL, and TL1A). Indeed, the RIPK1 kinase activity functions as a bifurcation point to caspase-8-mediated apoptosis in complex IIb, while in conditions of caspase-8 deficiency, it leads to necroptosis in the necrosome complex [65, 75, 212]. Of course, this type of switch can only function in cells that express the necroptotic machinery, viz., the RIPK3-MLKL axis.

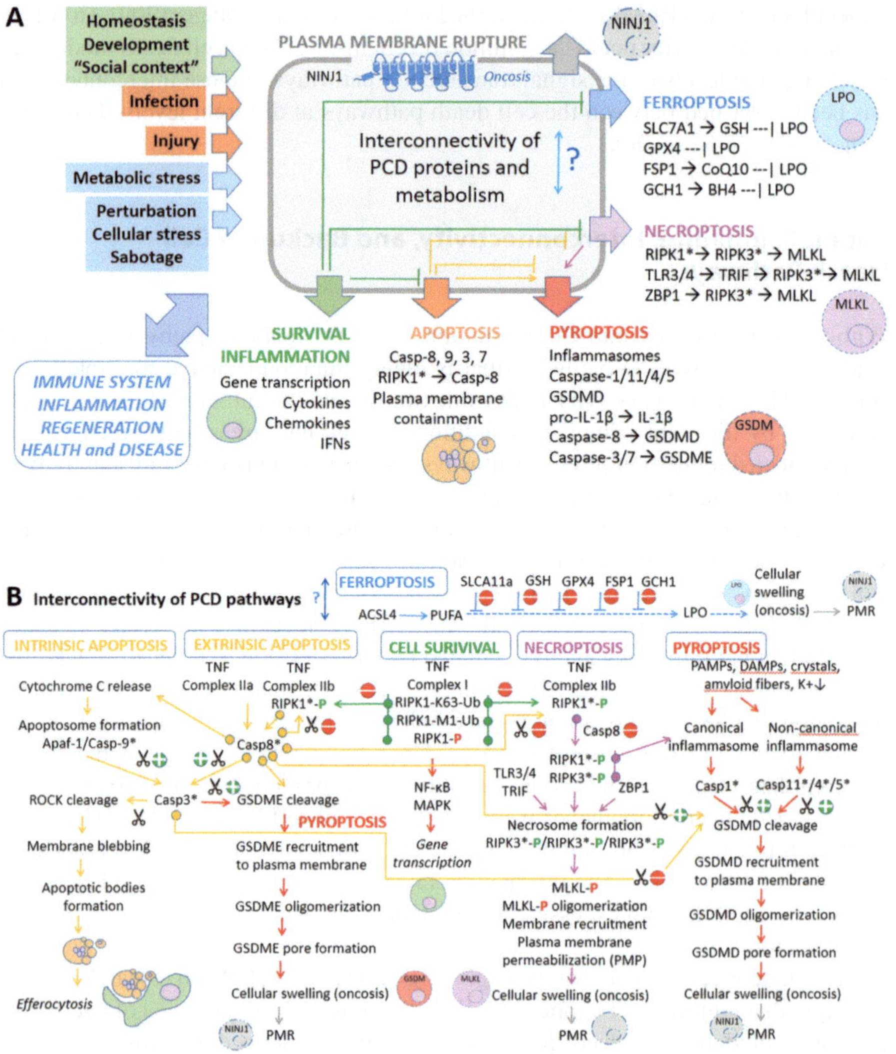

Fig. 6 Interconnectivity of cell death modalities. (**a**) Survival pathways modulate cell death pathways in multiple ways, including gene transcription (e.g., induction of anti-apoptotic genes and NRF2-mediated stress adaptive responses protecting against ferroptosis) and inducing PTMs (RIPK1 ubiquitylation and phosphorylation). Signaling of the three major cell death modalities (apoptosis, pyroptosis, and necroptosis) are highly interconnected. How ferroptosis is interconnected with apoptosis, pyroptosis, and necroptosis is still elusive. Since lipid peroxidation during ferroptosis not only happens at the plasma membrane but also affects other organelles, such as the endoplasmic reticulum [267], it can be expected that the associated cellular stress may also relate to the engagement of other cell death modalities. (**b**) This scheme represents the interconnection between cell survival (green), apoptosis (orange), necroptosis (pink), and pyroptosis (red) pathways. The scheme emphasizes two central hubs in the interconnectivity: caspase-8 and RIPK1. Caspase-8 activation bridges the extrinsic and the intrinsic apoptotic pathways and can directly activate the executioner caspase-3/7, induce pyroptosis by GSDME and GSDMD proteolytic

Caspase-8 does not only function as an initiator caspase of extrinsic apoptosis or as an inhibitor of complex II-mediated apoptosis and necrosome-mediated necroptosis, but in conditions where caspase-8 is inactive or MLKL is deleted, caspase-8 apparently also modulates pyroptosis. Indeed, genetic studies revealed mutant catalytically inactive *Casp8*$^{C362A/C362A}$ and *Mlkl*$^{-/-}$ deficient mice showed perinatal lethality involving ASC (PYCARD) speck formation, caspase-1, caspase-11, caspase-3/7 activation, IL-1β production, and a non-necroptotic function of RIPK3 [211, 213, 214]. These findings imply that the caspase-8 catalytic activity is not only required to keep in check RIPK1-mediated apoptosis and necroptosis [98, 99] but also prevents excessive pyroptosis by preventing ASC speck formation. On the other hand, the catalytically inactive caspase-8 platform in the absence of apoptosis and necroptosis pathways engages both canonical and non-canonical inflammasome activation, resulting in pyroptosis, in line with the concept of plasticity in the cell death outcome.

Apparently, in contrast to the inhibitory function of catalytically active caspase-8 on pyroptosis, as revealed by genetic studies in non-challenged animals, studies involving infection and challenges revealed that catalytically active caspase-8 positively affects inflammasome activation at multiple levels. TAK1 inhibition during *Yersinia* infection results in caspase-8-mediated pyroptosis by directly cleaving GSDMD and GSDME [167, 168]. As a platform, caspase-8 is required for efficient Toll-like-receptor-induced NLRP3 inflammasome priming and NF-κB-mediated cytokine production [215, 216]. Altogether, this suggests that both catalytic and platform functions of caspase-8 restrain and stimulate inflammasome-associated functions depending on the cellular context. However, it should be noted that conclusions derived from genetic studies using mutant signaling molecules often lead to imbalanced conditions and might differ from studies based on wild-type signaling molecules under challenging conditions, such as infection or TLR stimulation.

Finally, RIPK1 and RIPK3 are also intertwined in pathways involving apoptosis, necroptosis, and pyroptosis [217]. Catalytically inactive RIPK3 expression enhances complex IIb formation and RIPK1 kinase activity-mediated apoptosis [68, 218]. In the absence of IAPs and caspase-8 (again an imbalanced condition even without challenge), catalytically inactive RIPK3 promotes LPS-mediated NLRP3 inflammasome activation and IL-1β inflammatory responses independent of MLKL and necroptotic cell death [219]. Interestingly, the inflammasome-mediated activation of caspase-1 in cells that lack GSDMD cell death also proceeds to Bid-caspase-9-caspase-3-mediated cleavage of GSDME, leading to secondary necrosis [220]. In

Fig. 6 (continued) activation, and keep in check RIPK1*-mediated apoptosis and necroptosis. Caspase-3 inactivates GSDMD by proteolysis, suppressing the inflammasome-mediated GSDMD activation. RIPK1 K63 and M1 ubiquitylation engage survival pathways involving TBK1, TAK1, and IKKs (see Fig. 1). RIPK1 transubiquitylation by IKK and TBK1 puts RIPK1 in a survival mode, avoiding its activation and autophosphorylation, which would lead to cell death by apoptosis and necroptosis. RIPK1* and RIPK3* promote the activation of the canonical inflammasome under conditions of caspase-8 inactivation. All cell death pathways are eventually interconnected by induction of plasma membrane rupture

line with the possible connection between caspase-1 and apoptosis, the apoptotic executioner caspase-7 has been shown to become activated by caspase-1 [221, 222]. This offers the intriguing possibility that inflammasome activation may eventually result in immunosilent apoptotic cell death in certain conditions.

As can be noticed from the limited number of examples discussed above, the interconnectivity between apoptotic, necroptotic, and pyroptotic pathways is very complex. It depends on the availability, platform function, and catalytic activity of the signaling hubs, including caspase-8, RIPK1, and RIPK3. The question is how infection or cellular stress conditions regulate the two faces of these signaling hubs. Genetic studies that prevent specific cell death modalities by removing a crucial player reveal imbalances and remarkable plasticity. However, the essential question remains: which cellular stress or infection conditions interfere with these critical mediators of cell death modalities in wild-type constellations? In the case of bacterial and viral infection, many gene products have been identified as apoptosis, necroptosis, or pyroptosis [150]. The interplay between PCD modalities and ferroptosis remains to be explored. Since ferroptosis results from imbalances in lipid, redox, and iron metabolism, the associated cellular stress may also engage classical PCD modalities. However, until now, ferroptosis has been mainly studied by blocking the protective mechanisms against lipid peroxidation [44], which may reflect more a death by sabotage [201]. However, an indication that during ferroptosis also apoptotic pathways might be engaged is provided by the observation that extracellular vesicles (EVs) produced by ferroptotically dying cells contain molecules of the apoptotic and necroptotic pathways (such as activated caspase-3 and phosphorylated MLKL), suggesting that during ferroptosis induction molecules of the apoptotic and necroptotic pathways might be recognized and exported in EVs. This was not observed in total cell lysates from living, apoptotic, or necroptotic dying cells [223].

Finally, the intertwining of cell death pathways also raises the question of how this may determine the eventual outcome of either individual cells as a cell death modality with mixed morphological and biochemical features or as a mix of cell death modalities in a population. The mixed cell death modalities in a challenged population may have significant immunoregulatory and pathophysiological consequences. Whether a single cell shows mixed cell death features or a population of cells shows different cellular outcomes is an essential item, especially for PANoptosis. The latter has been described as a lytic cell death pathway initiated by innate immune sensors and driven by caspases and receptor-interacting protein kinases (RIPKs). It involves the formation of multiprotein PANoptosome complexes, with standard features of complexes associated with pyroptosis, apoptosis, and necroptosis [60]. The PANoptosis concept emphasizes the interconnectivity of the significant cell death modalities (apoptosis, necroptosis, and pyroptosis). However, the crucial question remains whether the outcome is determined as an individual cell death modality with mixed features or whether the mixed features result from different responses in a population of cells. Single-cell time-lapse analysis of morphological and biochemical features following different cell death-inducing conditions will resolve this important question with far-reaching therapeutic implications.

NINJ1: The Final Cataclysm of Necrotic Cell Death

Different cell death modalities, including secondary necrosis, pyroptosis, and ferroptosis, all end by eventual rupture of the plasma membrane (PMR), an oncosis-sensing process involving NINJ1 [51, 53, 162]. Since NINJ1 deficiency apparently does not affect PMR following necroptosis, the precise mechanism remains elusive [51, 53]. However, one could wonder what the reason for this potentially distinct mechanism for necroptosis could be. PMR represents a final burst of the cell, releasing intracellular content, including DAMPs and PAMPs (in case of infection). Interestingly, the cytokine release by GSDMD-mediated pyroptosis is unaffected by the absence of NINJ1, while the release of large DAMPs requires NINJ1 [51]. This differential timing provides two waves of factors released by pyroptotic dying cells: a first one with cytokines like IL-1β and IL-18, and a second one with additional DAMPs, creating a time kinetics and gradient of released factors. How NINJ1 activation occurs at the mechanistic level remains unclear, but cells that lack NINJ1 are stuck temporarily in the ballooning phase of cell death ("oncosis"), demonstrating that NINJ1 is not required for pyroptosis and ferroptosis per se [51, 162], and that the swelling by itself is somehow sensed and transcribed in active PMR induction. This idea is supported by the observation that swelling by hypotonic shock is sufficient to induce NINJ1 oligomerization [53]. Recently, the structure of homodimeric [54] and oligomerized NINJ1 [55, 224, 225], and comparison with non-PMR-causing NINJ2 [55, 225] provided molecular insights into how NINJ1 executes PMR.

NINJ1 is predicted to contain an N-terminal amphipathic helix (AH) and two transmembrane helices (TM1 and TM2) (Fig. 7). The AH domain is critical for the cytotoxic activity of NINJ1. During the resting state of non-stimulated cells, NINJ1 is inserted in the membrane and forms dimers. Following cell death induction by secondary necrosis, pyroptosis, and ferroptosis, NINJ1 forms oligomers [51]. A kinked AH helix is a crucial structural feature of active-state NINJ1 prior to oligomerization [55, 224, 225]. The presence of a kinked AH1 can be detected using a nanobody as a biosensor [54]. In the basal state, NINJ1 forms dimers where the straight (or unkinked) AHs of each monomer are buried in the dimer interaction surface, resulting in autoinhibition. In response to cell death stimuli, the AH straight α helix becomes interrupted by a kink, resulting in dimer dissociation, enabling oligomerization by interaction between the first kinked AH domain and the TM2 domain of the adjacent NINJ1 molecule. In such a model, the NINJ1 dimer would operate as a sensor of plasma membrane tension caused by the swelling of the dying cell ("oncosis"). Merely increased tension in the membrane might be sufficient to activate the dimer and the AH1 kink formation, leading to dimer dissociation [54]. The amphipathic nature of the AH domain contributes to the concave hydrophobic and convex hydrophilic surfaces of the NINJ1 oligomers, allowing the filament or ring structures to interact on one side with membrane lipids and on the other side

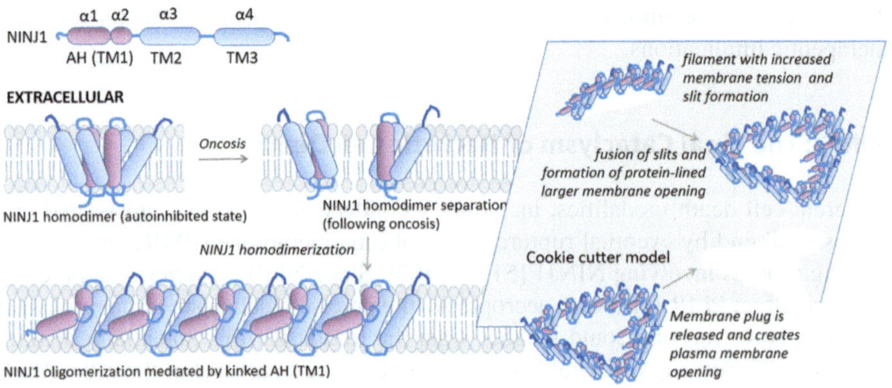

Fig. 7 Plasma membrane rupture by NINJ1. NINJ1 induces plasma membrane rupture following secondary necrosis, pyroptosis, and ferroptosis. It contains three transmembrane regions (TM1, TM2, and TM3) organized as α1-α2 (pink), α3 (blue), and α4 (blue). The TM1, also called AH1, has amphoteric features required for the cytotoxicity of NINJ1. In non-stressed cells, NINJ1 is inserted in the plasma membrane as a dimer in an auto-inhibited state. In this dimer, AH1 (TM1 or α1-α2) forms one straight α-helix with the hydrophilic face hidden in the dimeric interface. Following cell swelling ("oncosis"), membrane thinning occurs, and the dimers may get separated, resulting in a conformational change of AH1 and the induction of a kink between α1 and α2. This "kinked" α1 contacts the α4 of the adjacent NINJ1 molecule, resulting in oligomerization and filament formation that cause membrane slits. These slits can either fuse, resulting in an enlarged rupture lined by NINJ1, or may form irregularly shaped disks bordering membrane areas. The latter are released as NINJ1-surrounded membrane plugs ("cookie cutter model"), creating large openings

with the aqueous extracellular environment. Different interpretations and models exist on how these filaments or rings eventually cause PMR. The first model involves the formation of large membrane protein lines with a hydrophilic duct [224]. A second possibility is described as a "cookie cutter model" in which amphipathic NINJ1 forms membrane disk rings that extract the surrounded membrane plug [55, 225]. Comparison and mutational analysis between the cytotoxic NINJ1 and the non-cytotoxic NINJ2 revealed that NINJ1 forms straight fragments that induce positive curvature at the membrane ridge, causing slit formation in the plasma membrane. Two adjacent NINJ1-bordered slits can fuse and form a growing ring that surrounds membrane fragments that are punched out [55]. NINJ2, on the other hand, forms fragments with many opposing curves and does not induce slits in the plasma membrane due to lateral tension that works in opposing directions by the curved shape [55].

Though the cells still do die in the absence of *Ninj1* [51, 53, 162], it is possible that cells with limited gasdermin pore formation or ferroptotic damage could still remove the pores via ESCRT III function or repair the areas containing oxidized lipids and survive [27, 201, 226]. Whether plasma membrane permeabilization (PMP) following pyroptosis or ferroptosis is not the actual endpoint but still can be reverted, or whether NINJ1-mediated PMR is a "deader than dead" process, may look like a semantic discussion. However, suppose PMR is merely driven by a swelling-induced conformational change of NINJ1 dimer and oligomerization without needing ATP. In that case, it might be a last flash of an oncotic cell corpse functioning as an alarm in the organism by the massive release of DAMPs and PAMPs (in case of infection). In that scenario, what could be the ultimate cell death-related function of NINJ in vivo? Inhibition of NINJ1 or *Ninj1*-deficiency improved the release of PMR parameters in several hepatotoxicity models (TNF plus D-gal, TNF plus ConA, anti-Fas) as detected by reduced serum levels of ALT, AST, HMGB1, LDH, and IL-18. Furthermore, in a model of liver ischemia-reperfusion injury, *Ninj1*$^{-/-}$ mice showed reduced neutrophil infiltration [227]. This argues for the therapeutic targeting of NINJ1 in degenerative and inflammatory diseases.

Necroptosis, Pyroptosis, and Ferroptosis in Cancer

Many metastatic cancers remain difficult to treat due to tumor evolution and resistance to apoptosis. Common therapies such as chemotherapy or radiation primarily induce apoptosis, which is an immunologically silent process of cell death—at least in most cases. In contrast, necroptosis, pyroptosis, and ferroptosis—forms of cell death that can stimulate immune cells by releasing intracellular contents, serve as danger signals, or enhance antigen presentation—boost CD8+ T cell activities. Here, we will explore how immune activation through these non-apoptotic cell death modalities could revolutionize cancer treatments, identifying opportunities and challenges [64]..

Necroptosis

Necroptosis exhibits varying roles in cancer progression and suppression. As a tumor suppressor, necroptosis is a critical alternative in eliminating apoptosis-resistant cancer cells, primarily by stimulating anti-tumor immune responses by releasing DAMPs [228]. On the other hand, necroptosis can promote tumor growth by causing chronic inflammation, which fosters a pro-tumorigenic microenvironment that boosts cancer cell proliferation and aids immune evasion. It plays a significant role in shaping resistance mechanisms to cancer treatments. Inducing necroptosis offers a promising approach to tackle apoptosis resistance in chemotherapy, thereby improving the effectiveness of chemotherapeutic drugs like cisplatin. Nevertheless, cancer cells can adaptively escape necroptosis by upregulating inhibitory proteins and silencing the *Ripk3* and *Mlkl* genes, which contributes to resistance against targeted therapies [228, 229].

Therapeutic strategies that target necroptosis include both induction and inhibition. Necroptosis inducers include SMAC mimetics [230], combined TNF-α and caspase inhibitors [64], and natural compounds like shikonin and curcumin. These agents specifically trigger necroptosis in tumors that are resistant to apoptosis. Inhibitors of necroptosis, such as necrostatins targeting RIPK1 and MLKL inhibitors, are included [18], and mitigate necrosis-driven inflammation within the tumor microenvironment, potentially limiting tumor progression. Necroptosis holds clinical relevance across various cancer types. In colorectal cancer, lower RIPK3 expression is associated with poorer patient prognosis. Pancreatic cancer shows that necrotic inflammation creates immunosuppressive microenvironments [231], promoting tumor survival and growth. In glioblastoma, necrotic inflammation can enhance susceptibility to therapies. Lung and breast cancers exhibit varied impacts of necroptotic inflammation, heavily influenced by the specific tumor microenvironmental context [232]. Another prominent example is acute myeloid leukemia (AML), where RIPK3-dependent cell death and inflammation act as tumor-suppressive mechanisms [233]. Current clinical trials are actively investigating the therapeutic potential of necroptosis inducers, such as birinapant (a SMAC mimetic), in combination with existing chemotherapy or immunotherapies, aiming to manage resistant tumor types effectively [64].

Pyroptosis

Pyroptosis is a form of programmed cell death characterized by inflammation, which manifests through cell swelling, membrane rupture, and the release of pro-inflammatory cytokines and DAMPs (see above). In contrast to apoptosis, pyroptosis activates the immune system, primarily triggered by inflammasomes, caspase-1 activation, and gasdermin family proteins. Its role in cancer varies depending on the context, showing both tumor suppression and promotion. Mechanistically, pyroptosis involves inflammasome assembly, caspase-1 activation, and cleavage of GSDMD [42], forming pores in cell membranes. These pores facilitate the release of interleukins like IL-1β and IL-18, contributing to inflammation and significantly influencing tumor biology. Pyroptosis plays a dual role in cancer. As a tumor-suppressive mechanism, the induction of pyroptosis may stop tumor growth and initiate immune activation and inflammatory responses that draw immune cells to the tumor, effectively eradicating cancer cells.

Furthermore, pyroptosis impacts resistance to cancer therapies through its inflammatory signaling pathways. However, in chronic inflammation, pyroptosis might also lead to resistance to therapy. Therapeutic strategies for pyroptosis include employing chemotherapeutic drugs or immune modulators to trigger inflammasome activation alongside small molecules targeting gasdermin pore formation disruption. Promising preclinical findings suggest considerable anti-cancer potential using these methods. The role of pyroptosis has been confirmed across various cancer types. In colorectal cancer [234], inflammasome-mediated pyroptosis primarily functions to suppress tumors, while in breast cancer, pyroptosis influences therapeutic response differently across subtypes. In lung cancer, pyroptotic cell death impacts immune responses and treatment outcomes, highlighting its clinical importance. Recent studies have proposed the concept of targeting pyroptosis in cancer treatments, focusing on gasdermin modulators, inflammasome activators, and combination strategies to boost immunotherapeutic effectiveness.

Ferroptosis

Investigating the indirect effects of ferroptosis involves analyzing the release of intracellular molecules, such as DAMPs, which trigger inflammation in surrounding tissues. Early work assessed whether cancer cells undergoing ferroptosis could function as prophylactic vaccines to elicit antitumor immune responses. However, the results were not conclusive, with opposing experimental outcomes favoring either ferroptosis as an immunogenic cell death modality or not. More detailed analysis using genetically inducible systems of synchronized ferroptosis revealed that ferroptosis occurs in distinct phases—initial, intermediate, and terminal—each associated with specific cellular processes, such as very early lipid peroxidation and late release of DAMPs [235]. Prophylactic vaccination by subcutaneous injection with these synchronized ferroptotic cancer cells resulted in impaired antigen cross-presentation compared to mitoxantrone-induced immunogenic cell death, as witnessed by decreased protection against a tumor challenge. This interference in antigen cross-presentation might be explained by oxidized lipids competing for MHC-I loading with antigen [236]. Another immunosuppressive mechanism is also associated with ferroptosis induction. Indeed, during ferroptosis, tumor cells release prostaglandin E2 (PGE2) [237, 238], which hinders anti-tumor immunity by inhibiting the recruitment and activation of conventional dendritic cells (cDC1s) and natural killer (NK) cells.

This situation paradoxically suggests that ferroptosis may allow for immune evasion, especially in tumors with elevated PGE2 levels. Additional research indicates that polymorphonuclear myeloid-derived suppressor cells (PMN-MDSCs)—known for suppressing anti-tumor immunity—also undergo spontaneous ferroptosis within tumors [239]. While ferroptosis reduces PMN-MDSC numbers, it simultaneously releases oxidized lipids that inhibit T-cell function. Blocking ferroptosis in PMN-MDSCs reduced their immunosuppressive impacts, slowed tumor progression, and improved the effectiveness of immune checkpoint blockade therapies, particularly during early tumor development. Conversely, inducing ferroptosis in PMN-MDSCs might be an immunosuppressive tactic and a potential treatment target [239, 240].

Furthermore, research has highlighted aconitate decarboxylase 1 (ACOD1) as a vital protective enzyme in tumor-infiltrating neutrophils (TINs), activated by the GM-CSF-JAK/STAT5-C/EBPβ signaling pathway [241]. ACOD1 shields TINs from being sensitive to ferroptosis and promotes metastasis. Thus, inhibiting ACOD1 could decrease neutrophil infiltration, enhance anti-tumor immunity, and boost treatment effectiveness, particularly in metastatic breast cancers. Ferroptosis also plays a crucial role in regulating metastasis. In an experimental model analyzing two pathways of melanoma cell metastasis, it was observed that traversing the lymph node microenvironment triggered an anti-ferroptotic state reliant on oleic acid in an ACSL3-dependent manner, facilitating increased metastasis. This underscores the induction of ferroptosis resistance in lymph nodes as a mechanism that promotes metastasis [242].

One key discovery has linked ferroptosis to adaptive immune responses, indicating that activated CD8+ T-cells during immunotherapy prompt lipid peroxidation in tumor cells, thereby enhancing treatment efficacy. Interferon-gamma (IFNγ) produced by these T-cells decreases the expression of both subunits of the cystine glutamate antiporter system x_c^-, SLC3A2 and SLC7A11, in tumor cells, which impairs cystine uptake and elevates lipid peroxidation. Studies in mice demonstrated that the combination of cystine deprivation and checkpoint inhibitors significantly enhanced T-cell-mediated tumor immunity by promoting ferroptosis. In line with this, cancer patients with lower system x_c^- levels displayed stronger immune responses and better clinical outcomes. Additionally, IFN-γ and arachidonic acid (AA) released by T-cells triggered ferroptosis in tumor cells through the enzyme ACSL4, thereby bolstering anti-tumor immune activity.

Conversely, a deficiency in ACSL4—which limits sensitivity to ferroptosis—resulted in accelerated tumor growth, suggesting that enhancing ACSL4 could improve immune-mediated tumor control. Another crucial enzyme, GPX4, aids regulatory T-cells (Tregs) by preventing lipid peroxidation and ferroptosis. The targeted deletion of GPX4 in Tregs initiated ferroptosis, disturbed immune equilibrium, elevated mitochondrial superoxide levels, and increased IL-1β secretion, which amplified T helper 17 (TH17)-mediated anti-tumor responses. Nonetheless, generalized GPX4 inhibition could negatively impact conventional T-effector cells, thereby reducing overall anti-cancer immunity. Additionally, recent research highlights the role of lipid metabolism in immune responses, particularly how lipid accumulation induced by the CD36 receptor on tumor-infiltrating CD8+ T-cells leads to dysfunction via lipid peroxidation and p38 kinase activation. Focusing on this oxidized lipid-CD36 pathway may restore T-cell functionality and boost the therapeutic effectiveness of immunotherapy [243].

Necroptosis, Pyroptosis, and Ferroptosis in Pathological Immunity

Neuroinflammation: Multiple Sclerosis

Multiple sclerosis (MS) is a chronic autoimmune condition marked by inflammation and neurodegeneration. Inflammation damages the myelin sheath that covers nerve fibers in the central nervous system, resulting in various neurological symptoms. While inflammation-driven neurodegeneration is a key feature of MS, the exact mechanisms behind it remain unclear.

Early work explored the role of necroptosis in the development of multiple sclerosis (MS) to determine if necroptosis plays a key role in oligodendrocyte death in MS and if this pathway could serve as a potential therapeutic target [244]. Necroptosis was found to be active in the cortical white matter lesions of MS patients, characterized by defective caspase-8 activation. Typically, caspase-8 inhibits necroptosis; therefore, its impaired function correlates with increased activity of receptor-interacting protein kinases RIPK1 and RIPK3, along with mixed lineage kinase domain-like protein (MLKL), which collectively mediate necroptotic cell death. MS lesions also showed an elevated level of insoluble proteins, a common hallmark in other neurodegenerative diseases like Alzheimer's, Parkinson's, and Huntington's disease, suggesting a potential shared mechanism involving protein aggregation and necroptosis. Further in vitro experiments confirmed that TNF, a cytokine previously linked to inflammation in MS, directly triggers necroptosis in oligodendrocytes through RIPK1 and RIPK3. Notably, the pharmacological inhibition of RIPK1 using the inhibitor 7 N-1 effectively prevented TNF-induced necroptosis in oligodendrocytes in vitro. Consistent with these findings, in vivo treatment with 7 N-1 reduced oligodendrocyte death and demyelination in two animal models of MS: the cuprizone-induced demyelination model and the experimental autoimmune encephalomyelitis (EAE) model. Moreover, RIPK3-deficient mice showed resistance to demyelination in response to cuprizone, further validating the involvement of necroptosis.

Similar work was carried out to challenge the notion of the involvement of pyroptosis and, in particular, NLRP3 inflammasomes in the pathogenesis of MS [245] concerning the clinical question of nonresponders to the commonly used fingolimod treatment. Interestingly, nonresponders showed increased NLRP3 gene expression, enhanced ASC oligomerization, and elevated galectin-3 levels (a marker of cell damage) compared to responders all of which suggest that the effectiveness of fingolimod may depend on its ability to modulate inflammasome activity. Differences in inflammasome activation between responders and nonresponders could serve as biomarkers to predict treatment response and highlight potential new therapeutic targets for patients who do not respond to fingolimod therapy by targeting inflammasome-related pathways. Underscoring these results, similar work scrutinized the role of GSDMD-mediated pyroptosis [246] in the pathogenesis of MS, providing evidence of inflammasome activation and pyroptosis in both microglia (macrophages) and oligodendrocytes within the central nervous system (CNS) of MS patients and in experimental autoimmune encephalomyelitis (EAE), an animal model of MS. In vitro studies demonstrated that exposure to inflammatory stimuli induced pyroptosis in human microglia and oligodendrocytes, which the caspase-1 inhibitor VX-765 could inhibit. In vivo, VX-765 treatment in EAE models reduced the expression of inflammasome and pyroptosis-associated proteins, prevented axonal injury, and improved neurological function.

These findings suggest that GSDMD-mediated pyroptosis in glial cells contributes to inflammatory demyelination in MS and that targeting this pathway through caspase-1 inhibition may offer a therapeutic approach for MS. Lastly, work centered around neuronal inflammatory stress response revealed that neurons in MS and its mouse model activate the stimulator of interferon genes (STING). However, neuronal STING activation necessitates separating from the stromal interaction molecule 1 (STIM1), which is driven by glutamate excitotoxicity. This in turn triggers noncanonical STING signaling, leading to the autophagic degradation of GPX4, normally crucial for preventing ferroptosis. Both genetic and pharmacological approaches targeting STING in neurons offer protection against inflammation-induced neurodegeneration [247].

Dermatological Inflammation and Cell Death

A symbiotic interaction among keratinocytes, immune cells, and microbial communities is essential for skin homeostasis. Disruption in this balance may result in aberrant activation of cell death pathways frequently resulting in inflammatory skin disorders. One insight is that while cell death is vital for tissue remodeling and defense, uncontrolled or excessive cell death can trigger inflammation and disease. Apoptosis plays a key role in skin renewal, hair follicle cycling, and immune tolerance. Under steady state conditions the intrinsic and extrinsic apoptotic pathways promote organized cell elimination without causing inflammation. Nevertheless, apoptosis may become pro-inflammatory if phagocytic clearance is delayed, resulting in late apoptosis and possible tissue damage. This complexity is demonstrated in models of excessive keratinocyte apoptosis, which compromises barrier integrity and fosters inflammatory responses in conditions such as lupus and chronic dermatitis.

Necroptosis, a type of lytic cell death, plays a critical role in inflammation when apoptosis is inhibited. Driven by RIPK1, RIPK3, and MLKL, necroptosis leads to membrane rupture and the release of damage-associated molecular patterns (DAMPs), which enhance inflammatory responses. Research using mouse models shows that genetic alterations in necroptotic pathways exacerbate inflammatory skin conditions.

Pyroptosis, another pathway of inflammatory cell death, is primarily mediated by inflammasomes and caspase-1, resulting in the secretion of IL-1β and IL-18. While pyroptosis is involved in pathogen defense, its activation in keratinocytes and immune cells can also contribute to inflammatory skin diseases.

Ferroptosis has been proposed as a contributing factor in the pathogenesis of systemic lupus erythematosus (SLE) [248]. Mechanistically, the serum's autoantibodies and interferon-α trigger neutrophil ferroptosis by facilitating the interaction of the transcriptional repressor CREMα with the promoter of glutathione peroxidase 4 (*Gpx4*), a crucial ferroptosis regulator. This increased binding diminishes Gpx4 expression, resulting in higher levels of lipid-reactive oxygen species. Additionally, the finding that mice with neutrophil-specific *Gpx4* haploinsufficiency exhibit significant clinical characteristics of human SLE—including autoantibodies, neutropenia, skin lesions, and proteinuria—and that administering a specific ferroptosis inhibitor notably alleviates disease severity in lupus-prone mice underscoring the importance of neutrophil ferroptosis in lupus pathogenesis.

Inflammatory Bowel Disease (IBD)

Crohn's disease and ulcerative colitis, the two prevalent types of inflammatory bowel disease (IBD), impact over 2.5 million people of European descent, and their prevalence is increasing in other populations [249]. These diseases originate from a complex pathophysiology wherein, the complex symbiotic interplay between host and micrbiome is disrupted, triggering chronic intestinal inflammation [250].

A crucial gene series not identified in GWAS encodes proteins that control inflammatory cell death, known as regulated necrosis or necroptosis. The significance of this pathway for intestinal homeostasis was first demonstrated in murine studies involving caspase-8 [210] and RIPK1 [251]. Epithelial deletion of caspase-8 led to severe ileal inflammation marked by Paneth cell death, heightened immune cell infiltration, and significantly increased RIPK3 expression. Cell death occurred via aberrant TNF-signalling. Notably, the key roles of TNF, RIPK1, and RIPK3 in intestinal epithelial cell death were also illustrated in a murine model of ATG16L1 deficiency, an IBD susceptibility gene mentioned earlier [252]. Another prominent example illustrating the critical role of restraining cell death in the gut is provided by studies investigating a primary immunodeficiency termed XLP-2 syndrome, which is driven by the loss of X-linked inhibitor of apoptosis protein (XIAP, *BIRC4*) [252]. Here, loss of this key protein restraining aberrant cell death cues is sufficient to trigger very early onset Crohn's-like disease in affected boys.

Additionally, epithelial deletion of *Ripk1* resulted in increased TNF-dependent apoptotic cell death [253]. Subsequent mutations in RIPK1 and CASP8 have been found in patients with severe immune deficiencies and IBD [110, 254], supporting these murine findings. However, aside from inducing cell death, RIPK3 has also been shown to aid in epithelial tissue repair in the murine DSS colitis model [255]. Consequently, questions remain about the exact roles of these proteins in disease initiation and progression, particularly in non-Mendelian IBD [252].

The rising incidence of inflammatory bowel disease (IBD) has become a global concern, potentially linked to the adoption of a Western lifestyle. This lifestyle includes dietary habits enriched with the ω-6 polyunsaturated fatty acid (PUFA) arachidonic acid (AA), associated with an increased risk of developing IBD. Small intestinal epithelial cells (IECs) in Crohn's disease (CD) show reduced GPX4 activity and evidence of LPO [256]. PUFAs, particularly AA, induce a cytokine response in IECs, which is restricted by the presence of GPX4. Although GPX4 does not regulate AA metabolism, cytokine production is controlled by mechanisms like ferroptosis. A PUFA-enriched Western diet induces focal granuloma-like neutrophilic enteritis in mice with a single *Gpx4* allele deficiency in IECs. In conclusion, increased dietary PUFAs override the GPX4-restricted mucosal inflammation, mirroring aspects of human CD [256].

Conclusions

The immune system constantly encounters dead cells during regular cell turnover, injuries, and infections. Therefore, mechanisms have evolved to distinguish between various types of cell death to properly eliminate pathogens following infection and promote wound healing during injury while avoiding reactions against the body's cells that could lead to autoimmunity. However, an immune response targeting host tissue is also necessary to eradicate tumors, particularly after treatment with radio- and/or chemotherapy, which aims to induce tumor cell death. Therefore, a critical challenge in immunology is understanding how the immune system assesses whether cell death is immunogenic, tolerogenic, or 'silent' [257]. Eukaryotic cells can succumb to lytic necrotic cell death due to physical trauma or undergo programmed cell death in response to the activation of specific signaling cues [258]. The immune system's response to non-lytically or lytically dying cells can determine whether it maintains or returns to adequate homeostasis or triggers and worsens pathological outcomes.

In recent years, significant advances have been made in understanding various forms of cell death, including apoptosis, necroptosis, pyroptosis, NETosis, immunogenic cell death, and activation-induced cell death, and their implications for immunity. Recently, metabolic cell death, such as ferroptosis, entered the stage resulting from cellular stress following irreparable perturbation and imbalances of essential processes in redox, lipid, iron, recycling, and energy metabolism. These cell death pathways are intricately connected with innate immune pathways in the dying cell, leading to the concurrent execution of cell death and activation of inflammatory responses within the dying cells [259]. The immunogenicity of cell death is determined by a combination of the cytokines released, the adjuvanticity, and the antigenicity. Cells infected by pathogens, as well as cancer cells, exhibit increased antigenicity. Stress responses in dying cells trigger the release of adjuvant-like danger signals. Different types of immunogenic cell death are associated with distinct danger signals. Restoring the immunogenicity of cell death shows excellent potential for anticancer therapy [260].

Cellular stress, infections, and inflammation activate a series of regulated cell death mechanisms, including apoptosis, necroptosis, pyroptosis, and ferroptosis. Some of these mechanisms may share characteristics like plasma membrane permeabilization. However, their activation pathways and consequences differ; apoptosis is defined explicitly by caspase-mediated cell fragmentation without an inflammatory response. In contrast, necroptosis, pyroptosis, and ferroptosis involve lytic cell death that disrupts the membrane, resulting in inflammation by releasing intracellular contents.

Necroptosis depends on RIPK3-mediated phosphorylation and the activation of MLKL, which disrupts the membrane by creating protein-lipid pores. In pyroptosis, the activation of inflammasome sensors leads to the caspase-mediated cleavage of gasdermin proteins (primarily GSDMD), producing large protein-lined pores that discharge inflammatory cytokines. Ferroptosis results from metabolic imbalances involving iron and lipid peroxidation alongside diminished antioxidant defenses (GPX4, FSP1, GCH1), leading to lipid pore formation and eventual membrane damage. Recently discovered, NINJ1 is a critical mediator of plasma membrane rupture during secondary necrosis, pyroptosis, and ferroptosis. When cells undergo swelling (oncosis), NINJ1 oligomerizes, disrupting membrane integrity and releasing large molecules that trigger inflammation.

Understanding these unique pore-forming and plasma membrane rupture mechanisms holds therapeutic promise for conditions like cancer, inflammatory bowel diseases, degenerative disorders, and ischemic injuries. Targeting specific cell death processes and adjusting membrane repair pathways (e.g., ESCRT-III) might enhance or inhibit cell death in clinical contexts. Cell death pathways are interconnected, providing redundancy and flexibility, enabling cells to respond adaptively to stress or infections. This interconnected nature enables plasticity and diversity in cell death responses, both of which are vital for effective and well-balanced immune reactions and eventually returning to tissue homeostasis. Understanding the diversity and interconnectivity of cell death responses at the cellular and intercellular levels is essential for conceiving targeted treatments for inflammatory, malignant, and degenerative diseases.

Acknowledgments Research in the Vandenabeele unit is supported by the FWO research projects (3G033120, 3G0A9322, G0AA824N), EOS research consortium (EOS 30826052, 3G0I5722), UGent Special Research Fund Methusalem (01 M00709), iBOF (01IB3920), grants from the Foundation against Cancer (365A08921), CRIG and GGIG consortia, and VIB. We are grateful to Savvas Savvides for the MLKL structural modeling. This work was supported by the Deutsche Forschungsgemeinschaft (DFG) (CO 291/7-1), the DFG Priority Program SPP 2306 (CO 291/9-1, #461385412; CO 291/10-1, #461507177; CO 291/9-2; CO 291/10-2; CO 291/14-1), the CRC TRR 353 (CO 291/11-1, #471011418), the German Federal Ministry of Education and Research (BMBF) FERROPATH (01EJ2205B), and the European Research Council (ERC) under the European Union's Horizon 2020 research and innovation program (grant agreement No. GA 884754) to M.C.

References

1. Virchow, R., Chance, F., Goodsir, J., & King's College London. & pathological Institute of Berlin. (1860). *Cellular pathology as based upon physiological and pathological histology; twenty lectures delivered in the pathological Institute of Berlin during the months of February, March, and April, 1858*. John Churchill.
2. Gerschenson, L. E., & Geske, F. J. (2001). Virchow and apoptosis. *The American Journal of Pathology, 158*, 1543.
3. Clarke, P. G., & Clarke, S. (1996). Nineteenth century research on naturally occurring cell death and related phenomena. *Anatomy and Embryology (Berl.), 193*, 81–99.
4. Kerr, J. F., Wyllie, A. H., & Currie, A. R. (1972). Apoptosis: A basic biological phenomenon with wide-ranging implications in tissue kinetics. *British Journal of Cancer, 26*, 239–257.
5. Wyllie, A. H., Kerr, J. F., & Currie, A. R. (1980). Cell death: The significance of apoptosis. *International Review of Cytology, 68*, 251–306.
6. Ellis, H. M., & Horvitz, H. R. (1986). Genetic control of programmed cell death in the nematode C. elegans. *Cell, 44*, 817–829.
7. Hengartner, M. O., Ellis, R. E., & Horvitz, H. R. (1992). Caenorhabditis elegans gene ced-9 protects cells from programmed cell death. *Nature, 356*, 494–499.
8. Conradt, B., & Horvitz, H. R. (1998). The C. elegans protein EGL-1 is required for programmed cell death and interacts with the Bcl-2-like protein CED-9. *Cell, 93*, 519–529.
9. Knudson, C. M., Tung, K. S., Tourtellotte, W. G., Brown, G. A., & Korsmeyer, S. J. (1995). Bax-deficient mice with lymphoid hyperplasia and male germ cell death. *Science, 270*, 96–99.
10. Yuan, J., Shaham, S., Ledoux, S., Ellis, H. M., & Horvitz, H. R. (1993). The C. elegans cell death gene ced-3 encodes a protein similar to mammalian interleukin-1 beta-converting enzyme. *Cell, 75*, 641–652.
11. Zou, H., Henzel, W. J., Liu, X., Lutschg, A., & Wang, X. (1997). Apaf-1, a human protein homologous to C. elegans CED-4, participates in cytochrome c-dependent activation of caspase-3. *Cell, 90*, 405–413.

12. Hengartner, M. O., & Horvitz, H. R. (1994). C. elegans cell survival gene ced-9 encodes a functional homolog of the mammalian proto-oncogene bcl-2. *Cell, 76*, 665–676.
13. Ellis, R. E., Jacobson, D. M., & Horvitz, H. R. (1991). Genes required for the engulfment of cell corpses during programmed cell death in Caenorhabditis elegans. *Genetics, 129*, 79–94.
14. Liu, Q. A., & Hengartner, M. O. (1999). The molecular mechanism of programmed cell death in C. elegans. *Annals of the New York Academy of Sciences, 887*, 92–104.
15. Henson, P. M., Bratton, D. L., & Fadok, V. A. (2001). Apoptotic cell removal. *Current Biology, 11*, R795–R805.
16. Thorp, E., & Tabas, I. (2009). Mechanisms and consequences of efferocytosis in advanced atherosclerosis. *Journal of Leukocyte Biology, 86*, 1089–1095.
17. Fadok, V. A., et al. (1998). Macrophages that have ingested apoptotic cells in vitro inhibit proinflammatory cytokine production through autocrine/paracrine mechanisms involving TGF-beta, PGE2, and PAF. *The Journal of Clinical Investigation, 101*, 890–898.
18. Degterev, A., et al. (2008). Identification of RIP1 kinase as a specific cellular target of necrostatins. *Nature Chemical Biology, 4*, 313–321.
19. Zhang, D.-W., et al. (2009). RIP3, an energy metabolism regulator that switches TNF-induced cell death from apoptosis to necrosis. *Science, 325*, 332–336.
20. He, S., et al. (2009). Receptor interacting protein kinase-3 determines cellular necrotic response to TNF-alpha. *Cell, 137*, 1100–1111.
21. Cho, Y. S., et al. (2009). Phosphorylation-driven assembly of the RIP1-RIP3 complex regulates programmed necrosis and virus-induced inflammation. *Cell, 137*, 1112–1123.
22. Cerqueira, D. M., Pereira, M. S. F., Silva, A. L. N., Cunha, L. D., & Zamboni, D. S. (2015). Caspase-1 but not caspase-11 is required for NLRC4-mediated pyroptosis and restriction of infection by flagellated legionella species in mouse macrophages and in vivo. *Journal of Immunology, 195*, 2303–2311.
23. Fink, S. L., & Cookson, B. T. (2006). Caspase-1-dependent pore formation during pyroptosis leads to osmotic lysis of infected host macrophages. *Cellular Microbiology, 8*, 1812–1825.
24. Nicotera, P., Leist, M., & Ferrando-May, E. (1998). Intracellular ATP, a switch in the decision between apoptosis and necrosis. *Toxicology Letters, 102–103*, 139–142.
25. Tsujimoto, Y. (1997). Apoptosis and necrosis: Intracellular ATP level as a determinant for cell death modes. *Cell Death and Differentiation, 4*, 429–434.
26. Ha, H. C., & Snyder, S. H. (1999). Poly(ADP-ribose) polymerase is a mediator of necrotic cell death by ATP depletion. *Proceedings of the National Academy of Sciences of the United States of America, 96*, 13978–13982.
27. Ammendolia, D. A., Bement, W. M., & Brumell, J. H. (2021). Plasma membrane integrity: Implications for health and disease. *BMC Biology, 19*, 71.
28. Zhang, Y., Chen, X., Gueydan, C., & Han, J. (2018). Plasma membrane changes during programmed cell deaths. *Cell Research, 28*, 9–21.
29. Dal Peraro, M., & van der Goot, F. G. (2016). Pore-forming toxins: Ancient, but never really out of fashion. *Nature Reviews. Microbiology, 14*, 77–92.
30. Voskoboinik, I., Whisstock, J. C., & Trapani, J. A. (2015). Perforin and granzymes: Function, dysfunction and human pathology. *Nature Reviews. Immunology, 15*, 388–400.
31. Ivanova, M. E., et al. (2022). The pore conformation of lymphocyte perforin. *Science Advances, 8*, eabk3147.
32. Ni, T., et al. (2020). Structure and mechanism of bactericidal mammalian perforin-2, an ancient agent of innate immunity. *Science Advances, 6*, eaax8286.
33. Yu, X., Ni, T., Munson, G., Zhang, P., & Gilbert, R. J. C. (2022). Cryo-EM structures of perforin-2 in isolation and assembled on a membrane suggest a mechanism for pore formation. *The EMBO Journal, 41*, e111857.
34. Rodríguez-Silvestre, P., et al. (2023). Perforin-2 is a pore-forming effector of endocytic escape in cross-presenting dendritic cells. *Science, 380*, 1258–1265.
35. Serna, M., Giles, J. L., Morgan, B. P., & Bubeck, D. (2016). Structural basis of complement membrane attack complex formation. *Nature Communications, 7*, 10587.

36. Menny, A., et al. (2018). CryoEM reveals how the complement membrane attack complex ruptures lipid bilayers. *Nature Communications, 9*, 5316.
37. Mastellos, D. C., Hajishengallis, G., & Lambris, J. D. (2024). A guide to complement biology, pathology and therapeutic opportunity. *Nature Reviews. Immunology, 24*, 118–141.
38. Zhao, J., et al. (2012). Mixed lineage kinase domain-like is a key receptor interacting protein 3 downstream component of TNF-induced necrosis. *Proceedings of the National Academy of Sciences of the United States of America, 109*, 5322–5327.
39. Sun, L., et al. (2012). Mixed lineage kinase domain-like protein mediates necrosis signaling downstream of RIP3 kinase. *Cell, 148*, 213–227.
40. Rogers, C., et al. (2017). Cleavage of DFNA5 by caspase-3 during apoptosis mediates progression to secondary necrotic/pyroptotic cell death. *Nature Communications, 8*, 14128.
41. Wang, Y., et al. (2017). Chemotherapy drugs induce pyroptosis through caspase-3 cleavage of a gasdermin. *Nature, 547*, 99–103.
42. Shi, J., et al. (2015). Cleavage of GSDMD by inflammatory caspases determines pyroptotic cell death. *Nature, 526*, 660–665.
43. Kayagaki, N., et al. (2015). Caspase-11 cleaves gasdermin D for non-canonical inflammasome signalling. *Nature, 526*, 666–671.
44. Dixon, S. J., & Olzmann, J. A. (2024). The cell biology of ferroptosis. *Nature Reviews. Molecular Cell Biology, 25*, 424–442.
45. Friedmann Angeli, J. P., et al. (2014). Inactivation of the ferroptosis regulator Gpx4 triggers acute renal failure in mice. *Nature Cell Biology, 16*, 1180–1191.
46. Yang, W. S., et al. (2014). Regulation of ferroptotic cancer cell death by GPX4. *Cell, 156*, 317–331.
47. Dixon, S. J., et al. (2012). Ferroptosis: An iron-dependent form of nonapoptotic cell death. *Cell, 149*, 1060–1072.
48. Runas, K. A., & Malmstadt, N. (2015). Low levels of lipid oxidation radically increase the passive permeability of lipid bilayers. *Soft Matter, 11*, 499–505.
49. Wiczew, D., Szulc, N., & Tarek, M. (2021). Molecular dynamics simulations of the effects of lipid oxidation on the permeability of cell membranes. *Bioelectrochemistry, 141*, 107869.
50. Agmon, E., Solon, J., Bassereau, P., & Stockwell, B. R. (2018). Modeling the effects of lipid peroxidation during ferroptosis on membrane properties. *Scientific Reports, 8*, 5155.
51. Kayagaki, N., et al. (2021). NINJ1 mediates plasma membrane rupture during lytic cell death. *Nature, 591*, 131–136.
52. Majno, G., & Joris, I. (1995). Apoptosis, oncosis, and necrosis. An overview of cell death. *The American Journal of Pathology, 146*, 3–15.
53. Dondelinger, Y., et al. (2023). NINJ1 is activated by cell swelling to regulate plasma membrane permeabilization during regulated necrosis. *Cell Death & Disease, 14*, 755.
54. Pourmal, S., et al. (2025). Autoinhibition of dimeric NINJ1 prevents plasma membrane rupture. *Nature, 637*, 446–452.
55. Sahoo, B., Mou, Z., Liu, W., Dubyak, G., & Dai, X. (2025). How NINJ1 mediates plasma membrane rupture and why NINJ2 cannot. *Cell, 188*, 292–302.e11.
56. Zhu, Y., et al. (2024). NINJ1 regulates plasma membrane fragility under mechanical tension. *Res Square*. https://doi.org/10.21203/rs.3.rs-5237916/v1
57. Gong, Y.-N., et al. (2017). ESCRT-III acts downstream of MLKL to regulate necroptotic cell death and its consequences. *Cell, 169*, 286–300.e16.
58. Rühl, S., et al. (2018). ESCRT-dependent membrane repair negatively regulates pyroptosis downstream of GSDMD activation. *Science, 362*, 956–960.
59. Doerflinger, M., et al. (2020). Flexible usage and interconnectivity of diverse cell death pathways protect against intracellular infection. *Immunity, 53*, 533–547.e7.
60. Pandeya, A., & Kanneganti, T.-D. (2024). Therapeutic potential of PANoptosis: Innate sensors, inflammasomes, and RIPKs in PANoptosomes. *Trends in Molecular Medicine, 30*, 74–88.

61. Kim, J., Kim, H.-S., & Chung, J. H. (2023). Molecular mechanisms of mitochondrial DNA release and activation of the cGAS-STING pathway. *Experimental & Molecular Medicine, 55*, 510–519.
62. Kroemer, G., Galassi, C., Zitvogel, L., & Galluzzi, L. (2022). Immunogenic cell stress and death. *Nature Immunology, 23*, 487–500.
63. Clucas, J., & Meier, P. (2023). Roles of RIPK1 as a stress sentinel coordinating cell survival and immunogenic cell death. *Nature Reviews. Molecular Cell Biology, 24*, 835–852.
64. Meier, P., Legrand, A. J., Adam, D., & Silke, J. (2024). Immunogenic cell death in cancer: Targeting necroptosis to induce antitumour immunity. *Nature Reviews. Cancer, 24*, 299–315.
65. Pasparakis, M., & Vandenabeele, P. (2015). Necroptosis and its role in inflammation. *Nature, 517*, 311–320.
66. Berger, S. B., et al. (2014). Cutting edge: RIP1 kinase activity is dispensable for normal development but is a key regulator of inflammation in SHARPIN-deficient mice. *Journal of Immunology, 192*, 5476–5480.
67. Polykratis, A., et al. (2014). Cutting edge: RIPK1 kinase inactive mice are viable and protected from TNF-induced necroptosis in vivo. *Journal of Immunology, 193*, 1539–1543.
68. Newton, K., et al. (2014). Activity of protein kinase RIPK3 determines whether cells die by necroptosis or apoptosis. *Science, 343*, 1357–1360.
69. Wu, J., et al. (2013). Mlkl knockout mice demonstrate the indispensable role of Mlkl in necroptosis. *Cell Research, 23*, 994–1006.
70. Martens, S., Hofmans, S., Declercq, W., Augustyns, K., & Vandenabeele, P. (2020). Inhibitors targeting RIPK1/RIPK3: Old and new drugs. *Trends in Pharmacological Sciences, 41*, 209–224.
71. Degterev, A., Ofengeim, D., & Yuan, J. (2019). Targeting RIPK1 for the treatment of human diseases. *Proceedings of the National Academy of Sciences of the United States of America, 116*, 9714–9722.
72. Sun, X., Yin, J., Starovasnik, M. A., Fairbrother, W. J., & Dixit, V. M. (2002). Identification of a novel homotypic interaction motif required for the phosphorylation of receptor-interacting protein (RIP) by RIP3. *The Journal of Biological Chemistry, 277*, 9505–9511.
73. Li, J., et al. (2012). The RIP1/RIP3 necrosome forms a functional amyloid signaling complex required for programmed necrosis. *Cell, 150*, 339–350.
74. Dondelinger, Y., Hulpiau, P., Saeys, Y., Bertrand, M. J. M., & Vandenabeele, P. (2016). An evolutionary perspective on the necroptotic pathway. *Trends in Cell Biology, 26*, 721–732.
75. van Loo, G., & Bertrand, M. J. M. (2023). Death by TNF: A road to inflammation. *Nature Reviews Immunology, 23*, 289–303.
76. Delanghe, T., Dondelinger, Y., & Bertrand, M. J. M. (2020). RIPK1 kinase-dependent death: A symphony of phosphorylation events. *Trends in Cell Biology, 30*, 189–200.
77. Micheau, O., & Tschopp, J. (2003). Induction of TNF receptor I-mediated apoptosis via two sequential signaling complexes. *Cell, 114*, 181–190.
78. Dondelinger, Y., et al. (2019). Serine 25 phosphorylation inhibits RIPK1 kinase-dependent cell death in models of infection and inflammation. *Nature Communications, 10*, 1729.
79. Lafont, E., et al. (2018). TBK1 and IKKε prevent TNF-induced cell death by RIPK1 phosphorylation. *Nature Cell Biology, 20*, 1389–1399.
80. Xu, D., et al. (2018). TBK1 suppresses RIPK1-driven apoptosis and inflammation during development and in aging. *Cell, 174*, 1477–1491.e19.
81. Dondelinger, Y., et al. (2017). MK2 phosphorylation of RIPK1 regulates TNF-mediated cell death. *Nature Cell Biology, 19*, 1237–1247.
82. Menon, M. B., et al. (2017). p38MAPK/MK2-dependent phosphorylation controls cytotoxic RIPK1 signalling in inflammation and infection. *Nature Cell Biology, 19*, 1248–1259.
83. Jaco, I., et al. (2017). MK2 phosphorylates RIPK1 to prevent TNF-induced cell death. *Molecular Cell, 66*, 698–710.e5.
84. Wang, L., Du, F., & Wang, X. (2008). TNF-alpha induces two distinct caspase-8 activation pathways. *Cell, 133*, 693–703.

85. Laurien, L., et al. (2020). Autophosphorylation at serine 166 regulates RIP kinase 1-mediated cell death and inflammation. *Nature Communications, 11*, 1747.
86. Dondelinger, Y., et al. (2013). RIPK3 contributes to TNFR1-mediated RIPK1 kinase-dependent apoptosis in conditions of cIAP1/2 depletion or TAK1 kinase inhibition. *Cell Death and Differentiation, 20*, 1381–1392.
87. Declercq, W., Vanden Berghe, T., & Vandenabeele, P. (2009). RIP kinases at the crossroads of cell death and survival. *Cell, 138*, 229–232.
88. Meng, H., et al. (2018). Death-domain dimerization-mediated activation of RIPK1 controls necroptosis and RIPK1-dependent apoptosis. *Proceedings of the National Academy of Sciences of the United States of America, 115*, E2001–E2009.
89. Vercammen, D., et al. (1998). Inhibition of caspases increases the sensitivity of L929 cells to necrosis mediated by tumor necrosis factor. *The Journal of Experimental Medicine, 187*, 1477–1485.
90. Vercammen, D., et al. (1998). Dual signaling of the Fas receptor: Initiation of both apoptotic and necrotic cell death pathways. *The Journal of Experimental Medicine, 188*, 919–930.
91. Holler, N., et al. (2000). Fas triggers an alternative, caspase-8-independent cell death pathway using the kinase RIP as effector molecule. *Nature Immunology, 1*, 489–495.
92. Cauwels, A., Janssen, B., Waeytens, A., Cuvelier, C., & Brouckaert, P. (2003). Caspase inhibition causes hyperacute tumor necrosis factor-induced shock via oxidative stress and phospholipase A2. *Nature Immunology, 4*, 387–393.
93. Duprez, L., et al. (2011). RIP kinase-dependent necrosis drives lethal systemic inflammatory response syndrome. *Immunity, 35*, 908–918.
94. Feng, S., et al. (2007). Cleavage of RIP3 inactivates its caspase-independent apoptosis pathway by removal of kinase domain. *Cellular Signalling, 19*, 2056–2067.
95. Oberst, A., et al. (2011). Catalytic activity of the caspase-8-FLIP(L) complex inhibits RIPK3-dependent necrosis. *Nature, 471*, 363–367.
96. Kaiser, W. J., et al. (2011). RIP3 mediates the embryonic lethality of caspase-8-deficient mice. *Nature, 471*, 368–372.
97. Zhang, H., et al. (2011). Functional complementation between FADD and RIP1 in embryos and lymphocytes. *Nature, 471*, 373–376.
98. Newton, K., et al. (2019). Cleavage of RIPK1 by caspase-8 is crucial for limiting apoptosis and necroptosis. *Nature, 574*, 428–431.
99. Zhang, X., Dowling, J. P., & Zhang, J. (2019). RIPK1 can mediate apoptosis in addition to necroptosis during embryonic development. *Cell Death & Disease, 10*, 245.
100. Lalaoui, N., et al. (2020). Mutations that prevent caspase cleavage of RIPK1 cause autoinflammatory disease. *Nature, 577*, 103–108.
101. Tao, P., et al. (2020). A dominant autoinflammatory disease caused by non-cleavable variants of RIPK1. *Nature, 577*, 109–114.
102. Huyghe, J., et al. (2022). ATG9A prevents TNF cytotoxicity by an unconventional lysosomal targeting pathway. *Science, 378*, 1201–1207.
103. Liu, L., et al. (2022). Tankyrase-mediated ADP-ribosylation is a regulator of TNF-induced death. *Science Advances, 8*, eabh2332.
104. Dillon, C. P., et al. (2014). RIPK1 blocks early postnatal lethality mediated by caspase-8 and RIPK3. *Cell, 157*, 1189–1202.
105. Rickard, J. A., et al. (2014). RIPK1 regulates RIPK3-MLKL-driven systemic inflammation and emergency hematopoiesis. *Cell, 157*, 1175–1188.
106. Kaiser, W. J., et al. (2014). RIP1 suppresses innate immune necrotic as well as apoptotic cell death during mammalian parturition. *Proceedings of the National Academy of Sciences of the United States of America, 111*, 7753–7758.
107. Anderton, H., et al. (2019). RIPK1 prevents TRADD-driven, but TNFR1 independent, apoptosis during development. *Cell Death and Differentiation, 26*, 877–889.
108. Dowling, J. P., Alsabbagh, M., Del Casale, C., Liu, Z.-G., & Zhang, J. (2019). TRADD regulates perinatal development and adulthood survival in mice lacking RIPK1 and RIPK3. *Nature Communications, 10*, 705.

109. Newton, K., et al. (2016). RIPK3 deficiency or catalytically inactive RIPK1 provides greater benefit than MLKL deficiency in mouse models of inflammation and tissue injury. *Cell Death and Differentiation, 23*, 1565–1576.
110. Cuchet-Lourenço, D., et al. (2018). Biallelic RIPK1 mutations in humans cause severe immunodeficiency, arthritis, and intestinal inflammation. *Science, 361*, 810–813.
111. Uchiyama, Y., et al. (2019). Primary immunodeficiency with chronic enteropathy and developmental delay in a boy arising from a novel homozygous RIPK1 variant. *Journal of Human Genetics, 64*, 955–960.
112. Li, Y., et al. (2019). Human RIPK1 deficiency causes combined immunodeficiency and inflammatory bowel diseases. *Proceedings of the National Academy of Sciences of the United States of America, 116*, 970–975.
113. Kaiser, W. J., et al. (2013). Toll-like receptor 3-mediated necrosis via TRIF, RIP3, and MLKL. *The Journal of Biological Chemistry, 288*, 31268–31279.
114. He, S., Liang, Y., Shao, F., & Wang, X. (2011). Toll-like receptors activate programmed necrosis in macrophages through a receptor-interacting kinase-3-mediated pathway. *Proceedings of the National Academy of Sciences of the United States of America, 108*, 20054–20059.
115. Meylan, E., et al. (2004). RIP1 is an essential mediator of toll-like receptor 3-induced NF-kappa B activation. *Nature Immunology, 5*, 503–507.
116. Chen, N.-J., et al. (2008). Beyond tumor necrosis factor receptor: TRADD signaling in toll-like receptors. *Proceedings of the National Academy of Sciences of the United States of America, 105*, 12429–12434.
117. Gitlin, A. D., et al. (2020). Integration of innate immune signalling by caspase-8 cleavage of N4BP1. *Nature, 587*, 275–280.
118. Zinngrebe, J., et al. (2016). -LUBAC deficiency perturbs TLR3 signaling to cause immunodeficiency and autoinflammation. *The Journal of Experimental Medicine, 213*, 2671–2689.
119. Weber, A., et al. (2010). Proapoptotic signalling through toll-like receptor-3 involves TRIF-dependent activation of caspase-8 and is under the control of inhibitor of apoptosis proteins in melanoma cells. *Cell Death and Differentiation, 17*, 942–951.
120. Feoktistova, M., et al. (2011). cIAPs block Ripoptosome formation, a RIP1/caspase-8 containing intracellular cell death complex differentially regulated by cFLIP isoforms. *Molecular Cell, 43*, 449–463.
121. Thapa, R. J., et al. (2016). DAI senses influenza a virus genomic RNA and activates RIPK3-dependent cell death. *Cell Host & Microbe, 20*, 674–681.
122. Lin, J., et al. (2016). RIPK1 counteracts ZBP1-mediated necroptosis to inhibit inflammation. *Nature, 540*, 124–128.
123. Koerner, L., et al. (2024). ZBP1 causes inflammation by inducing RIPK3-mediated necroptosis and RIPK1 kinase activity-independent apoptosis. *Cell Death and Differentiation, 31*, 938–953.
124. Rodriguez, D. A., et al. (2022). Caspase-8 and FADD prevent spontaneous ZBP1 expression and necroptosis. *Proceedings of the National Academy of Sciences of the United States of America, 119*, e2207240119.
125. Rodriguez, D. A., et al. (2024). The interaction between RIPK1 and FADD controls perinatal lethality and inflammation. *Cell Reports, 43*, 114335.
126. Cai, Z.-Y., et al. (2024). A ZBP1 isoform blocks ZBP1-mediated cell death. *Cell Reports, 43*, 114221.
127. Jiao, H., et al. (2020). Z-nucleic-acid sensing triggers ZBP1-dependent necroptosis and inflammation. *Nature, 580*, 391–395.
128. Newton, K., et al. (2016). RIPK1 inhibits ZBP1-driven necroptosis during development. *Nature, 540*, 129–133.
129. Jang, T.-H., et al. (2014). Structural study of the RIPoptosome core reveals a helical assembly for kinase recruitment. *Biochemistry, 53*, 5424–5431.
130. Imai, T., et al. (2024). The RIPK1 death domain restrains ZBP1- and TRIF-mediated cell death and inflammation. *Immunity, 57*, 1497–1513.e6.

131. Wang, Z., et al. (2008). Regulation of innate immune responses by DAI (DLM-1/ZBP1) and other DNA-sensing molecules. *Proceedings of the National Academy of Sciences of the United States of America, 105*, 5477–5482.
132. Takaoka, A., et al. (2007). DAI (DLM-1/ZBP1) is a cytosolic DNA sensor and an activator of innate immune response. *Nature, 448*, 501–505.
133. Rebsamen, M., et al. (2009). DAI/ZBP1 recruits RIP1 and RIP3 through RIP homotypic interaction motifs to activate NF-kappaB. *EMBO Reports, 10*, 916–922.
134. Kaiser, W. J., Upton, J. W., & Mocarski, E. S. (2008). Receptor-interacting protein homotypic interaction motif-dependent control of NF-kappa B activation via the DNA-dependent activator of IFN regulatory factors. *Journal of Immunology, 181*, 6427–6434.
135. Chen, W., et al. (2013). Diverse sequence determinants control human and mouse receptor interacting protein 3 (RIP3) and mixed lineage kinase domain-like (MLKL) interaction in necroptotic signaling. *The Journal of Biological Chemistry, 288*, 16247–16261.
136. Murphy, J. M., et al. (2013). The pseudokinase MLKL mediates necroptosis via a molecular switch mechanism. *Immunity, 39*, 443–453.
137. Meng, Y., et al. (2021). Human RIPK3 maintains MLKL in an inactive conformation prior to cell death by necroptosis. *Nature Communications, 12*, 6783.
138. Petrie, E. J., et al. (2018). Conformational switching of the pseudokinase domain promotes human MLKL tetramerization and cell death by necroptosis. *Nature Communications, 9*, 2422.
139. Meng, Y., et al. (2023). Phosphorylation-dependent pseudokinase domain dimerization drives full-length MLKL oligomerization. *Nature Communications, 14*, 6804.
140. Zhang, Y., et al. (2021). The MLKL kinase-like domain dimerization is an indispensable step of mammalian MLKL activation in necroptosis signaling. *Cell Death & Disease, 12*, 638.
141. Davies, K. A., et al. (2020). Distinct pseudokinase domain conformations underlie divergent activation mechanisms among vertebrate MLKL orthologues. *Nature Communications, 11*, 3060.
142. Garnish, S. E., et al. (2021). Conformational interconversion of MLKL and disengagement from RIPK3 precede cell death by necroptosis. *Nature Communications, 12*, 2211.
143. Davies, K. A., Czabotar, P. E., & Murphy, J. M. (2024). Death at a funeral: Activation of the dead enzyme, MLKL, to kill cells by necroptosis. *Current Opinion in Structural Biology, 88*, 102891.
144. Garnish, S. E., et al. (2024). Inhibitors identify an auxiliary role for mTOR signalling in necroptosis execution downstream of MLKL activation. *The Biochemical Journal, 481*, 1125–1142.
145. Hildebrand, J. M., et al. (2014). Activation of the pseudokinase MLKL unleashes the four-helix bundle domain to induce membrane localization and necroptotic cell death. *Proceedings of the National Academy of Sciences of the United States of America, 111*, 15072–15077.
146. Cai, Z., et al. (2014). Plasma membrane translocation of trimerized MLKL protein is required for TNF-induced necroptosis. *Nature Cell Biology, 16*, 55–65.
147. Davies, K. A., et al. (2018). The brace helices of MLKL mediate interdomain communication and oligomerisation to regulate cell death by necroptosis. *Cell Death and Differentiation, 25*, 1567–1580.
148. Nailwal, H., & Chan, F. K.-M. (2019). Necroptosis in anti-viral inflammation. *Cell Death and Differentiation, 26*, 4–13.
149. Palmer, S. N., Chappidi, S., Pinkham, C., & Hancks, D. C. (2021). Evolutionary profile for (host and viral) MLKL indicates its activities as a battlefront for extensive counteradaptation. *Molecular Biology and Evolution, 38*, 5405–5422.
150. Verdonck, S., Nemegeer, J., Vandenabeele, P., & Maelfait, J. (2022). Viral manipulation of host cell necroptosis and pyroptosis. *Trends in Microbiology, 30*, 593–605.
151. Chen, X., et al. (2014). Translocation of mixed lineage kinase domain-like protein to plasma membrane leads to necrotic cell death. *Cell Research, 24*, 105–121.
152. Dondelinger, Y., et al. (2014). MLKL compromises plasma membrane integrity by binding to phosphatidylinositol phosphates. *Cell Reports, 7*, 971–981.

153. Wang, H., et al. (2014). Mixed lineage kinase domain-like protein MLKL causes necrotic membrane disruption upon phosphorylation by RIP3. *Molecular Cell, 54*, 133–146.
154. Quarato, G., et al. (2016). Sequential engagement of distinct MLKL phosphatidylinositol-binding sites executes necroptosis. *Molecular Cell, 61*, 589–601.
155. Su, L., et al. (2014). A plug release mechanism for membrane permeation by MLKL. *Structure, 22*, 1489–1500.
156. Dovey, C. M., et al. (2018). MLKL requires the inositol phosphate code to execute necroptosis. *Molecular Cell, 70*, 936–948.e7.
157. Hildebrand, J. M., et al. (2020). A missense mutation in the MLKL brace region promotes lethal neonatal inflammation and hematopoietic dysfunction. *Nature Communications, 11*, 3150.
158. Samson, A. L., et al. (2020). MLKL trafficking and accumulation at the plasma membrane control the kinetics and threshold for necroptosis. *Nature Communications, 11*, 3151.
159. Xia, B., et al. (2016). MLKL forms cation channels. *Cell Research, 26*, 517–528.
160. Flores-Romero, H., Ros, U., & Garcia-Saez, A. J. (2020). Pore formation in regulated cell death. *The EMBO Journal, 39*, e105753.
161. Yoon, S., Kovalenko, A., Bogdanov, K., & Wallach, D. (2017). MLKL, the protein that mediates necroptosis, also regulates endosomal trafficking and extracellular vesicle generation. *Immunity, 47*, 51–65.e7.
162. Ramos, S., Hartenian, E., Santos, J. C., Walch, P., & Broz, P. (2024). NINJ1 induces plasma membrane rupture and release of damage-associated molecular pattern molecules during ferroptosis. *The EMBO Journal, 43*, 1164–1186.
163. Martinon, F., Burns, K., & Tschopp, J. (2002). The inflammasome: A molecular platform triggering activation of inflammatory caspases and processing of proIL-beta. *Molecular Cell, 10*, 417–426.
164. Fink, S. L., & Cookson, B. T. (2005). Apoptosis, pyroptosis, and necrosis: Mechanistic description of dead and dying eukaryotic cells. *Infection and Immunity, 73*, 1907–1916.
165. Cejna, M., Fritsch, G., Printz, D., Schulte-Hermann, R., & Bursch, W. (1994). Kinetics of apoptosis and secondary necrosis in cultured rat thymocytes and S.49 mouse lymphoma and CEM human leukemia cells. *Biochemistry and Cell Biology, 72*, 677–685.
166. Payne, C. M., Bjore, C. G., Jr., & Schultz, D. A. (1992). Change in the frequency of apoptosis after low- and high-dose X-irradiation of human lymphocytes. *Journal of Leukocyte Biology, 52*, 433–440.
167. Orning, P., et al. (2018). Pathogen blockade of TAK1 triggers caspase-8-dependent cleavage of gasdermin D and cell death. *Science, 362*, 1064–1069.
168. Sarhan, J., et al. (2018). Caspase-8 induces cleavage of gasdermin D to elicit pyroptosis during Yersinia infection. *Proceedings of the National Academy of Sciences of the United States of America, 115*, E10888–E10897.
169. Chen, K. W., et al. (2019). Extrinsic and intrinsic apoptosis activate pannexin-1 to drive NLRP3 inflammasome assembly. *The EMBO Journal, 38*, e101638.
170. Shi, J., et al. (2014). Inflammatory caspases are innate immune receptors for intracellular LPS. *Nature, 514*, 187–192.
171. Shi, X., et al. (2023). Recognition and maturation of IL-18 by caspase-4 noncanonical inflammasome. *Nature, 624*, 442–450.
172. Devant, P., et al. (2023). Structural insights into cytokine cleavage by inflammatory caspase-4. *Nature, 624*, 451–459.
173. Vande Walle, L., & Lamkanfi, M. (2024). Drugging the NLRP3 inflammasome: from signalling mechanisms to therapeutic targets. *Nature Reviews. Drug Discovery, 23*, 43–66.
174. Coll, R. C., & Schroder, K. (2025). Inflammasome components as new therapeutic targets in inflammatory disease. *Nature Reviews. Immunology, 25*, 22–41.
175. De Schutter, E., et al. (2021). Punching holes in cellular membranes: Biology and evolution of gasdermins. *Trends in Cell Biology, 31*, 500–513.
176. Liu, X., Xia, S., Zhang, Z., Wu, H., & Lieberman, J. (2021). Channelling inflammation: Gasdermins in physiology and disease. *Nature Reviews. Drug Discovery, 20*, 384–405.

177. Broz, P., Pelegrín, P., & Shao, F. (2020). The gasdermins, a protein family executing cell death and inflammation. *Nature Reviews. Immunology, 20*, 143–157.
178. He, W.-T., et al. (2015). Gasdermin D is an executor of pyroptosis and required for interleukin-1β secretion. *Cell Research, 25*, 1285–1298.
179. Liu, Z., et al. (2020). Caspase-1 engages full-length gasdermin D through two distinct interfaces that mediate caspase recruitment and substrate cleavage. *Immunity, 53*, 106–114.e5.
180. Wang, K., et al. (2020). Structural mechanism for GSDMD targeting by autoprocessed caspases in pyroptosis. *Cell, 180*, 941–955.e20.
181. Devant, P., & Kagan, J. C. (2023). Molecular mechanisms of gasdermin D pore-forming activity. *Nature Immunology, 24*, 1064–1075.
182. Liu, X., et al. (2016). Inflammasome-activated gasdermin D causes pyroptosis by forming membrane pores. *Nature, 535*, 153–158.
183. Ding, J., et al. (2016). Pore-forming activity and structural autoinhibition of the gasdermin family. *Nature, 535*, 111–116.
184. Liu, Z., et al. (2019). Crystal structures of the full-length murine and human gasdermin D reveal mechanisms of autoinhibition, lipid binding, and oligomerization. *Immunity, 51*, 43–49.e4.
185. Xia, S., et al. (2021). Gasdermin D pore structure reveals preferential release of mature interleukin-1. *Nature, 593*, 607–611.
186. Mulvihill, E., et al. (2018). Mechanism of membrane pore formation by human gasdermin-D. *The EMBO Journal, 37*, e98321.
187. Schaefer, S. L., & Hummer, G. (2022). Sublytic gasdermin-D pores captured in atomistic molecular simulations. *eLife, 11*, e81432.
188. Margheritis, E., et al. (2024). Gasdermin D cysteine residues synergistically control its palmitoylation-mediated membrane targeting and assembly. *The EMBO Journal, 43*, 4274–4297.
189. Newton, K., Strasser, A., Kayagaki, N., & Dixit, V. M. (2024). Cell death. *Cell, 187*, 235–256.
190. Jiang, X., Stockwell, B. R., & Conrad, M. (2021). Ferroptosis: Mechanisms, biology and role in disease. *Nature Reviews. Molecular Cell Biology, 22*, 266–282.
191. Lewerenz, J., Ates, G., Methner, A., Conrad, M., & Maher, P. (2018). Oxytosis/ferroptosis-(re-) emerging roles for oxidative stress-dependent non-apoptotic cell death in diseases of the central nervous system. *Frontiers in Neuroscience, 12*, 214.
192. Murphy, T. H., Miyamoto, M., Sastre, A., Schnaar, R. L., & Coyle, J. T. (1989). Glutamate toxicity in a neuronal cell line involves inhibition of cystine transport leading to oxidative stress. *Neuron, 2*, 1547–1558.
193. Miyamoto, M., Murphy, T. H., Schnaar, R. L., & Coyle, J. T. (1989). Antioxidants protect against glutamate-induced cytotoxicity in a neuronal cell line. *The Journal of Pharmacology and Experimental Therapeutics, 250*, 1132–1140.
194. Kagan, V. E., et al. (2017). Oxidized arachidonic and adrenic PEs navigate cells to ferroptosis. *Nature Chemical Biology, 13*, 81–90.
195. Doll, S., et al. (2019). FSP1 is a glutathione-independent ferroptosis suppressor. *Nature, 575*, 693–698.
196. Bersuker, K., et al. (2019). The CoQ oxidoreductase FSP1 acts parallel to GPX4 to inhibit ferroptosis. *Nature, 575*, 688–692.
197. Mishima, E., et al. (2022). A non-canonical vitamin K cycle is a potent ferroptosis suppressor. *Nature, 608*, 778–783.
198. Zheng, J., & Conrad, M. (2025). Ferroptosis: When metabolism meets cell death. *Physiological Reviews, 105*, 651–706.
199. Hadian, K., & Stockwell, B. R. (2023). The therapeutic potential of targeting regulated non-apoptotic cell death. *Nature Reviews. Drug Discovery, 22*, 723–742.
200. Mao, C., Wang, M., Zhuang, L., & Gan, B. (2024). Metabolic cell death in cancer: Ferroptosis, cuproptosis, disulfidptosis, and beyond. *Protein & Cell, 15*, 642–660.
201. Green, D. R. (2024). Cell death: Revisiting the roads to ruin. *Developmental Cell, 59*, 2523–2531.

202. Seiler, A., et al. (2008). Glutathione peroxidase 4 senses and translates oxidative stress into 12/15-lipoxygenase dependent- and AIF-mediated cell death. *Cell Metabolism, 8*, 237–248.
203. Mishima, E., Wahida, A., Seibt, T., & Conrad, M. (2023). Diverse biological functions of vitamin K: From coagulation to ferroptosis. *Nature Metabolism, 5*, 924–932.
204. Soula, M., et al. (2020). Metabolic determinants of cancer cell sensitivity to canonical ferroptosis inducers. *Nature Chemical Biology, 16*, 1351–1360.
205. Kraft, V. A. N., et al. (2020). GTP cyclohydrolase 1/tetrahydrobiopterin counteract ferroptosis through lipid remodeling. *ACS Central Science, 6*, 41–53.
206. Freitas, F. P., et al. (2024). 7-Dehydrocholesterol is an endogenous suppressor of ferroptosis. *Nature, 626*, 401–410.
207. Dixon, S. J., et al. (2015). Human haploid cell genetics reveals roles for lipid metabolism genes in nonapoptotic cell death. *ACS Chemical Biology, 10*, 1604–1609.
208. Liang, D., et al. (2023). Ferroptosis surveillance independent of GPX4 and differentially regulated by sex hormones. *Cell, 186*, 2748–2764.e22.
209. Kono, H., & Rock, K. L. (2008). How dying cells alert the immune system to danger. *Nature Reviews. Immunology, 8*, 279–289.
210. Günther, C., et al. (2011). Caspase-8 regulates TNF-α-induced epithelial necroptosis and terminal ileitis. *Nature, 477*, 335–339.
211. Fritsch, M., et al. (2019). Caspase-8 is the molecular switch for apoptosis, necroptosis and pyroptosis. *Nature, 575*, 683–687.
212. Vandenabeele, P., Galluzzi, L., Vanden Berghe, T., & Kroemer, G. (2010). Molecular mechanisms of necroptosis: An ordered cellular explosion. *Nature Reviews. Molecular Cell Biology, 11*, 700–714.
213. Newton, K., et al. (2019). Activity of caspase-8 determines plasticity between cell death pathways. *Nature, 575*, 679–682.
214. Schwarzer, R., Jiao, H., Wachsmuth, L., Tresch, A., & Pasparakis, M. (2020). FADD and caspase-8 regulate gut homeostasis and inflammation by controlling MLKL- and GSDMD-mediated death of intestinal epithelial cells. *Immunity, 52*, 978–993.e6.
215. Allam, R., et al. (2014). Mitochondrial apoptosis is dispensable for NLRP3 inflammasome activation but non-apoptotic caspase-8 is required for inflammasome priming. *EMBO Reports, 15*, 982–990.
216. Maelfait, J., et al. (2008). Stimulation of toll-like receptor 3 and 4 induces interleukin-1beta maturation by caspase-8. *The Journal of Experimental Medicine, 205*, 1967–1973.
217. Speir, M., & Lawlor, K. E. (2021). RIP-roaring inflammation: RIPK1 and RIPK3 driven NLRP3 inflammasome activation and autoinflammatory disease. *Seminars in Cell & Developmental Biology, 109*, 114–124.
218. Mandal, P., et al. (2014). RIP3 induces apoptosis independent of pronecrotic kinase activity. *Molecular Cell, 56*, 481–495.
219. Lawlor, K. E., et al. (2015). RIPK3 promotes cell death and NLRP3 inflammasome activation in the absence of MLKL. *Nature Communications, 6*, 6282.
220. Tsuchiya, K., et al. (2019). Caspase-1 initiates apoptosis in the absence of gasdermin D. *Nature Communications, 10*, 2091.
221. Erener, S., et al. (2012). Inflammasome-activated caspase 7 cleaves PARP1 to enhance the expression of a subset of NF-κB target genes. *Molecular Cell, 46*, 200–211.
222. Lamkanfi, M., et al. (2008). Targeted peptidecentric proteomics reveals caspase-7 as a substrate of the caspase-1 inflammasomes. *Molecular & Cellular Proteomics, 7*, 2350–2363.
223. Cappe, B., et al. (2023). Systematic compositional analysis of exosomal extracellular vesicles produced by cells undergoing apoptosis, necroptosis and ferroptosis. *Journal of Extracellular Vesicles, 12*, e12365.
224. Degen, M., et al. (2023). Structural basis of NINJ1-mediated plasma membrane rupture in cell death. *Nature, 618*, 1065–1071.
225. David, L., et al. (2024). NINJ1 mediates plasma membrane rupture by cutting and releasing membrane disks. *Cell, 187*, 2224–2235.e16.

226. Cooper, S. T., & McNeil, P. L. (2015). Membrane repair: Mechanisms and pathophysiology. *Physiological Reviews, 95*, 1205–1240.
227. Kayagaki, N., et al. (2023). Inhibiting membrane rupture with NINJ1 antibodies limits tissue injury. *Nature, 618*, 1072–1077.
228. Najafov, A., Chen, H., & Yuan, J. (2017). Necroptosis and cancer. *Trends Cancer, 3*, 294–301.
229. Yin, J., Yu, Y., Huang, X., & Chan, F. K.-M. (2024). Necroptosis in immunity, tissue homeostasis, and cancer. *Current Opinion in Immunology, 89*, 102455.
230. Fulda, S. (2017). Smac mimetics to therapeutically target IAP proteins in cancer. *International Review of Cell and Molecular Biology, 330*, 157–169.
231. Seifert, L., et al. (2016). The necrosome promotes pancreatic oncogenesis via CXCL1 and Mincle-induced immune suppression. *Nature, 532*, 245–249.
232. Yatim, N., et al. (2015). RIPK1 and NF-κB signaling in dying cells determines cross-priming of $CD8^+$ T cells. *Science, 350*, 328–334.
233. Höckendorf, U., et al. (2016). RIPK3 restricts myeloid leukemogenesis by promoting cell death and differentiation of leukemia initiating cells. *Cancer Cell, 30*, 75–91.
234. Zaki, M. H., et al. (2011). The NOD-like receptor NLRP12 attenuates colon inflammation and tumorigenesis. *Cancer Cell, 20*, 649–660.
235. Wiernicki, B., et al. (2022). Cancer cells dying from ferroptosis impede dendritic cell-mediated anti-tumor immunity. *Nature Communications, 13*, 3676.
236. Veglia, F., et al. (2017). Lipid bodies containing oxidatively truncated lipids block antigen cross-presentation by dendritic cells in cancer. *Nature Communications, 8*, 2122.
237. Morotti, M., et al. (2024). PGE2 inhibits TIL expansion by disrupting IL-2 signalling and mitochondrial function. *Nature, 629*, 426–434.
238. Friedmann Angeli, J. P., Krysko, D. V., & Conrad, M. (2019). Ferroptosis at the crossroads of cancer-acquired drug resistance and immune evasion. *Nature Reviews. Cancer, 19*, 405–414.
239. Kim, R., et al. (2022). Ferroptosis of tumour neutrophils causes immune suppression in cancer. *Nature, 612*, 338–346.
240. Bell, H. N., Stockwell, B. R., & Zou, W. (2024). Ironing out the role of ferroptosis in immunity. *Immunity, 57*, 941–956.
241. Zhao, Y., et al. (2023). Neutrophils resist ferroptosis and promote breast cancer metastasis through aconitate decarboxylase 1. *Cell Metabolism, 35*, 1688–1703.e10.
242. Ubellacker, J. M., et al. (2020). Lymph protects metastasizing melanoma cells from ferroptosis. *Nature, 585*, 113–118.
243. Xu, S., et al. (2021). Uptake of oxidized lipids by the scavenger receptor CD36 promotes lipid peroxidation and dysfunction in CD8+ T cells in tumors. *Immunity, 54*, 1561–1577.e7.
244. Ofengeim, D., et al. (2015). Activation of necroptosis in multiple sclerosis. *Cell Reports, 10*, 1836–1849.
245. Malhotra, S., et al. (2023). Increased NLRP3 inflammasome activation and pyroptosis in patients with multiple sclerosis with fingolimod treatment failure. *Neurology Neuroimmunology & Neuroinflammation, 10*, e200100.
246. McKenzie, B. A., et al. (2018). Caspase-1 inhibition prevents glial inflammasome activation and pyroptosis in models of multiple sclerosis. *Proceedings of the National Academy of Sciences of the United States of America, 115*, E6065–E6074.
247. Woo, M. S., et al. (2024). STING orchestrates the neuronal inflammatory stress response in multiple sclerosis. *Cell, 187*, 4043–4060.e30.
248. Li, P., et al. (2021). Glutathione peroxidase 4-regulated neutrophil ferroptosis induces systemic autoimmunity. *Nature Immunology, 22*, 1107–1117.
249. Matsuzawa-Ishimoto, Y., et al. (2017). Autophagy protein ATG16L1 prevents necroptosis in the intestinal epithelium. *The Journal of Experimental Medicine, 214*, 3687–3705.
250. Jostins, L., et al. (2012). Host-microbe interactions have shaped the genetic architecture of inflammatory bowel disease. *Nature, 491*, 119–124.
251. Takahashi, N., et al. (2014). RIPK1 ensures intestinal homeostasis by protecting the epithelium against apoptosis. *Nature, 513*, 95–99.
252. Zeissig, Y., et al. (2015). XIAP variants in male Crohn's disease. *Gut, 64*, 66–76.

253. Lehle, A. S., et al. (2019). Intestinal inflammation and dysregulated immunity in patients with inherited caspase-8 deficiency. *Gastroenterology, 156*, 275–278.
254. Moriwaki, K., et al. (2014). The necroptosis adaptor RIPK3 promotes injury-induced cytokine expression and tissue repair. *Immunity, 41*, 567–578.
255. Mayr, L., et al. (2020). Dietary lipids fuel GPX4-restricted enteritis resembling Crohn's disease. *Nature Communications, 11*, 1775.
256. Green, D. R., Ferguson, T., Zitvogel, L., & Kroemer, G. (2009). Immunogenic and tolerogenic cell death. *Nature Reviews. Immunology, 9*, 353–363.
257. Jorgensen, I., Rayamajhi, M., & Miao, E. A. (2017). Programmed cell death as a defence against infection. *Nature Reviews. Immunology, 17*, 151–164.
258. Yatim, N., Cullen, S., & Albert, M. L. (2017). Dying cells actively regulate adaptive immune responses. *Nature Reviews. Immunology, 17*, 262–275.
259. Galluzzi, L., Buqué, A., Kepp, O., Zitvogel, L., & Kroemer, G. (2017). Immunogenic cell death in cancer and infectious disease. *Nature Reviews. Immunology, 17*, 97–111.
260. Demarco, B., et al. (2020). Caspase-8-dependent gasdermin D cleavage promotes antimicrobial defense but confers susceptibility to TNF-induced lethality. *Science Advances, 6*, eabc3465.
261. Doll, S., et al. (2017). ACSL4 dictates ferroptosis sensitivity by shaping cellular lipid composition. *Nature Chemical Biology, 13*, 91–98.
262. Yan, B., et al. (2021). Membrane damage during ferroptosis is caused by oxidation of phospholipids catalyzed by the oxidoreductases POR and CYB5R1. *Molecular Cell, 81*, 355–369.e10.
263. Xie, Y., et al. (2017). The tumor suppressor p53 limits ferroptosis by blocking DPP4 activity. *Cell Reports, 20*, 1692–1704.
264. Hassannia, B., et al. (2018). Nano-targeted induction of dual ferroptotic mechanisms eradicates high-risk neuroblastoma. *Journal Of Clinical Investigation, 128*(8), 3341–3355.
265. Fang, Y., et al. (2021). Inhibiting ferroptosis through disrupting the NCOA4-FTH1 interaction: A new mechanism of action. *ACS Central Science, 7*, 980–989.
266. Pedrera, L., et al. (2021). Ferroptotic pores induce Ca2+ fluxes and ESCRT-III activation to modulate cell death kinetics. *Cell Death and Differentiation, 28*, 1644–1657.
267. von Krusenstiern, A. N., et al. (2023). Identification of essential sites of lipid peroxidation in ferroptosis. *Nature Chemical Biology, 19*, 719–730.

Entosis in Health and Disease

Jyotirekha Das and Michael Overholtzer

Abstract

Entosis is a mechanism of cell-in-cell formation that resembles phagocytosis but is regulated by distinct processes related to cell-cell adhesion and actomyosin tension. Entosis results in the engulfment of cells with high levels of tension, which become internalized into cells with lower tension, following the formation of adherens junctions between the pair. The resulting "cell-in-cell" structures typically resolve in a manner that leads to the death of the internalized cells, although other fates, including internalized cell release, are also possible. Entosis is induced in response to a variety of stressors, sometimes in parallel to other forms of cell death, and has been observed to occur in cancer as well as normal tissues. Here we discuss the mechanism of entosis and consider its unique properties compared to other forms of cell death and engulfment.

Keywords

Entosis · Entotic cell death · Adherens junctions · E-cadherin · Rho · ROCK · Autophagy · Lysosome

---

J. Das
Cell Biology Program, Memorial Sloan Kettering Cancer Center, New York, NY, USA
e-mail: dasj3@mskcc.org

M. Overholtzer (✉)
Cell Biology Program, Memorial Sloan Kettering Cancer Center, New York, NY, USA

Louis V. Gerstner, Jr. Graduate School of Biomedical Sciences, Memorial Sloan Kettering Cancer Center, New York, NY, USA
e-mail: overhom1@mskcc.org

C. D. Gregory et al. (eds.), *Apoptosis and Other forms of Cell Death in the Regulation of Inflammation in Health and Disease*, Advances in Experimental Medicine and Biology 1481, https://doi.org/10.1007/978-3-031-92785-0_9

Abbreviations

p190RHOGAP	p190 Rho GTPase Activating Protein
AMPK	AMP-activated protein kinase
BCL-2	B-cell leukemia/lymphoma-2
C-MYC	Cellular myelocytomatosis oncogene
ECM	Extracellular matrix
JNK	c-Jun-N-terminal kinase
KRAS	Kirsten rat sarcoma virus
mTOR	Mechanistic target of rapamycin
PDZ-RHOGEF	PDZ-Rho guanine nucleotide exchange factor
RAC1	Ras-related C3 botulinum toxin substrate 1
RHOA	Ras homolog family member A
ROCK	Rho-associated coiled-coil containing protein kinase
TRAIL	Tumor necrosis factor (TNF)-related apoptosis-inducing ligand
UV	Ultraviolet radiation

Introduction to the Entosis Mechanism

"Entosis," named after the Greek entos, meaning inside or within, is a mechanism of cell death that was discovered in mammalian cells cultured in the absence of extracellular matrix (ECM) [1]. Detachment from ECM is a stress condition that is well known to induce cell death, for example, through an apoptotic mechanism called anoikis [2]. Entosis emerged in this context as a completely different way for cells to die. Rather than dying through cell suicide, like apoptosis and most other regulated forms of cell death [3], entotic cells are killed, non-cell-autonomously, by being engulfed alive.

Cell engulfment mechanisms are typically thought of as eating-like processes that allow cells to swallow passive targets. Apoptotic cells, for example, are engulfed after signaling to activate phagocytes, which utilize the polymerization of actin to deform their membranes and swallow dying cells [4]. Entotic cells, on the other hand, are not committed to death before they are engulfed, and they also actively control their own engulfment. Actin- and myosin-based tension localized within entotic cells provides the driving force necessary to deform neighboring cells through an invasion-like mechanism [5]. Whereas actin polymerization during phagocytosis typically involves signaling through the RAC1-GTPase (Ras-related C3 botulinum toxin substrate 1) within engulfing cells [6], entosis is instead regulated by RHOA (Ras homolog family member A) and its downstream effector ROCK (Rho-associated coiled-coil containing protein kinase), which activate myosin contraction within invading cells that are, as a result, stiffer than their hosts [5]. This mechanism is guided by adherens junctions that both anchor entotic cells to their hosts and also orient contractile myosin toward the back of the invading cells, generating a force vector that pushes entotic cells forward as junctions grow and ultimately swallow cells completely (Fig. 1) [7].

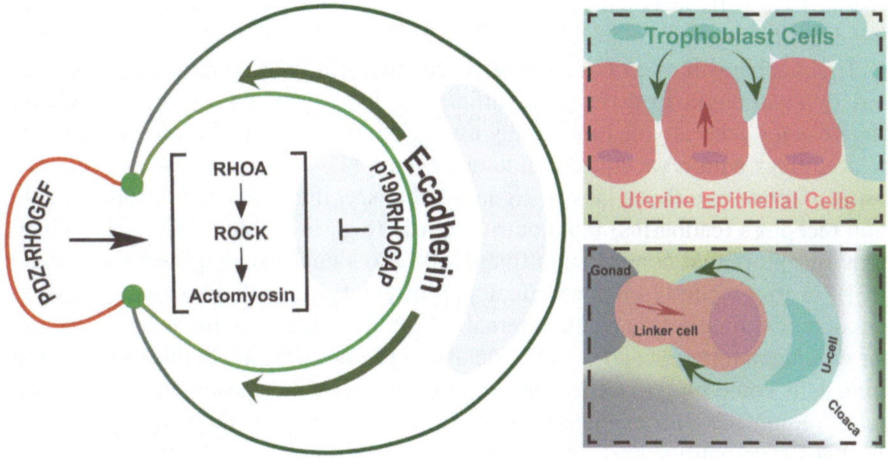

Fig. 1 The mechanism of entosis and its potential roles in development. Left: Entosis results in cell engulfment but is regulated by RHOA- and ROCK-dependent actomyosin tension within internalizing cells, which effectively invade their hosts. Adherens junctions, mediated by epithelial-type cadherins such as E-cadherin, anchor entotic cells and hosts together and also signal to polarize RHOA activity in internalizing cells away from the cell-cell adhesion interface (through the activity of p190RHOGAP) and toward the cortex at the back of invading cells (through the activity of PDZ-RHOGEF). Right top: Entosis may participate in embryo implantation, as trophectoderm cells can ingest uterine epithelial cells through this mechanism, which creates space and also implants the embryo deeper into the underlying tissue. Right bottom: Entosis may also participate in clearing the linker cell in *C. elegans* development, a mechanism that would reinforce the connection between the gonad and cloaca and also promote the formation of a contiguous lumen

The spatial orientation of contractile myosin that drives entotic cell uptake is thought to be established by at least two parallel mechanisms. One involves the recruitment of the RHOA activator PDZ-RHOGEF to the cell cortex opposite the junctional interface [8], and another involves the binding of an RHOA inhibitor, p190RHOGAP, to p120 catenin localized at cell junctions [9]. The latter mechanism is also a general component of cell adhesion-based signaling [10], suggesting that mechanical forces normally associated with cell-cell adhesion are utilized to drive the uptake of entotic cells. When completed, entosis leads to the full engulfment of viable cells, forming what are called "cell-in-cell" structures, named to denote the unusual appearance of whole cells engulfed inside large vacuole-like structures [11].

Introduction to Entotic Cell Death

Most cells engulfed by entosis die by a non-apoptotic mechanism referred to as "entotic cell death." Some manage to first divide inside their hosts, and a small percentage also find a way to escape, emerging unharmed and rejoining the cell population. However, most will die [1]. Entotic cell death results from the maturation of the membrane surrounding live-engulfed cells in a manner that leads to lysosome fusion. Live cells are then killed, non-cell-autonomously, by lysosomal enzymes and degraded by host cells [1, 12].

The mechanism of entotic cell death directly resembles phagosome maturation and is not thought to differ in significant ways, except for its timing. Whereas phagosomes generally mature rapidly upon closure, entotic cells can exist for hours or even days within vacuoles that do not mature [1]. The reason for this is unclear but may relate to either a lack of signaling that specifies maturation from the adhesion receptors (cadherins) that control entosis (e.g., in a manner similar to phagocytic receptors that control engulfment and also signal for phagosome maturation [13]); or to a recently observed final step of entotic vacuole closure that involves scission of a small piece of the internalizing cell [14], a step that is likely required for vacuole maturation and could be delayed in some contexts. In this case, internalizing cells could play a role in determining the timing of scission and, by causing a delay, remain alive for extended periods of time. In any case, the majority of entotic vacuoles eventually mature, and most entotic cells die as a result [1, 12].

That entotic cells can remain alive for extended periods within apparently enclosed vacuoles [1] raises the question of how they can manage to survive. It is clear that internalized cells have limited access to the extracellular environment and show a loss of growth factor and nutrient-dependent signaling, and that they upregulate autophagy [1, 12, 15]. The inhibition of autophagy within internalized cells, or the disruption of lysosome function, inhibits survival, as entotic cells are then observed to undergo apoptosis within their vacuoles, a fate that is clearly distinguishable by morphological criteria (condensed and fragmented cells and nuclei) and inhibition by overexpression of the anti-apoptotic protein BCL-2 (B-cell leukemia/lymphoma-2) [1, 12, 16]. These findings demonstrate that survival in the enclosed, nutrient-restricted environment of the entotic vacuole requires autophagy, presumably to provide essential nutrients through lysosome-mediated recycling. Intriguingly, entotic cells are also observed to undergo apoptosis upon withdrawal of amino acids from media or in response to inhibition of the nutrient-sensing kinase mTOR (mechanistic target of rapamycin) in host cells, suggesting that the external nutrient environment can also affect the viability of internalized cells through a poorly understood mechanism that may be linked to control over vacuole maturation by host cells [15].

The different possible fates of entotic cells highlight an emerging concept that individual cells within stressed populations can take different paths to death. As discussed, entosis was discovered in cell populations that were denied adhesion to matrix, where mixed fates, including entosis, apoptosis, necrotic forms of death, and even terminal differentiation (cornification), can occur in parallel [1, 2, 17, 18]. For entosis as well, although most internalized cells will die with a non-apoptotic morphology and in a manner unaffected by BCL-2, a small percentage can be observed to undergo apoptosis even in untreated cultures [1]. The full rescue of entotic cells requires blocking more than one path to death, as overexpression of BCL-2 inhibits apoptosis but not entotic cell death [1, 12], while inhibition of lysosome function blocks entotic cell death but leads to apoptosis [1, 16]. The complete rescue of entotic cells therefore requires the inhibition of both lysosome activity and apoptosis, which then allows internalized cells to escape unharmed, presumably having found enough time to find their way out.

Entosis in Normal Physiology?

The mechanism of entosis was discovered at a time when it was becoming apparent that different forms of cell death could be regulated in metazoan cells in addition to apoptosis [1, 19, 20]. To date, 20 or more different cell death mechanisms have now been described [3, 21–23]. So, why are there so many ways to die? It is possible that some mechanisms have features that are so unique that they underlie specialized utilization in particular contexts. Regulated forms of necrosis, for example, including necroptosis and pyroptosis, are induced in response to infection, and their necrotic modes of execution may be the most efficient ways to release inflammatory molecules that can alert immune responses [19, 20]. Apoptosis, on the other hand, may be largely anti-inflammatory, and this feature may make it particularly useful in normal tissues [24]. Other forms of death, such as ferroptosis, may not be active death mechanisms *per se* but result instead from catastrophic failures of processes that are essential for keeping cells alive [25].

So, what about entosis? There is accumulating evidence that entosis could occur in normal physiology in specific contexts where its cell adhesion-based mechanism might be useful [7]. The mechanism of embryo implantation, for example, which requires contact to be established between the blastocyst and uterine wall and then a deeper embedding of the embryo into the underlying tissue, has been shown to involve the entotic ingestion and removal of uterine epithelial cells by trophectoderm cells that line the exterior of the embryo [26]. Through entosis, trophectoderm cells can utilize the adhesions that first establish contact with the uterine wall to also engulf and clear epithelial cells, a process that both creates space for implantation and embeds the embryo deeper into the underlying mesenchyme. The occurrence of entosis in this context is supported by imaging and inhibitor-based studies in mice and by an in vitro assay that shows the ability of trophectoderm cells to ingest uterine epithelial cells, but further studies are needed to continue to explore this mode of regulation.

Entosis has also been shown to occur in *C. elegans,* where it can clear a specialized cell that is involved in the development of the male gonad called the linker cell [14]. The linker cell leads a collective migration that shapes the developing gonad and then forms the initial junctions that link the gonad to the digestive opening called the cloaca. Through entosis, the same junctions that establish the gonad-to-cloaca connection can also engulf and clear the linker cell, a coordinated mechanism that may reinforce the inter-organ linkage while at the same time eliminating this cell that otherwise stands in the way of forming a contiguous lumen. Removal of the linker cell is critical for development, as its continued presence can render animals sterile [27]. Entosis was shown to be involved in linker cell clearance largely through imaging-based approaches, and further studies are needed to continue to investigate this mechanism [14].

Entosis in Disease

Since its discovery, entosis has been linked to cancer [1]. Structures resembling those formed by entosis, involving the engulfment of whole cells into cell-in-cell structures that are decorated with cell adhesion proteins, are found in numerous cancers, including many carcinomas (breast, pancreatic, colon, lung, prostate, liver, stomach, esophageal, rectal, and head and neck cancers), consistent, perhaps, with the regulation of entosis by epithelial-type cadherins [1, 22, 28–36]. Many other types of cancer also exhibit cell-in-cell structures that could result from entosis or from other forms of engulfment that can target live cells through eating-like mechanisms that are referred to as cell cannibalism, reviewed extensively elsewhere [22].

In some cancers, such as breast, lung, pancreatic and rectal carcinoma, and head and neck cancers, entotic cell structures are found most frequently in particularly aggressive subsets and can correlate with negative clinical outcomes, suggesting that this mechanism could be associated with disease progression [28, 30–32, 35–38]. While entosis leads to cell death and might therefore be predicted to have tumor-suppressive activity [1], this process also has significant effects on host cells, which, for example, can scavenge large amounts of metabolites through the digestion of internalized cells [39]. Nutrient scavenging can support the survival and even the proliferation of host cells under conditions of nutrient deprivation, which are common in cancer tissues [39]. Host cells can therefore benefit significantly from killing their neighbors through this mechanism. Moreover, when host cells divide, the presence of internalized cells can disrupt the completion of division by acting as a barrier to the formation of proper contractile rings during cytokinesis. As a result, host cells fail division at a high rate in culture and exhibit a high frequency of multinucleation in cancerous tissues [37]. Entosis therefore not only supplies nutrients to hosts to support their survival and proliferation but also generates host cell-derived lineages that transmit gross aneuploidy, the result of a failure of cytokinesis [37]. These two linked and potentially potent effects may underlie, at least in part, the promotion of cancer progression by entosis.

The engulfment of live cells by their neighbors also has the intrinsic potential to be competitive between the individual cells within a population. Entosis has indeed been shown to promote competition between cells in culture-based models, selecting for cells with lower setpoints of tension, a known feature of cells expressing particular oncogenes such as activated KRAS (Kirsten rat sarcoma virus) [5, 40]. Cancer cells have also been shown to internalize non-cancer cells preferentially by this mechanism, an effect potentially linked to lowered tension levels that is a general property of malignant cells [5]. In developing tissues, individual cells are known to engage in "cell competition" based on relative setpoints of fitness. Cells that are less fit, called "losers," become eliminated, likely through apoptosis, in response to proximity to cells that are more fit, or "winners." Cell fitness, in this case, is a relative property that can be driven at least in part by differential expression of oncogenes such as KRAS and C-MYC [41]. In a similar manner, entosis can promote the elimination of loser cells that have higher levels of mechanical tension by winner cells that have lower tension and, as discussed, high levels of aneuploidy and internal nutrient stores derived from entotic cell digestion.

Entosis in Cancer Therapy?

Entosis has now been shown to occur in response to numerous stress conditions in addition to matrix detachment, including starvation for glucose [40], exposure to ultraviolet radiation (UV) [42], and treatment with anti-mitotic drugs [43, 44]. In each case, entosis induction results from changes within internalizing cells that control the process of cell uptake through either stress kinase signaling in UV-treated cells (JNK (c-Jun-N-terminal kinase)); low energy-based signaling in glucose-starved cells (AMPK (AMP-activated protein kinase)); or mechanical changes in cells treated with mitotic poisons. These conditions induce entosis in adherent cultures alongside other forms of death, including apoptosis and necrosis, which occur together with entosis in complex mixtures [40, 42], in a manner perhaps similar to cells cultured in suspension where entosis was originally discovered [1]. Inhibiting entosis in glucose-starved or UV-treated cell populations leads to compensatory increases in other forms of cell death, and vice versa [40, 42].

As nutrient starvation, UV exposure, and treatment with anti-mitotic drugs are well known to cause other forms of cell death, the presence of death mixtures is perhaps not surprising, but the findings raise at least two important questions. First, what determines the path for individual cells, entotic or otherwise? And second, do death mixtures occur in clinical contexts, for example, in cancer therapy?

It is possible that cells experiencing high levels of stress are destined to die, and whether or not they undergo entosis is based on the presence of neighboring cells that are suitable to act as hosts. For example, in UV-treated cells, it was observed that JNK signaling was initially activated to high levels in every cell within a population, but over time, some cells showed reduced signaling, and these acted as hosts to internalize entotic cells with sustained high levels of signaling [42]. In this case, it is possible that cells with high levels of stress may be destined to die, and their entotic path is related to population heterogeneity and proximity to cells with reduced signaling, perhaps a sign of relative fitness. For entosis occurring in response to starvation, AMPK signaling was instead observed to be induced in a heterogeneous manner, leading to differences between individual cells, where those with high levels of signaling were internalized into cells with lower levels [40]. Here as well, high levels of stress signaling may ultimately lead to cell death, but the path to entosis may rely on the manifestation of population-scale heterogeneities.

Whether entosis is induced by cancer therapy remains an open question, although several reports have now shown that this process is induced by treatment with therapeutic, or potentially therapeutic, compounds, including drugs that inhibit mitosis [43, 44] or kinase signaling [45], affect redox potential [46], or induce cell death (TRAIL (tumor necrosis factor (TNF)-related apoptosis-inducing ligand)) [30]. Entosis could be induced in cancers as part of mixed death responses, in which case the percentages of different death types are likely to have an impact on therapeutic outcomes. High levels of entosis would be predicted to contribute to a cell death response and regression in the short term but, in the longer term, promote the development of gross aneuploidy and selection for cells with highly malignant characteristics that could ultimately drive the development of more aggressive disease. Understanding whether entosis is induced therapeutically and how its presence, in addition to other forms of cell death, might correlate with patient outcomes are important areas for future study. The quantification of death mixtures in clinical contexts is broadly an open area and likely an important avenue of investigation in cell death research as we seek to understand how different death mechanisms interface with human disease.

Concluding Remarks

Here, we have highlighted entosis and entotic cell death to discuss the basic mechanisms that drive these processes and consider their possible functions in normal physiology and disease. While we focused on particular examples to illustrate key discussion points, many papers have now been published in the entosis field, and other recent reviews may better capture the full scope of this research (e.g., please see refs. (7, 22)).

Entosis has been argued to occur in particular contexts in normal physiology, and here we discussed two examples that may help to illustrate why: A common theme between embryo implantation and linker cell clearance is the need to establish adhesion, and adhesion is perhaps the single most defining feature of entosis. In both cases, one cell population critically needs to adhere to another, and at the same time, cells that stand in the way need to be killed and cleared. Entosis can conceivably promote each of these activities through one mechanism, a simple model that may allow for the clearance of particular cells in development while also reinforcing adhesion strength between cells or cell populations (Fig. 1). Clearly, these mechanisms need further exploration, but this common thread could underscore the specialized utility of entosis in normal tissues.

We have also discussed the potential functions of entosis in disease by focusing on cancer, where entotic cell structures are clearly observable, although this mechanism could contribute to additional diseases, including degenerative conditions [47]. On balance, entosis appears to promote cancer progression, and the clinical correlations and cellular mechanisms discussed herein support this conclusion. However, this process is complex, and there may be instances where entotic cell death could instead be a favorable indicator. The tumor-suppressive activity of entosis was noted in its first description [1], and since this time, favorable prognostic indicators have emerged in some contexts [33, 48, 49]. Like other forms of cell death, entosis may have complex effects on the dynamics of cell populations, although on balance it likely promotes cancer progression and should probably be avoided in therapeutic contexts, if possible.

One additional emerging area relating to cancer therapy is the potential of cell-in-cell structures, like those formed by entosis and other mechanisms of engulfment

[22], to shield internalized cells from targeted therapies, for example, cytotoxic T lymphocyte-based anti-cancer treatments [50]. Here, the transient ability of internalized cells to survive within entotic vacuoles could allow such cells to hide from targeted therapeutics that need direct plasma membrane access, for example, cytotoxic cell-based therapies, or even antibodies, nanotherapeutics, or some chemotherapeutic drugs. Whether entotic cell shielding represents a significant mechanism of resistance to targeted therapies is an important subject for future research.

Acknowledgments The research underlying this work is funded by a grant from the NCI (R35CA263846 to M.O.).

Competing Interests The authors declare no competing interests.

References

1. Overholtzer, M., Mailleux, A. A., Mouneimne, G., Normand, G., Schnitt, S. J., King, R. W., et al. (2007). A nonapoptotic cell death process, entosis, that occurs by cell-in-cell invasion. *Cell, 131*(5), 966–979.
2. Frisch, S. M., & Francis, H. (1994). Disruption of epithelial cell-matrix interactions induces apoptosis. *The Journal of Cell Biology, 124*(4), 619–626.
3. Galluzzi, L., Vitale, I., Aaronson, S. A., Abrams, J. M., Adam, D., Agostinis, P., et al. (2018). Molecular mechanisms of cell death: Recommendations of the nomenclature committee on cell death 2018. *Cell Death and Differentiation, 25*(3), 486–541.
4. Ravichandran, K. S. (2011). Beginnings of a good apoptotic meal: The find-me and eat-me signaling pathways. *Immunity, 35*(4), 445–455.
5. Sun, Q., Luo, T., Ren, Y., Florey, O., Shirasawa, S., Sasazuki, T., et al. (2014). Competition between human cells by entosis. *Cell Research, 24*(11), 1299–1310.
6. Elliott, M. R., & Ravichandran, K. S. (2016). The dynamics of apoptotic cell clearance. *Developmental Cell, 38*(2), 147–160.
7. Kim, S., Lee, D., Kim, S. E., & Overholtzer, M. (2024). Entosis: The core mechanism and crosstalk with other cell death programs. *Experimental & Molecular Medicine, 56*(4), 870–876.
8. Purvanov, V., Holst, M., Khan, J., Baarlink, C., & Grosse, R. (2014). G-protein-coupled receptor signaling and polarized actin dynamics drive cell-in-cell invasion. *eLife, 3*, 3.
9. Sun, Q., Cibas, E. S., Huang, H., Hodgson, L., & Overholtzer, M. (2014). Induction of entosis by epithelial cadherin expression. *Cell Research, 24*(11), 1288–1298.
10. Wildenberg, G. A., Dohn, M. R., Carnahan, R. H., Davis, M. A., Lobdell, N. A., Settleman, J., & Reynolds, A. B. (2006). p120-catenin and p190RhoGAP regulate cell-cell adhesion by coordinating antagonism between Rac and Rho. *Cell, 127*(5), 1027–1039.
11. Overholtzer, M., & Brugge, J. S. (2008). The cell biology of cell-in-cell structures. *Nature Reviews. Molecular Cell Biology, 9*(10), 796–809.
12. Florey, O., Kim, S. E., Sandoval, C. P., Haynes, C. M., & Overholtzer, M. (2011). Autophagy machinery mediates macroendocytic processing and entotic cell death by targeting single membranes. *Nature Cell Biology, 13*(11), 1335–1343.
13. Kinchen, J. M., & Ravichandran, K. S. (2008). Phagosome maturation: Going through the acid test. *Nature Reviews. Molecular Cell Biology, 9*(10), 781–795.
14. Lee, Y., Hamann, J. C., Pellegrino, M., Durgan, J., Domart, M. C., Collinson, L. M., et al. (2019). Entosis controls a developmental cell Clearance in C. elegans. *Cell Reports, 26*(12), 3212–20 e4.
15. Kim, S. E., Zhang, J., Jiang, E., & Overholtzer, M. (2021). Amino acids and mechanistic target of rapamycin regulate the fate of live engulfed cells. *The FASEB Journal, 35*(10), e21909.

16. Su, Y., Ren, H., Tang, M., Zheng, Y., Zhang, B., Wang, C., et al. (2021). Role and dynamics of vacuolar pH during cell-in-cell mediated death. *Cell Death & Disease, 12*(1), 119.
17. Hawk, M. A., Gorsuch, C. L., Fagan, P., Lee, C., Kim, S. E., Hamann, J. C., et al. (2018). RIPK1-mediated induction of mitophagy compromises the viability of extracellular-matrix-detached cells. *Nature Cell Biology, 20*(3), 272–284.
18. Mailleux, A. A., Overholtzer, M., Schmelzle, T., Bouillet, P., Strasser, A., & Brugge, J. S. (2007). BIM regulates apoptosis during mammary ductal morphogenesis, and its absence reveals alternative cell death mechanisms. *Developmental Cell, 12*(2), 221–234.
19. Degterev, A., Huang, Z., Boyce, M., Li, Y., Jagtap, P., Mizushima, N., et al. (2005). Chemical inhibitor of nonapoptotic cell death with therapeutic potential for ischemic brain injury. *Nature Chemical Biology, 1*(2), 112–119.
20. Fink, S. L., & Cookson, B. T. (2005). Apoptosis, pyroptosis, and necrosis: Mechanistic description of dead and dying eukaryotic cells. *Infection and Immunity, 73*(4), 1907–1916.
21. Mao, C., Wang, M., Zhuang, L., & Gan, B. (2024). Metabolic cell death in cancer: Ferroptosis, cuproptosis, disulfidptosis, and beyond. *Protein Cell, 15*, 642.
22. Fais, S., & Overholtzer, M. (2018). Cell-in-cell phenomena in cancer. *Nature Reviews. Cancer, 18*(12), 758–766.
23. Malireddi, R. K. S., Kesavardhana, S., & Kanneganti, T. D. (2019). ZBP1 and TAK1: Master regulators of NLRP3 inflammasome/pyroptosis, apoptosis, and necroptosis (PAN-optosis). *Frontiers in Cellular and Infection Microbiology, 9*, 406.
24. Perry, J. S. A., Morioka, S., Medina, C. B., Iker Etchegaray, J., Barron, B., Raymond, M. H., et al. (2019). Interpreting an apoptotic corpse as anti-inflammatory involves a chloride sensing pathway. *Nature Cell Biology, 21*(12), 1532–1543.
25. Dixon, S. J., Lemberg, K. M., Lamprecht, M. R., Skouta, R., Zaitsev, E. M., Gleason, C. E., et al. (2012). Ferroptosis: An iron-dependent form of nonapoptotic cell death. *Cell, 149*(5), 1060–1072.
26. Li, Y., Sun, X., & Dey, S. K. (2015). Entosis allows timely elimination of the luminal epithelial barrier for embryo implantation. *Cell Reports, 11*(3), 358–365.
27. Abraham, M. C., Lu, Y., & Shaham, S. (2007). A morphologically conserved nonapoptotic program promotes linker cell death in Caenorhabditis elegans. *Developmental Cell, 12*(1), 73–86.
28. Hayashi, A., Fan, J., Chen, R., Ho, Y. J., Makohon-Moore, A. P., Lecomte, N., et al. (2020). A unifying paradigm for transcriptional heterogeneity and squamous features in pancreatic ductal adenocarcinoma. *Nature Cancer, 1*(1), 59–74.
29. Wen, S., Shang, Z., Zhu, S., Chang, C., & Niu, Y. (2013). Androgen receptor enhances entosis, a non-apoptotic cell death, through modulation of rho/ROCK pathway in prostate cancer cells. *The Prostate, 73*(12), 1306–1315.
30. Bozkurt, E., Dussmann, H., Salvucci, M., Cavanagh, B. L., Van Schaeybroeck, S., Longley, D. B., et al. (2021). TRAIL signaling promotes entosis in colorectal cancer. *The Journal of Cell Biology, 220*(11), e202010030.
31. Mackay, H. L., Moore, D., Hall, C., Birkbak, N. J., Jamal-Hanjani, M., Karim, S. A., et al. (2018). Genomic instability in mutant p53 cancer cells upon entotic engulfment. *Nature Communications, 9*(1), 3070.
32. Wang, R., Zhu, Y., Zhong, H., Gao, X., Sun, Q., & He, M. (2022). Homotypic cell-in-cell structures as an adverse prognostic predictor of hepatocellular carcinoma. *Frontiers in Oncology, 12*, 1007305.
33. Wang, Y., Niu, Z., Zhou, L., Zhou, Y., Ma, Q., Zhu, Y., et al. (2021). Subtype-based analysis of cell-in-cell structures in esophageal squamous cell carcinoma. *Frontiers in Oncology, 11*, 670051.
34. Huang, H., Chen, A., Wang, T., Wang, M., Ning, X., He, M., et al. (2015). Detecting cell-in-cell structures in human tumor samples by E-cadherin/CD68/CD45 triple staining. *Oncotarget, 6*(24), 20278–20287.
35. Schenker, H., Buttner-Herold, M., Fietkau, R., & Distel, L. V. (2017). Cell-in-cell structures are more potent predictors of outcome than senescence or apoptosis in head and neck squamous cell carcinomas. *Radiation Oncology, 12*(1), 21.

36. Schwegler, M., Wirsing, A. M., Schenker, H. M., Ott, L., Ries, J. M., Buttner-Herold, M., et al. (2015). Prognostic value of homotypic cell internalization by nonprofessional phagocytic cancer cells. *BioMed Research International, 2015*, 359392.
37. Krajcovic, M., Johnson, N. B., Sun, Q., Normand, G., Hoover, N., Yao, E., et al. (2011). A nongenetic route to aneuploidy in human cancers. *Nature Cell Biology, 13*(3), 324–330.
38. Song, J., Xu, R., Zhang, H., Xue, X., Ruze, R., Chen, Y., et al. (2023). Cell-in-cell-mediated entosis reveals a progressive mechanism in pancreatic cancer. *Gastroenterology, 165*(6), 1505–21 e20.
39. Krajcovic, M., Krishna, S., Akkari, L., Joyce, J. A., & Overholtzer, M. (2013). mTOR regulates phagosome and entotic vacuole fission. *Molecular Biology of the Cell, 24*(23), 3736–3745.
40. Hamann, J. C., Surcel, A., Chen, R., Teragawa, C., Albeck, J. G., Robinson, D. N., & Overholtzer, M. (2017). Entosis is induced by glucose starvation. *Cell Reports, 20*(1), 201–210.
41. Cong, B., & Cagan, R. L. (2024). Cell competition and cancer from Drosophila to mammals. *Oncogene, 13*(1), 1.
42. Chen, R., Ram, A., Albeck, J. G., & Overholtzer, M. (2021). Entosis is induced by ultraviolet radiation. *iScience., 24*(8), 102902.
43. Durgan, J., Tseng, Y. Y., Hamann, J. C., Domart, M. C., Collinson, L., Hall, A., et al. (2017). Mitosis can drive cell cannibalism through entosis. *eLife, 6*, 6.
44. Liang, J., Niu, Z., Zhang, B., Yu, X., Zheng, Y., Wang, C., et al. (2021). p53-dependent elimination of aneuploid mitotic offspring by entosis. *Cell Death and Differentiation, 28*(2), 799–813.
45. Liu, J., Wang, L., Zhang, Y., Li, S., Sun, F., Wang, G., et al. (2019). Induction of entosis in prostate cancer cells by nintedanib and its therapeutic implications. *Oncology Letters, 17*(3), 3151–3162.
46. Khalkar, P., Diaz-Argelich, N., Antonio Palop, J., Sanmartin, C., & Fernandes, A. P. (2018). Novel methylselenoesters induce programed cell death via entosis in pancreatic cancer cells. *International Journal of Molecular Sciences, 19*(10), 2849.
47. Sterling, N. A., Park, J. Y., Park, R., Cho, S. H., & Kim, S. (2023). An entosis-like process induces mitotic disruption in Pals1 microcephaly pathogenesis. *Nature Communications, 14*(1), 82.
48. Zhang, X., Niu, Z., Qin, H., Fan, J., Wang, M., Zhang, B., et al. (2019). Subtype-based prognostic analysis of cell-in-cell structures in early breast cancer. *Frontiers in Oncology, 9*, 895.
49. Bauer, M. F., Hildebrand, L. S., Rosahl, M. C., Erber, R., Schnellhardt, S., Buttner-Herold, M., et al. (2022). Cell-in-cell structures in early breast cancer are prognostically valuable. *Cells, 12*(1), 81.
50. Gutwillig, A., Santana-Magal, N., Farhat-Younis, L., Rasoulouniriana, D., Madi, A., Luxenburg, C., et al. (2022). Transient cell-in-cell formation underlies tumor relapse and resistance to immunotherapy. *eLife, 11*, 11.

Index

C. D. Gregory et al. (eds.), *Apoptosis and Other forms of Cell Death in the Regulation of Inflammation in Health and Disease*, Advances in Experimental Medicine and Biology 1481, https://doi.org/10.1007/978-3-031-92785-0

MIX
Papier aus verantwortungsvollen Quellen
Paper from responsible sources
FSC® C105338

If you have any concerns about our products,
you can contact us on
ProductSafety@springernature.com

In case Publisher is established outside the EU,
the EU authorized representative is:
Springer Nature Customer Service Center GmbH
Europaplatz 3, 69115 Heidelberg, Germany

Printed by Libri Plureos GmbH
in Hamburg, Germany